Lecture Notes in Networks and Systems

Volume 1201

The series "Lecture Notes in Networks and Systems" publishes the latest developments in Networks and Systems—quickly, informally and with high quality. Original research reported in proceedings and post-proceedings represents the core of LNNS.

Volumes published in LNNS embrace all aspects and subfields of, as well as new challenges in, Networks and Systems.

The series contains proceedings and edited volumes in systems and networks, spanning the areas of Cyber-Physical Systems, Autonomous Systems, Sensor Networks, Control Systems, Energy Systems, Automotive Systems, Biological Systems, Vehicular Networking and Connected Vehicles, Aerospace Systems, Automation, Manufacturing, Smart Grids, Nonlinear Systems, Power Systems, Robotics, Social Systems, Economic Systems and other. Of particular value to both the contributors and the readership are the short publication timeframe and the world-wide distribution and exposure which enable both a wide and rapid dissemination of research output.

The series covers the theory, applications, and perspectives on the state of the art and future developments relevant to systems and networks, decision making, control, complex processes and related areas, as embedded in the fields of interdisciplinary and applied sciences, engineering, computer science, physics, economics, social, and life sciences, as well as the paradigms and methodologies behind them.

Indexed by SCOPUS, EI Compendex, INSPEC, WTI Frankfurt eG, zbMATH, SCImago.

All books published in the series are submitted for consideration in Web of Science.

For proposals from Asia please contact Aninda Bose (aninda.bose@springer.com).

Anand J. Kulkarni · Mostafa Hajiaghaei-Keshteli
Editors

Advances in Intelligent Systems for Sustainable Agriculture

Proceedings of the 3rd International Conference, ICISA 2024

Springer

Editors
Anand J. Kulkarni
Institute of Artificial Intelligence
Dr. Vishwanath Karad MIT World Peace University
Pune, Maharashtra, India

Mostafa Hajiaghaei-Keshteli
School of Engineering and Sciences
Monterrey Institute of Technology and Higher Education (ITESM)
Monterrey, Mexico

ISSN 2367-3370 ISSN 2367-3389 (electronic)
Lecture Notes in Networks and Systems
ISBN 978-981-97-9838-4 ISBN 978-981-97-9839-1 (eBook)
https://doi.org/10.1007/978-981-97-9839-1

This Springer imprint is published by the registered company Springer Nature Singapore Pte Ltd.
The registered company address is: 152 Beach Road, #21-01/04 Gateway East, Singapore 189721, Singapore

Preface

The agricultural domain is one of the oldest professions of the human civilization. It significantly contributes to the GDP of all the countries in the world. Since the first industrial revolution, the agricultural system processes are being automated for increasing the crop yield, logistics and supply chain management, utilization of the resources such as land, workforce, etc.; however, the evergrowing global population is still demanding significant improvement in these domains. As the state-of-the-art Artificial Intelligence (AI) and Machine Learning (ML) tools and techniques have demonstrated their usability and effectiveness in almost all kinds of industrial and business processes, their application is very much needed for the agricultural domain and its associated areas.

In order to be a driver and contributor in this direction, the Institute of Artificial Intelligence of the Dr. Vishwanath Karad MIT World Peace University, Pune, India, successfully organized the 3rd International Conference on Intelligent Systems and Applications (ICISA) on April 26 and 27, 2024 with a focus 'Intelligent Systems for Sustainable Agriculture'. Several academicians, researchers, scientists, professionals, and students shared their findings, knowledge, and expertise in the fields associated with intelligent systems in plant disease detection, food and grain monitoring, digitalization and automation of agriculture equipment, soil health analysis, remote sensing, crop predictions, recommendations and price forecasting, resource management system, decision-making system, agricultural waste supply chain network design, crop monitoring system, etc.

Two elite keynote talks were delivered by Prof. Hossein Bonakdari, University of Ottawa, ON, Canada, and Prof. Anand Nayyar, Duy Tan University, Da Nang, Vietnam. We would like to extend our special thanks to Pro-Vice-Chancellor Dr. Milind Pande and Dean R&D Dr. Murthy Chavali of the MIT World Peace University for their support and motivation in seamless organization of the conference. Thanks are due to very distinguished reviewers for their timely support and competent team members Dr. Ishaan R. Kale and Dr. Apoorva S. Shastri, Dr. Rohit Ghadge, Dr. Ratnakar Ghorpade and Dr. Sushil Kumar Gupta, Dr. Sumant Shinde and Dr. Puskaraj D. Sonawwanay for successfully organizing the event. Importantly, we thank Ms. Kamiya Khatter, Senior Editor, Springer, and Divya Meiyazhagan, Production Administrator, Springer, for all the help they provided.

Pune, India Anand J. Kulkarni
Puebla, Mexico Mostafa Hajiaghaei-Keshteli

From Fields to Data: Agriculture Transformation with Intelligent Systems

Abstract

In the face of rapid technological advancements and climate change, agriculture is transforming by integrating intelligent systems. The impact of climate change on farming underscores the need for predictive and prescriptive insights that intelligent systems provide. This note explores the transformative potential of intelligent systems in sustainable agriculture, from traditional farming practices to adopting data-driven, adaptive solutions. Four key aspects of AI in action demonstrate the practical application of analytics in agriculture. Descriptive analytics identifies historical patterns in crop yields, soil health, and pest infestations, while diagnostic analytics delves into the causes behind these patterns. Predictive analytics anticipates future scenarios like crop yields and pest outbreaks using machine learning models trained on historical data. Finally, prescriptive analytics offers tailored recommendations to optimize irrigation, fertilization, and pest management, ensuring precise resource use and maximum profitability. By coupling the power of analytics with advanced sensing technologies, machine learning models, and data-driven decision-making, farmers can navigate the complexities of modern agriculture with precision, resilience, and innovation. Intelligent systems empower farmers to predict, adapt, and optimize their practices amidst ever-changing environmental conditions, paving the way for sustainable and resilient farming solutions.

Introduction

Agriculture has undergone a remarkable evolution, transitioning from traditional farming practices to adopting intelligent systems (Montalvo-Romero et al., 2023). IoT, AI, and robotics advancements have led to data-driven and automated agriculture, revolutionizing farming practices (Charania and Li, 2020). This marks a new phase in digital agriculture, enabling predictive analytics, autonomous machinery, and big data analytics (Mishra and Kim, 2024). Climate change presents a significant challenge to modern agriculture, necessitating adaptive, data-driven farming solutions and the reengineering of all sectors related to product production, processing, and delivery. (Karataeva et al., 2020). Studies have shown that climate change can lead to increased volumes of runoff (Soltani et al., 2021) and more significant losses of soil and nutrients due to rising temperatures and intense precipitation with a negative impact on crop yields (Mason et al., 2021; Szejgis et al., 2024). Furthermore, drought can significantly reduce crop yield and productivity due to its adverse effects on nutrient uptake and assimilation and radiation use efficiency (Ullah et al., 2019). To address these challenges, predictive analytics based on weather patterns, historical data, and crop growth stages can forecast yield outcomes and potential threats, helping farmers prepare for extreme weather events (Mall et al., 2017; Zeynoddin et al., 2020; Feng et al. 2020; Kumari et al., 2023). Precision agriculture ensures optimal resource use with variable rate irrigation and nutrient application (Grenon et al., 2023; Zeynoddin et al., 2023). Meanwhile, autonomous machinery can work continuously, adjusting operations based on real-time data, and data-driven carbon farming practices like no-till and cover cropping can reduce greenhouse gas emissions (Bai et al., 2019; Gong et al., 2021; Contasti et al., 2023). Intelligent systems in agriculture integrate sensing, learning, and autonomous decision-making technologies to manage agricultural processes efficiently (Savita et al., 2023; Araujo et al., 2023; Khanna et al., 2023; Zhu, 2024).

Climate change presents a significant challenge to modern agriculture, necessitating adaptive, data-driven farming solutions and the reengineering of all sectors related to product production, processing, and delivery. Intelligent systems address these challenges through key components like sensing technologies, which integrate Internet of Things (IoT) sensors and devices for real-time data collection on crucial environmental parameters such as soil moisture, temperature, and humidity. This empowers farmers to make informed irrigation, fertilization, and pest control decisions. Learning technologies, including machine learning algorithms and data analytics, extract valuable insights from the collected data, enabling proactive decision-making, early disease detection, and accurate yield prediction. Autonomous decision-making, supported by artificial intelligence and machine learning, provides valuable insights to farmers by optimizing resource utilization and enhancing agricultural productivity through tailored recommendations. These technologies bring several benefits to agriculture, including efficient resource management by enabling real-time monitoring of soil conditions and planning agronomic inputs, maximizing resource use efficiency, and minimizing environmental impact. Additionally, they increase productivity by addressing water scarcity, reducing water consumption, and boosting crop production compared to traditional methods. Moreover, intelligent systems promote sustainable agriculture by optimizing farming practices, increasing productivity, and mitigating environmental impact. Thus, these systems empower farmers to navigate the complexities of modern agriculture with precision, resilience, and innovation. Together, these technologies offer a more resilient, sustainable future for farming amidst the uncertainties of climate change.

Understanding Climate Adaptation in Agriculture

Climate adaptation refers to adjusting practices, systems, and infrastructure to reduce the adverse impacts of climate change while capitalizing on potential opportunities. In the agricultural sector, climate adaptation is critical due to the significant sensitivity of crops, livestock, and farming practices to changing weather patterns and environmental conditions. As climate change intensifies, farmers face rising temperatures, shifting precipitation patterns, more frequent extreme weather events, and altered pest and disease dynamics, all of which threaten food production and security. Adaptive strategies include the development of drought-tolerant crop varieties, implementing precision irrigation to conserve water, diversifying cropping systems to improve resilience, and using predictive analytics to anticipate and respond to climate risks. Moreover, adopting agroecological practices like cover cropping, no-till farming, and agroforestry enhances soil health, sequesters carbon, and creates more resilient ecosystems. By integrating intelligent systems, such as IoT-based sensors, machine learning models, and automated machinery, farmers can monitor climate conditions in real-time, refine predictive models, and optimize resource use. This proactive approach to climate adaptation not only safeguards agricultural productivity but also contributes to global efforts to mitigate the impacts of climate change.

Farmers today are confronted with various challenges exacerbated by climate change, among them being unpredictable weather patterns, shifting growing seasons, and pest migration. Unpredictable weather events, such as prolonged droughts, intense rainfall, and heatwaves, disrupt traditional farming practices and significantly impact crop yields. For instance, extended droughts deplete soil moisture and reduce crop growth, while heavy rains can lead to soil erosion, nutrient leaching, and flooding of low-lying fields. Shifting growing seasons further complicate farming as the onset of spring and autumn becomes less predictable, affecting planting and harvesting times. Crops that rely on specific temperature and daylight cues for growth and maturation now face altered timelines, resulting in mismatched growth stages and reduced yields. Simultaneously, climate change has triggered the migration of pests and diseases into new regions, posing a significant threat to crop health. Diseases like late blight and rust fungi are becoming more prevalent due to changing humidity and temperature conditions. The unpredictability of weather, changing seasons, and pest migration pose challenges for farmers in maintaining profitable production. Farmers increasingly turn to data-driven solutions to address these issues, such as predictive analytics for climate forecasting, integrated pest management strategies, and intelligent systems that provide real-time monitoring and adaptive responses. However, adopting these technologies requires significant investment and training, posing further challenges for small-scale and resource-constrained farmers.

Intelligent Systems in Agriculture: An Overview

Intelligent systems in agriculture integrate advanced technologies to enhance efficiency and sustainability. As shown in Table 1, intelligent systems bring together sensing technologies, satellite imagery, predictive analytics, and robotics to provide comprehensive data-driven solutions in agriculture. Table 1 provides examples of the key components that constitute intelligent systems in agriculture.

The Role of Data in Agriculture: From Field Sensors to Satellite Imagery

Data have emerged as a cornerstone of modern agriculture, empowering farmers to make informed decisions and optimize their practices through precise, real-time insights. Field sensors play a pivotal role by providing granular data on soil and crop health. Soil sensors measure moisture levels, pH, temperature, and nutrient content, enabling farmers to tailor irrigation and fertilization practices to the specific needs of different field zones. Plant sensors, such as spectral reflectance devices and chlorophyll meters, detect nutrient deficiencies, water stress, and disease onset early, helping farmers apply corrective measures before significant yield losses occur. Environmental sensors monitor key climate variables like temperature, humidity, wind speed, and precipitation, offering valuable data for forecasting weather impacts on crops.

Satellite imagery extends the reach of data collection, providing a bird's-eye view of large agricultural areas. High-resolution multispectral and hyperspectral imagery enables the calculation of vegetation indices like the Normalized Difference Vegetation Index (NDVI), which assesses crop health and vigor across fields. Thermal imaging identifies areas of water stress by detecting temperature variations in plant canopies, while microwave radar imagery penetrates clouds to estimate soil moisture. Moreover, satellite data, combined with advanced predictive analytics, aids in yield prediction, early disease detection, and efficient resource allocation.

Together, data from field sensors and satellite imagery underpin precision agriculture practices, allowing farmers to implement site-specific management strategies. Variable rate technology (VRT) uses soil and crop data to optimize input application, ensuring that fertilizers, pesticides, and water are used efficiently. Predictive models, fueled by historical and real-time data, provide insights into yield forecasts, pest outbreaks, and weather anomalies, enabling proactive decision-making. As a result, data-driven agriculture not only improves productivity and sustainability but also enhances the resilience of farming systems to climate change and market volatility.

Table 1 Components of intelligent systems in agriculture

Sensors (IoT devices)	Soil sensors	Soil moisture
		Tensiometers
		pH and nutrient sensors
	Environmental sensors	Thermocouples and thermistors
		Anemometers
		Rain gauges and hygrometers
	Plant sensors	Spectral reflectance sensors
		Thermal imaging cameras
		Dendrometers and phytomonitoring systems
Satellite imagery	Remote sensing data	Landsat and Sentinel-2
		MODIS
	Thermal and microwave imaging	Thermal imaging
		Microwave imaging
	Vegetation indices	NDVI (Normalized Difference Vegetation Index)
		EVI (Enhanced Vegetation Index)
		NDWI (Normalized Difference Water Index)
Predictive	Historical data analysis and real-time data gathering	Time-series analysis
		Machine learning models
		Predictive models: Yield prediction models (process-based models and statistical models) Pest and disease forecasting models (degree-day models and spatio-temporal models)
Robotics	Automated machinery	Autonomous tractors and harvesters
		Drones
		Thermal cameras
		Spraying drones
		Weeding robots

Machine Learning and Predictive Analytics in Crop Management

Machine learning and predictive analytics have revolutionized crop management by providing farmers with actionable insights and enabling data-driven decision-making. These technologies leverage vast datasets from historical records, field sensors, and satellite imagery to build predictive models that enhance yield estimation, optimize input usage, and anticipate risks. Supervised learning algorithms like Random Forests, Support Vector Machines (SVMs), and Gradient Boosting Machines are employed to predict crop yields based on variables such as soil properties, weather conditions, and crop health indices. For instance, yield prediction models can integrate multispectral imagery data (e.g., NDVI), soil moisture levels, and historical climate information to accurately forecast crop productivity. Predictive analytics also extends to disease and pest forecasting, providing early warnings that empower farmers to implement timely and targeted interventions. Degree-day models, which track insect life cycles based on temperature data, are combined with spatial models to predict pest emergence and spread. Machine learning models identify correlations between environmental factors and disease outbreaks, helping farmers anticipate risks like rust, blight, and root rot. Furthermore, unsupervised learning techniques, such as clustering algorithms, analyze field variability to delineate management zones for site-specific input applications. This ensures that fertilizers, pesticides, and irrigation are precisely tailored to the unique needs of each zone, reducing waste and minimizing environmental impact. Time-series analysis and recurrent neural networks (RNNs) enable predictive analytics in climate forecasting, providing short- and long-term weather predictions crucial for crop planning. For instance, predictive models can estimate rainfall patterns, frost events, and drought conditions, allowing farmers to adjust sowing and harvesting schedules. Additionally, reinforcement learning is used in automated machinery to refine decision-making in real-time, enabling tractors, drones, and robotic implements to adapt their actions based on changing field conditions.

IoT and Drones for Real-Time Data Collection

The integration of IoT (Internet of Things) devices and drones in agriculture has fundamentally transformed real-time data collection, providing farmers with unprecedented insights into their crops and fields. IoT devices, such as soil and plant sensors, environmental monitors, and automated machinery, are interconnected to form smart networks that continuously monitor and transmit crucial agricultural data. Soil sensors measure moisture levels, temperature, pH, and nutrient content, enabling farmers to implement precision irrigation and fertilization based on real-time soil conditions. For instance, capacitive soil moisture sensors detect water availability across different soil depths, while nitrate-selective ion sensors measure nutrient concentrations, providing a comprehensive picture of soil health. Environmental sensors monitor key climate variables like temperature, humidity, wind speed, and rainfall, offering predictive insights into crop weather impacts. Automated weather stations with IoT connectivity provide localized climate data, complementing broader regional forecasts. Plant sensors, including chlorophyll meters and spectral reflectance devices, assess leaf water content, chlorophyll levels, and growth stages, helping farmers identify early nutrient deficiencies and water stress. All these data are transmitted to cloud platforms, where advanced analytics provide actionable insights through user-friendly dashboards and alerts. Drones, or unmanned aerial vehicles (UAVs), enhance real-time data collection by providing high-resolution, multispectral imagery that reveals spatial variability across fields. Equipped with multispectral cameras, drones capture images in visible, near-infrared, and thermal bands, enabling the calculation of vegetation indices like NDVI (Normalized Difference Vegetation Index) and GNDVI (Green NDVI). These indices detect early signs of crop stress, disease outbreaks, and growth anomalies. Thermal drone cameras identify temperature variations in plant canopies, pinpointing areas of water stress and irrigation inefficiencies. Moreover, LiDAR-equipped drones create high-resolution elevation maps, aiding in field drainage planning and erosion control. Spraying drones further optimize crop management by precisely applying pesticides and fertilizers based on real-time imagery analysis. This ensures targeted treatment of affected zones, reducing chemical usage and environmental impact. Drones also facilitate early detection of pest infestations by identifying hotspots and guiding ground-based scouting efforts. IoT and drones have enabled a paradigm shift in agricultural data collection, delivering real-time, high-resolution insights that empower farmers to make informed decisions. By integrating these technologies with predictive analytics and intelligent systems, farmers can achieve precise resource management, mitigate risks, and enhance the overall sustainability of their operations.

From Patterns to Outputs: AI in Action

1. Descriptive Analytics: Identifying Patterns from Historical Data

Descriptive analytics is the foundation of data analysis, focused on summarizing and interpreting historical data to uncover patterns and trends. In agriculture, descriptive analytics helps farmers understand their past performance by analyzing variables such as crop yields, soil health, and weather conditions over multiple growing seasons. For instance, farmers can use descriptive analytics to identify patterns like consistent yield declines in certain field zones, recurring pest infestations, or trends in soil nutrient depletion. By integrating data from multiple sources such as IoT field sensors, satellite imagery, and farm management software, farmers can gain a comprehensive view of their historical performance. Crop yield data, collected through combine harvesters equipped with yield monitors, can be mapped spatially to identify low-yielding zones within fields. Soil sensor data provide insights into long-term changes in soil moisture, pH levels, and nutrient concentrations, highlighting areas of persistent nutrient depletion. Environmental sensors record temperature, humidity, and precipitation patterns, revealing climate trends that affect crop growth. Moreover, descriptive analytics leverages pest and disease monitoring data collected through pheromone traps, scouting records, and remote sensing to identify recurring infestations. For example, heatmaps and temporal charts can visualize the spread of pests like potato beetles or aphids, helping farmers pinpoint areas that require targeted interventions. Similarly, disease prevalence can be mapped to specific microclimates within fields, offering clues to conditions that favor disease development. Farmers can also analyze irrigation data to assess water usage efficiency and identify patterns of over- or under-irrigation. By correlating irrigation schedules with soil moisture data, farmers can evaluate the effectiveness of their irrigation systems and adjust schedules accordingly. Additionally, dashboards that visualize historical fertilizer and pesticide application rates against yield outcomes enable farmers to fine-tune input usage for maximum productivity. Ultimately, farmers gain insights into historical crop health, irrigation efficiency, and input usage by visualizing this data through graphs, heatmaps, and dashboards. These visualizations highlight the areas of improvement and enable farmers to implement evidence-based practices for enhanced productivity, such as adopting variable-rate technology, optimizing irrigation schedules, or adjusting planting densities. Descriptive analytics thus provides a crucial first step toward data-driven decision-making in agriculture.

2. Diagnostic Analytics: Finding Causes and Effects Behind Observed Patterns

While descriptive analytics answers "What happened?" diagnostic analytics delves deeper into "Why did this happen?" This involves uncovering the underlying causes behind observed patterns by analyzing relationships between variables. In crop management, diagnostic analytics can help explain why certain fields consistently underperform by correlating yield data with soil quality, weather patterns, and pest prevalence. To identify the root causes of underperformance, diagnostic analytics starts with collecting and integrating a wide range of data. Yield maps, soil health records, satellite imagery, and weather data are combined to analyze spatial and temporal relationships. For instance, a diagnostic analysis might reveal that declining yields in a particular zone are due to soil nutrient depletion and increased pest pressure. Soil nutrient depletion can be identified through historical soil testing data, while pest pressure is assessed by analyzing scouting records, remote sensing imagery, and pheromone trap data.

Advanced diagnostic models, such as regression analysis and decision tree algorithms, help uncover the complex interactions between multiple factors affecting crop yields. For example, regression models can correlate yield data with historical precipitation patterns, soil moisture levels, and fertilizer application rates to determine the impact of water stress or nutrient deficiencies on crop productivity. Decision tree algorithms can identify specific conditions under which crops are more susceptible to disease outbreaks, such as certain humidity ranges, soil temperatures, and planting densities. Similarly, diagnostic analytics can explain why certain crops are more susceptible to disease outbreaks based on historical planting schedules and environmental factors. By analyzing planting dates alongside climate data (e.g., temperature, humidity), and crop health indices, farmers can identify planting windows that minimize disease risk. For instance, late planting might coincide with peak humidity levels that favor fungal disease outbreaks. Furthermore, spectral imagery from drones or satellites can detect early signs of disease in crops, enabling diagnostic models to correlate these symptoms with specific microclimatic conditions within the field. Pest prevalence and migration patterns can also be analyzed using diagnostic analytics. Temporal analysis of scouting and pheromone trap data, combined with weather models, can reveal how pest populations respond to changing climate conditions. For instance, a sudden increase in aphid populations might be linked to an unusually warm spring, while the movement of certain beetles could correlate with shifts in wind patterns. Farmers can make informed decisions to rectify underlying issues and optimize their practices by understanding these cause-and-effect relationships. Corrective actions might include adjusting fertilizer application rates, modifying planting schedules, or implementing targeted pest control measures. Diagnostic analytics thus empowers farmers to move beyond symptomatic management to proactive and strategic interventions that enhance productivity and sustainability.

3. Predictive Analytics: Forecasting Future Scenarios Based on Current Data

Predictive analytics employs statistical models and machine learning techniques to anticipate future scenarios based on current and historical data. In agriculture, predictive analytics enables farmers to forecast crop yields, anticipate pest outbreaks, and predict the impacts of climate variability on production. Farmers can implement proactive strategies that reduce risks and optimize productivity by harnessing these insights.

Crop Yield Forecasting

Machine learning models such as Random Forests, Gradient Boosting Machines, and Artificial Neural Networks (ANNs) are widely used to forecast crop yields. These models are trained on historical data, including multispectral satellite imagery, soil sensor data, weather forecasts, and agronomic practices like planting density, irrigation schedules, and fertilizer application rates. For instance, satellite-based vegetation indices like NDVI (Normalized Difference Vegetation Index) and EVI (Enhanced Vegetation Index) provide insights into crop health, while soil moisture sensors track water availability. By incorporating weather forecasts and climate projections, machine learning models accurately predict yields. For example, a model can forecast wheat yields by analyzing soil moisture levels, NDVI values, and historical rainfall data, helping farmers make informed decisions about harvest timing and market positioning.

Pest and Disease Outbreak Prediction

Predictive analytics also plays a crucial role in forecasting pest and disease outbreaks. Machine learning models trained on historical pest population data, weather patterns, and crop phenology can anticipate pest emergence and migration. For example, degree-day models track insect development stages based on temperature data, predicting when pests like potato beetles or armyworms will reach damaging life stages. Similarly, disease models integrate weather forecasts with multispectral imagery to predict the likelihood of diseases like late blight or rust. By analyzing leaf wetness duration, humidity, and temperature data, these models provide early warnings to farmers, enabling timely pesticide applications or crop rotations to minimize damage.

Climate Impact Prediction

Understanding the impacts of climate variability on crop production is critical for long-term planning. Predictive models use historical climate data and future projections to estimate the effects of temperature changes, rainfall patterns, and extreme weather events on crops. For instance, time-series analysis and autoregressive models forecast drought conditions and frost events, allowing farmers to adjust sowing and irrigation schedules. Additionally, machine learning models can simulate the impact of rising temperatures on different crop varieties, identifying those that are more resilient to heat stress. These predictions help farmers choose suitable crop varieties and implement climate adaptation strategies, such as changing planting dates or investing in drought-tolerant seeds.

Resource Optimization and Market Positioning

Predictive analytics provides valuable insights into optimal resource management. Models that predict future yields and crop health can suggest precise irrigation and fertilization schedules, ensuring efficient resource use. For instance, predictive models can determine nitrogen fertilizer requirements based on predicted yield outcomes and soil nutrient status. Additionally, market forecasting models analyze global supply and demand trends, price movements, and trade policies to predict future commodity prices. By combining yield forecasts with price predictions, farmers can decide the best time to sell their produce and negotiate favorable contracts. In summary, predictive analytics allows farmers to anticipate challenges and opportunities, enabling proactive resource management, risk reduction, and strategic planning. Whether it's forecasting yields, preventing pest outbreaks, or adapting to climate change, predictive models are transforming agriculture into a data-driven, resilient, and sustainable industry.

Prescriptive Analytics: Suggesting Optimal Courses of Action

Prescriptive analytics takes predictive insights a step further by recommending specific actions to achieve desired outcomes. This involves using optimization algorithms and machine learning models to analyze current data, predict future events, and prescribe the best course of action. Prescriptive analytics helps farmers refine their irrigation schedules, fertilizer application rates, and pest management strategies in agriculture. Prescriptive analytics helps farmers refine their irrigation schedules, fertilizer application rates, and pest management strategies in agriculture. These models optimize irrigation by using soil moisture data, weather forecasts, and crop growth stages to generate precise schedules, while variable-rate fertilization recommendations are made by analyzing soil nutrient maps and predicting crop demands. They guide pest and disease management by suggesting optimal intervention timing and dosage based on predictive insights. Moreover, they assist in crop selection and rotation planning by evaluating soil health, market trends, and climate forecasts, and support supply chain optimization and market positioning through predictive yield forecasts and commodity price projections. Prescriptive analytics also improves machinery usage and labor management by analyzing field data and equipment capacity, providing tailored recommendations that empower farmers to navigate the complexities of modern agriculture with precision and confidence.

Conclusion

The integration of intelligent systems in sustainable agriculture marks a pivotal transformation in the sector, offering unprecedented opportunities to navigate the complexities of modern farming. Enhanced Prediction, driven by machine learning and predictive analytics, enables accurate forecasting of crop yields, pest outbreaks, and climate impacts. This predictive capability is instrumental in proactive resource management, reducing risks, and enhancing productivity. Evolutionary Learning ensures continuous improvement in farming practices by analyzing historical data to uncover cause-and-effect relationships, enabling farmers to refine their pest control, irrigation, and fertilization strategies. Intelligent systems are not just tools but strategic partners in agriculture's evolution. They provide farmers with the ability to predict, adapt, and optimize their practices amidst ever-changing environmental conditions, ensuring a future of sustainable, resilient, and innovative farming. By embracing these technologies, the agricultural sector can address global food security challenges and pave the way for a more sustainable relationship between humanity and the land.

Hossein Bonakdari
Department of Civil Engineering
University of Ottawa
Ottawa, ON, Canada
hbonakda@uottawa.ca

References

1. Araujo SO, Peres RS, Filipe L, Manta-Costa A, Lidon F, Ramalho JC, Barata J (2023) Intelligent data-driven decision support for agricultural systems-ID3SAS. IEEE Access, 11:115798–115815. 10.1109/ACCESS.2023.3324813
2. Bai X, Huang Y, Ren W, Coyne M, Jacinthe P-A, Tao B, Hui D, Yang J, Matocha C (2019) Responses of soil carbon sequestration to climate-smart agriculture practices: A meta-analysis. Global change biology, 25(8):2591–2606. 10.1111/gcb.14658
3. Charania I, Li X (2020) Smart farming: Agriculture's shift from a labor intensive to technology native industry, Internet of Things (Netherlands), 9, art. no. 100142. 10.1016/j.iot.2019.100142
4. Contasti AL, Firth AG, Baker BH, Brooks JP, Locke MA, Morin DJ (2023) Balancing tradeoffs in climate-smart agriculture: will selling carbon credits offset potential losses in the net yield income of small-scale soybean (Glycine max L.) producers in the mid-southern United States? Decision analysis, 20(4):252–275. 10.1287/deca.2023.0478
5. Feng P, Wang B, Liu DL, Waters C, Xiao D, Shi L, Yu Q (2020) Dynamic wheat yield forecasts are improved by a hybrid approach using a biophysical model and machine learning technique. Agricultural and forest meteorology, 285–286, 107922. 10.1016/j.agrformet.2020.107922
6. Gong Y, Li P, Sakagami N, Komatsuzaki M (2021) No-tillage with rye cover crop can reduce net global warming potential and yield-scaled global warming potential in the long-term organic soybean field. Soil and tillage research, 205, 104747. 10.1016/j.still.2020.104747
7. Grenon G, Madramootoo CA, von Sperber C, Ebtehaj I, Bonakdari H, Singh B Nutrient release in drainage discharge from organic soils under two different agricultural water management systems. (2023) Hydrological Processes, 37, pp e14953. 10.1002/hyp.14953
8. Karataeva OG, Pekalski MI, Pekalski DI (2021) Development of intellectual agriculture in modern economic conditions, Lecture Notes in Networks and Systems, 205, pp 149–153. 10.1007/978-3-030-73097-0_18
9. Khanna R, Yadav AL (2023) DMS: A Necessity in Agricultural Planning. 14th International Conference on Computing Communication and Networking Technologies, ICCCNT 2023. 10.1109/ICCCNT56998.2023.10306573
10. Kumari S, Muthulakshmi P (2024) Predicting weather conditions using machine learning for improving crop production. AI Applications for Business, Medical, and Agricultural Sustainability, pp 267–302. 10.4018/979-8-3693-5266-3.ch010
11. Mall RK, Gupta A, Sonkar G (2017) Effect of Climate Change on Agricultural Crops. Current Developments in Biotechnology and Bioengineering: Crop Modification, Nutrition, and Food Production, pp 23–46. 10.1016/B978-0-444-63661-4.00002-5

12. Mason RE, Merrill SC, Görres J, Faulkner J, Niles MT (2021) Agronomic and environmental performance of dairy farms in a warmer, wetter climate. Journal of Soil and Water Conservation, 76, pp 76–88. 10.2489/jswc.2021.00169
13. Mishra A, Kim S (2021) A Comprehensive Survey on AgriTech to Pioneer the HCI-Based Future of Farming. Lecture Notes in Computer Science (including subseries Lecture Notes in Artificial Intelligence and Lecture Notes in Bioinformatics), 14531 LNCS, pp 310–325. 10.1007/978-3-031-53827-8_28
14. Montalvo-Romero N, Montiel-Rosales A, Purroy-Vasquez R, Quechulpa-Perez P (2023) Agro-Technological Systems in Traditional Agriculture Assistance: A Systematic Review. IEEE Access, 11, pp 123047–123069, 10.1109/ACCESS.2023.3329087
15. Savita, Vimal V (2023) Integrating IoT-Based Environmental Monitoring and Data Analytics for Crop-Specific Smart Agriculture Management: A Multivariate Analysis. Proceedings—International Conference on Technological Advancements in Computational Sciences, ICTACS 2023, pp 368–373. 10.1109/ICTACS59847.2023.10390277
16. Soltani K, Ebtehaj I, Amiri A, Azari A, Gharabaghi B, Bonakdari H (2021) Mapping the spatial and temporal variability of flood susceptibility using remotely sensed normalized difference vegetation index and the forecasted changes in the future. Science of the Total Environment, 770, pp 145288. 10.1016/j.scitotenv.2021.145288
17. Szejgis J, Nielsen UN, Dijkstra FA, Carrillo Y (2024) Prolonged drought moderates flood effects on soil nutrient pools across a rainfall gradient. Soil Biology and Biochemistry, 193, 109404. 10.1016/j.soilbio.2024.109404
18. Ullah H, Santiago-Arenas R, Ferdous Z, Attia A, Datta A (2019) Improving water use efficiency, nitrogen use efficiency, and radiation use efficiency in field crops under drought stress: A review. Advances in Agronomy, 156, pp 109–157. 10.1016/bs.agron.2019.02.002
19. Zeynoddin M, Ebtehaj I, Bonakdari H (2020) Development of a linear based stochastic model for daily soil temperature prediction: One step forward to sustainable agriculture. Computers and Electronics in Agriculture, 176, pp 105636. 10.1016/j.compag.2020.105636
20. Zeynoddin M, Gumiere SG, Bonakdari H (2023) Enhancing water use efficiency in precision irrigation: data-driven approaches for addressing data gaps in time series. Frontiers in Water, 5, pp 1237592. 10.3389/frwa.2023.1237592
21. Zhu C (2024) Research On Data Analysis and Decision Support of Artificial Intelligence Technology in Intelligent Agricultural System, Journal of Logistics, Informatics and Service Science, 11, pp 429–443. 10.33168/JLISS.2024.0128

Contents

Intelligent Systems for Disease Prediction and Assessment-I

Disease Detection in Grape Leaves Using Convolutional Neural Networks

Varsha Naik, Trupti Baraskar, Viren Khatri, Parimal Kolhe, and Mitheelesh Katyarmal

Abstract Grape cultivation is vital to the global economy, but various diseases threaten grapevine health and yield. Early and accurate disease detection is crucial for implementing effective control measures. Traditional methods rely on visual inspection by experts, which is subjective and time-consuming. Automated approaches using deep learning offer a promising alternative. The goal of this study was to create a convolutional neural network (CNN) model that can identify grapevine leaf diseases reliably. The aim of the study was to train the model for distinguishing between healthy and infected leaves by three diseases that are common in them such as black rot, esca, and leaf blight. We collected a dataset of images containing healthy and diseased grape leaves. The dataset was pre-processed and augmented to increase data diversity and ensure consistency. The CNN model was developed using TensorFlow and Keras and consisted of convolutional, pooling, dense, and activation layers. Optimizations were implemented using cross-entropy loss function, adam optimizer, and Dropout layers. Model performance was evaluated using metrics such as accuracy of the model, the precision at which it cannot wrongly label a negative sample, the recall at which it can find all the positive outputs, and F1-score which is the ratio between precision and recall. The developed CNN model achieved high accuracy in disease identification. We observed significant accuracy, precision, F1-score, and recall values across all disease categories, demonstrating the model's effectiveness in distinguishing between healthy and diseased leaves. This study successfully demonstrates the potential of CNNs for automated grape leaf disease identification. The developed model offers a promising tool for viticulturists to improve early disease detection and ultimately enhance grape yield and quality.

V. Naik (✉) · T. Baraskar · V. Khatri · P. Kolhe · M. Katyarmal
School of Computer Engineering and Technology, MIT World Peace University, Dr. Vishwanath Karad, Pune, Maharashtra 411038, India
e-mail: varsha.powar@mitwpu.edu.in

T. Baraskar
e-mail: trupti.baraskar@mitwpu.edu.in

V. Khatri
e-mail: vk102002@gmail.com

A. J. Kulkarni and M. Hajiaghaei-Keshteli (eds.), *Advances in Intelligent Systems for Sustainable Agriculture*, Lecture Notes in Networks and Systems 1201,
https://doi.org/10.1007/978-981-97-9839-1_1

Future research could focus on expanding the model's capabilities to identify a wider range of diseases and exploring its potential for real-time applications in vineyards.

Keywords Grape leaf disease · CNN · Deep learning · Disease identification · Viticulture

1 Introduction

Grapevines are a vital agricultural crop worldwide, contributing significantly to the global economy and cultural landscape. The fruit they produce is not only enjoyed fresh but also transformed into various products such as wine, raisins, and juice. However, the health and productivity of grapevines are constantly threatened by various diseases that can significantly impact yield and quality. These diseases can cause significant damage to leaves, stems, and fruits, leading to reduced production, economic losses, and even vineyard abandonment in severe cases.

Traditionally, grapevine disease detection has relied heavily on visual inspection by experienced professionals. However, this approach suffers from several limitations. It is often subjective, inconsistent, and time-consuming, especially when dealing with large vineyards and numerous vines. Additionally, accurate diagnosis requires expertise and familiarity with various disease symptoms, which can lead to misdiagnoses and delayed responses. Furthermore, visual inspection becomes increasingly difficult in early stages of infection when symptoms might be subtle or absent.

The recent advancements in the artificial intelligence sector, particularly deep learning, have ushered in a new era for precision agriculture. Deep learning algorithms [1] have demonstrated remarkable capabilities in various image recognition tasks, including plant disease identification. Convolutional neural networks (CNNs) [2], a specific type of deep learning architecture, have proven particularly successful in this domain due to their ability to extract relevant features from images and learn complex relationships between features and disease labels. Utilizing deep learning for grapevine disease detection offers several advantages over traditional methods. Deep learning models can potentially:

- Achieve high accuracy and consistency: By analysing large datasets of diseased and healthy leaf images, deep learning models can learn to identify subtle disease symptoms with high accuracy and consistency, overcoming the subjectivity and limitations of human-based inspection.
- Enable early and rapid disease detection: Deep learning models can be trained to detect diseases at early stages, even before visible symptoms appear, allowing for prompt implementation of control measures and minimizing potential losses.
- Automate the detection process: Deep learning models can automate the disease detection process, significantly reducing the time and labour required compared to manual inspection, freeing up resources for other vineyard management tasks.

- Improve disease management strategies: By providing early and accurate disease information, deep learning models can assist growers in implementing targeted disease management strategies, optimizing resource allocation, and maximizing vineyard productivity.
- Reduce the need for expert knowledge: Deep learning models eliminate the need for extensive expertise in disease diagnosis, making disease detection accessible to a wider range of individuals and organizations involved in grapevine cultivation.

This study investigates the development and evaluation of a deep learning-based model for automated grapevine disease identification. The model is designed to identify and classify three common grapevine diseases: black rot, esca, and leaf blight, based on images of affected leaves. This study explores the potential of deep learning to revolutionize grapevine disease detection and contribute to improved vineyard management practices.

2 Literature Survey

This [3] paper presents a comprehensive solution for the detection and classification of grape leaf diseases in India, a critical issue affecting fruit productivity. Leveraging image processing techniques and Support Vector Machine (SVM) classification [4], the study employs K-means clustering [5] for segmentation, extracting colour and texture features to accurately identify and classify diseases. The proposed system achieves an overall accuracy of 88.89%, with notable success in identifying Downy-class grape leaves. Comparative analysis with a previous system demonstrates improved performance, emphasizing the efficacy of the proposed approach in addressing the challenges of grape disease management.

This [6] paper aims to improve the classification of grape leaf diseases using multiclass SVM. The system uses image pre-processing techniques such as resizing, enhancement, and smoothing, and transforms the image into both LAB and HSI colour models. Firstly, the disease area is segmented using K-mean clustering and subsequent extraction of texture and colour features with GLCM method and standard MATLAB functions respectively. Afterwards, the multiclass SVM classifiers are trained with these features into either healthy or diseased classes. The system achieves 90% average accuracy for all three-diseased leaf and healthy leaf compared to 82.5% in the case of taking only LAB colour model for feature extraction.

Four modified deep learning models are developed in this [7] research project for grape leaf disease detection and classification on a developed grape leaf dataset. Targeted diseases include black rot, black measles, leaf blight and phylloxera. Transfer learning has been done with three pre-trained machine learning models (VGG16 [8], MobileNet [9], and AlexNet [10]). According to the reported comparative evaluation results, the accuracy is better compared to the pre-trained models as well as performance improvement. An ensemble model that combines these four developed models enhances final detection and classification accuracy achieving

100% accuracy for grape disease and bug detection. This proposed model can be easily extended to be used widely in the grapevine industry. The authors will further develop a real-time smart farming system based on camera and drone data plus remote sensing for supporting real-time multi-dimensional disease detection and classification by using their developed models, and ensemble model for future smart agriculture/farming.

The proposed system in [11] aims to effectively detect grape diseases through leaf feature inspection, utilizing photographic images for disease recognition. The algorithm is tested on three diseases affecting grape plants: Downy Mildew, Black rot, and Powdery mildew. The experimental outcomes demonstrate the effectiveness of the approach in accurately detecting leaf illnesses with minimal computational effort. The authors plan to extend this work by developing hybrid algorithms and automatically estimating the severity of the detected disease in the future. The research project focuses on the development of an improved CNN model for diagnosing grape leaf diseases. The study collected a large dataset of grape leaf images, including various diseases and healthy leaves, and utilized deep learning techniques for disease detection and classification. The results obtained in this research show that real time diagnosis or detection of grape leaves diseases is possible. In addition, the accuracy and performance of the developed model is quite high. Farmers suffer from reduced profits and low yields due to the occurrence of leaf diseases in grapes. The aim of this paper is to identify and categorize grape leaf diseases using a Convolutional Neural Network (CNN). In the identification and classification of grape leaf diseases, the accuracy rate of the CNN model was 91.37%, indicating its suitability for effective disease diagnosis and management.

For detecting diseases in grapevines, an automatic system is proposed in [12]. It uses image processing and machine learning techniques in segmenting and identifying diseased regions on grape leaves. The system achieves a testing accuracy of 93% using global thresholding and Support Vector Machine (SVM) for classification. By effectively extracting texture and colour features from the identified diseased parts, the system demonstrates the potential for accurate disease detection with minimal computational effort. The use of machine learning techniques such as SVM, random forest, and Adaboost for classification proves to be successful, with the system showing promise for practical application in the detection and control of grape leaf diseases.

This paper presents a real-time detector for grape leaf diseases using an improved deep convolutional neural network (CNN) [13] called Faster DR-IACNN [14]. The system utilizes digital image processing technology to expand the grape leaf disease dataset (GLDD) and constructs a high-performance detection model for grape leaf diseases. The proposed Enhanced DR-IACNN model employs the Inception-v1 module, Inception-ResNet-v2 module, and SE-blocks to enhance feature extraction capabilities. The detection model achieves a precision of 81.1% mAP on GLDD and a detection speed of 15.01 FPS, demonstrating the potential of the Faster DR-IACNN model for real-time grape leaf disease detection [15].

```
Count of Images in each Class of Training dataset -

Total number of images in Healthy class: 1692
Total number of images in Leaf Blight class: 1722
Total number of images in Black Rot class: 1888
Total number of images in ESCA class: 1920
```

Fig. 1 Data visualization

3 Methodology

3.1 Data Collection

In this research, we utilized a dataset readily available on Kaggle [16], a renowned platform for data science enthusiasts and researchers. This dataset encompasses a wealth of grape leaf images categorized into four distinct classes: healthy leaves, those infected with black rot, leaves suffering from leaf blight, and those affected by esca disease.

Figure 1 provides an example distribution of the images across the four classes within the dataset. As evident, the number of images per class is relatively balanced, minimizing potential biases and promoting fair learning during the training process.

3.2 Data Pre-Processing

To effectively train our CNN model for grape leaf disease detection, we established a robust data pipeline capable of handling a large number of images. This pipeline involved meticulous pre-processing steps to ensure the model received high-quality and standardized data for optimal learning.

One crucial step was image scaling and resizing. We applied a scaling factor of 1/255 to normalize pixel values between 0 and 1, allowing the model to efficiently process and learn from the numeric data. Additionally, we resized all images to a uniform size of 256 × 256 pixels, ensuring consistency in input dimension and facilitating efficient processing by the CNN.

To further enhance the training process and prevent overfitting, we implemented data augmentation [17]. This technique artificially increases the size and diversity of the dataset by applying various transformations to the existing images. In our pipeline, we employed techniques such as:

- Width and height shifts: Randomly shifting the image horizontally and vertically to introduce spatial variation.
- Shearing: Tilting the image slightly to add perspective distortion.
- Zooming: Scaling the image up or down to simulate different viewing distances.
- Horizontal flipping: Mirroring the image to increase variability.
- Brightness adjustment: Modifying image brightness to introduce lighting variations.
- Channel shifting: Slightly altering the colour channels to increase robustness to colour variations.

These augmentation techniques provided the model with a broader spectrum of image features to learn from, enhancing its generalizability and reducing its susceptibility to overfitting on the specific training data.

Furthermore, we incorporated a background removal step. This step involved utilizing a pre-trained model [18] to automatically remove the background from the images, leaving only the relevant leaf area for analysis. This process focused the model's attention on the diseased regions and reduced the impact of irrelevant background information, thereby improving the accuracy and efficiency of disease detection.

To optimize the training process and ensure unbiased evaluation, we split our data into three distinct partitions:

- Training set (80%): Used for training the model and learn the disease recognition patterns.
- Validation set (10%): Used for monitoring the model's performance during training and fine-tuning its hyperparameters for optimal results.
- Testing set (10%): Used to evaluate the model's final performance on unseen data and assess its generalizability to real-world scenarios.

Finally, we performed label encoding to convert the class labels (healthy, black rot, leaf blight, esca) into numerical representations understandable by the machine learning model. Additionally, we normalized the pixel values of the images to ensure numerical stability and facilitate efficient computation within the model.

By implementing these comprehensive data pre-processing steps, we prepared a high-quality and standardized dataset for training our CNN model. This meticulous approach ensures the model receives the optimal data it needs to learn and perform accurate grape leaf disease detection. Figure 2: Visualization of the data pre-processing pipeline.

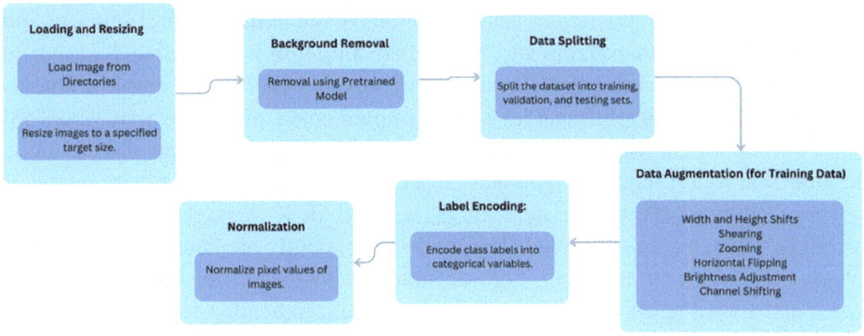

Fig. 2 Data Pre-processing pipeline

3.3 Model Development

This section discusses the model development and refinement of the Convolutional Neural Network (CNN) model for automated grape leaf disease detection. Our methodology was to distinguish healthy grape leaves from those affected by the diseases black rot, leaf blight, and esca.

The initial CNN architecture (Iteration 1) consisted of 3 convolutional layers with ReLU activation followed by max pooling layer trained on a dataset of 7222 images with balanced representation of healthy and diseased leaves. Numerical results of the model appeared promising, achieving initial accuracy of 97.89%, however, when tasked with classifying individual leaf images, it consistently failed. This discrepancy between training and real-world performance highlighted the model's overfitting and lack of generalizability.

To address the limitations and shortcomings of the model in iteration 1, two key improvements were done in iteration 2. Firstly, Keras built-in hyper tuning tool was utilized to automate the selection of optimal parameters for the CNN model. This eliminated human bias and ensured a more comprehensive search for the best configuration. Secondly, the model complexity was increased with a deeper architecture with six convolutional layers, allowing the model to extract richer features and improve its discrimination capabilities.

Despite these improvements in iteration 2, the model still struggled with generalizability, exhibiting similar limitations to the initial model. Although the test accuracy was increased to 99% with a low loss of 2.8% still the classification report revealed low F1-score, recall, and precision values, indicating poor performance on unseen data. These findings suggest that further modifications are necessary to overcome the model's overfitting and enhance its generalizability to correctly identify different diseases. The use of data augmentation to broaden the diversity of datasets, dropout layers to avoid overfitting, early stopping to end training when validation accuracy reaches a plateau, transfer learning to extract features, and regularization methods like L1 or L2 regularization were all noted as potential directions for future development.

Recognizing the impracticality and limitations of the initial models for real-world applications due to their failure to generalize, Iteration 3 was more focused on reducing model complexity and incorporating domain expertise. A simplified architecture with fewer convolutional layers was implemented, alongside active incorporation of domain knowledge about grape leaf diseases. Emphasis was placed on the classification report rather than numerical accuracy, guiding model refinement.

These adjustments in the iteration 3 resulted in a marked improvement in overall performance of the model. The model consistently predicted correct disease categories for new images with high accuracy of 99.44%. Metrics like accuracy, recall, and F1-score all showed significant progress, indicating robust and reliable performance. Minimal misclassifications further solidified the model's effectiveness.

As shown in Fig. 3, the model architecture described here shows a convolutional neural network (CNN) intended for automated grape leaf disease detection. After the model receives an image of a grape leaf as input, it uses several convolutional layers to extract important information. These layers make use of filters that move over the picture, picking out patterns like edges, textures, and shapes that are important for classifying diseases. A max pooling layer lowers the spatial dimensionality of the retrieved features after every convolutional operation. After being flattened, the feature vector moves through fully connected dense layers, where complex feature interactions are discovered. The last layer ultimately produces three values, which are likely to represent the three disease categories that are being targeted (black rot, esca, and leaf blight). Throughout the network, activation functions such as ReLU are used to create non-linearity and improve the model's ability to learn complex patterns. The model's architecture suggests that it places a high priority on extracting spatial features from the images of grape leaves, specifically concentrating on texture and colour distribution to distinguish between leaves that are healthy and those that are diseased.

4 Results

This study aimed to develop a reliable and automated method for grape leaf disease detection using convolutional neural networks (CNNs). In Iteration 1, the model achieved high numerical accuracy (97.89%) during training but encountered challenges in generalization to unseen data, displaying poor performance, particularly in single image predictions. The classification report from this iteration revealed consistently low scores across all metrics, indicating significant room for improvement.

Iteration 2 included various improvements in response to the shortcomings noted in the previous iteration. Six convolutional layers were added to the model, increasing its complexity, and Keras hyperparameter tweaking was used to optimize the model's parameters. The model continued to have issues with generalizability and correct image classification even though it maintained high accuracy on the training set.

Fig. 3 Model architecture

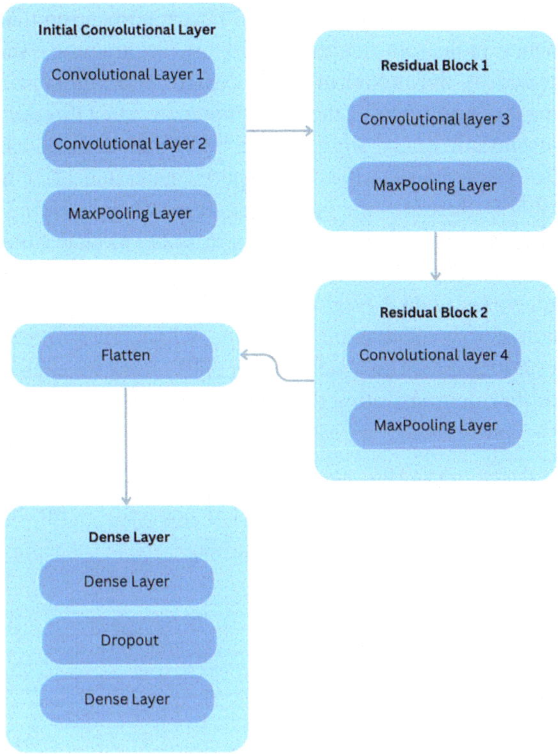

Table 1 illustrates the classification reports generated after model training of iteration 2.

Iteration 3 concentrated on streamlining the architecture and lowering model complexity in response to the ongoing difficulties, actively incorporating domain knowledge into the development process. Significant improvements in performance

Table 1 Classification report of iteration 2

	Precision	Recall	F1-score	Support
Black rot	0.28	0.27	0.27	235
ESCA	0.25	0.25	0.25	240
Healthy	0.20	0.20	0.20	212
Leaf blight	0.22	0.23	0.22	215
Accuracy			0.24	902
Macro Avg	0.24	0.24	0.24	902
Weighted Avg	0.24	0.24	0.24	902

and generalization resulted from these modifications. A number of noteworthy performance indicators, including support, accuracy, F1-score, and recall, were demonstrated by the model in its successful and highly accurate classification of single images. Additionally, the confusion matrix that this iteration created showed very few misclassifications, suggesting a good degree of disease detection efficacy. Table 2 illustrates the classification reports generated after model training of iteration 3. Table 3 illustrates the parameters used while building the model. After the training of the model a confusion matrix was also plotted which can be seen in Fig. 4.

Promising results are seen in the confusion matrix Fig. 4 for the three disease categories of leaf blight, esca, and black rot. There is a high degree of accuracy indicated by the diagonal elements, which represent accurately anticipated cases and are noticeable. The True Positive (TP) scores for each disease provide more evidence for this, demonstrating the model's accuracy in identifying infected leaves. Detailed Analysis:

- Black rot—The model shows good accuracy for black rot identification, with 231 cases accurately predicted and only 4 misclassifications. Notably, TN =

Table 2 Classification report of iteration 3

	Precision	Recall	F1-score	Support
Black rot	1.00	0.92	0.96	472
ESCA	0.93	1.00	0.96	480
Healthy	1.00	1.00	1.00	423
Leaf blight	1.00	1.00	1.00	430
Accuracy			0.98	1805
Macro Avg	0.98	0.98	0.98	1805
Weighted Avg	0.98	0.98	0.98	1805

Table 3 Best model parameters

Units (Input layer)	32
Units (Conv2D_1)	32
Units (Conv2D_2)	64
Units (Conv2D_3)	128
Activation	ReLU
MaxPooling	MaxPooling (2 × 2)
Units (Output layer)	4
Optimizer	Adam
Loss function	Categorical crossentropy
Batch size	32
Epochs	16
Activation (Output Layer)	Softmax

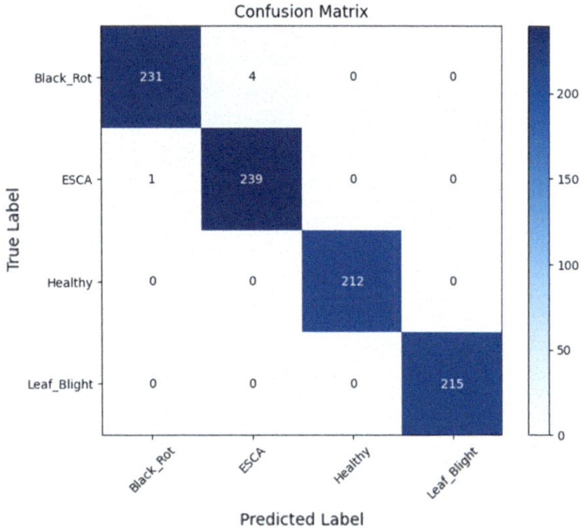

Fig. 4 Confusion matrix of iteration 3

200 revealed that no healthy leaves were mistakenly diagnosed as black rot, demonstrating the disease's great specificity.

- Esca—With 239 accurate predictions and just one incorrectly identified healthy leaf, the algorithm also performs admirably for esca. The TP score of 239 indicates that the model well captures the unique characteristics of leaves infected with esca.
- Leaf blight—As with the other diseases, there are 215 accurate predictions for leaf blight and just 0 incorrect classifications of healthy leaves, indicating good detection accuracy. This strengthens the model's capacity to distinguish between leaves that are diseased and those that are healthy.

While there are a few instances of misclassification, the model misidentified 4 black rot cases as esca and 1 esca case as black rot. Similarly, it mistook 1 healthy leaf for leaf blight. These minor errors could be due to subtle variations in disease presentation or limitations in the training data. But the overall performance of the model is encouraging.

5 Conclusion

This study has successfully developed a Convolutional Neural Network (CNN) model for the automatic detection of diseases in grape leaves. Notably, the model achieved an impressive 99.45% validation accuracy, showcasing its efficacy in accurately distinguishing between ESCA disease, Black Rot disease, Leaf Blight disease, and Healthy Leaves. These results emphasize the potential of CNNs in categorising grape leaf diseases, offering substantial promise for enhanced disease management and optimised yield in grape vineyards.

Emphasis is placed on achieving a balanced model, as a mere increase in the number of layers does not always translate to improved classification results. The difficulties lie in finding an equilibrium between model complexity and accuracy. Therefore, the construction of an ideal CNN model necessitates meticulous architectural consideration. Superiority in CNN models are highly dependent on not solely on the quantity of layers but by skilfully producing models which navigate the fine lines between under-fitting and overfitting.

6 Future Scope

This research proposes a multifaceted approach to empower farmers with a robust and informative crop disease detection tool. By integrating the model with Internet of Things (IoT) devices and mobile applications, real-time disease monitoring becomes a reality, enabling farmers to receive immediate notifications and take timely action against outbreaks. Expanding the model's capabilities to simultaneously identify multiple diseases addresses the crucial challenge of complex disease combinations, often faced by farmers. Further refinement to assess disease severity would provide invaluable insights for targeted treatment strategies, optimizing resource allocation and minimizing yield losses.

Enhancing the model's generalizability and robustness is crucial for real-world applicability. This can be achieved through leveraging transfer learning from pre-trained models and investigating model ensembles. Additionally, active learning strategies and exploring advanced data augmentation techniques can optimize data utilization and enhance model performance, addressing potential data scarcity and bias.

Building trust and transparency is paramount for widespread adoption. Developing methods to explain the model's predictions (Explainable AI) empowers agricultural professionals to understand its decision-making process. This fosters trust and enables them to tailor its use for their specific needs, ultimately leading to better informed decision-making. By implementing these synergistic advancements, this research strives to equip farmers with a powerful tool for proactive disease management, ultimately contributing to improved agricultural productivity and sustainability.

References

1. Sarker IH (2021) Deep learning: A comprehensive overview on techniques, taxonomy, applications and research directions. SN Comput Sci 2:420. https://doi.org/10.1007/s42979-021-00815-1
2. Yamashita R, Nishio M, Do RKG et al (2018) Convolutional neural networks: an overview and application in radiology. Insights Imaging 9:611–629. https://doi.org/10.1007/s13244-018-0639-9

3. Padol PB, Yadav AA (2016) SVM classifier based grape leaf disease detection. In: 2016 Conference on Advances in Signal Processing (CASP), Pune, India, pp 175–179. https://doi.org/10.1109/CASP.2016.7746160.
4. Theodoros E, Massimiliano P (2001) Support vector machines: Theory and applications. 2049:249–257. https://doi.org/10.1007/3-540-44673-7_12
5. Li Y, Wu H (2012) A clustering method based on k-means algorithm. Phys Procedia 25:1104–1109. https://doi.org/10.1016/j.phpro.2012.03.206
6. Agrawal N, Singhai J, Agarwal DK (2017) Grape leaf disease detection and classification using multi-class support vector machine. In: 2017 International Conference on Recent Innovations in Signal processing and Embedded Systems (RISE), Bhopal, India, pp 238–244. https://doi.org/10.1109/RISE.2017.8378160.
7. Huang Z, Qin A, Lu J, Menon A, Gao J (2020) Grape leaf disease detection and classification using machine learning. In: 2020 International Conferences on Internet of Things (iThings) and IEEE Green Computing and Communications (GreenCom) and IEEE Cyber, Physical and Social Computing (CPSCom) and IEEE Smart Data (SmartData) and IEEE Congress on Cybermatics (Cybermatics), Rhodes, Greece, pp 870-877. https://doi.org/10.1109/iThings-GreenCom-CPSCom-SmartData-Cybermatics50389.2020.00150
8. Srikanth T (2019) Transfer learning using VGG-16 with deep convolutional neural network for classifying images. Int J Sci Res Publ (IJSRP) 9:9420. https://doi.org/10.29322/IJSRP.9.10.2019.p9420
9. Andrew H, Menglong Z, Bo C, Dmitry K, Weijun W, Tobias W, Marco A, Hartwig A (2017) MobileNets: Efficient convolutional neural networks for mobile vision applications
10. Singh I, Goyal G, Chandel A (2022) AlexNet architecture based convolutional neural network for toxic comments classification. J King Saud University—Computer Inf Sci-Es 34(9):7547–7558. ISSN 1319–1578. https://doi.org/10.1016/j.jksuci.2022.06.007
11. Narvekar PR, Kumbhar MM, Patil SN (2014) Grape leaf diseases detection & analysis using SGDM matrix method. Int J Innov Res Comput Com-Munication Eng 2(3):3365
12. Jaisakthi SM, Mirunalini P, Thenmozhi D, Vatsala (2019) Grape leaf disease identification using machine learning techniques. In: 2019 International Conference on Computational Intelligence in Data Science (ICCIDS), Chennai, India, pp 1–6. https://doi.org/10.1109/ICCIDS.2019.8862084.
13. Zhaohua H, Ally Q, Jingshu L, Aparna M, Jerry G (2020) Grape leaf disease classification
14. Rahul C, Ghanshala KK, Joshi RC (2018) Convolutional neural network (CNN) for image detection and recognition. In: 2018 first international conference on secure cyber computing and communication (ICSCCC). IEEE, 2018.
15. https://www.kaggle.com/datasets/rm1000/grape-disease-dataset-original
16. Shorten C, Khoshgoftaar TM (2019) A survey on image data augmentation for deep learning. J Big Data 6:60. https://doi.org/10.1186/s40537-019-0197-0
17. https://github.com/danielgatis/rembg
18. Rune B, Pål H, Michael R, Håkon S (2018) Automatic Hyperparameter Optimization in Keras for the MediaEval 2018 Medico Multimedia Task

Leaf Disease Detection Using Fuzzy Edge Detection and Clustering

Vrishani Shah, Mustafa Poonawala, Apoorva Shastri, and Sujit Naik

Abstract Sustainable agriculture plays an important role in the economy and for the environment. The introduction of natural methods to maintain plant health and prevent it from harm from differing climate and weather has become imperative. Due to the diseases spreading in plants, chemical fertilizers and pesticides are used which deplete soil nutrients which cause the later crop yield to deteriorate. We propose the utilization of Fuzzy Logic in Edge Detection and Clustering using K-Means and Density Based clustering of image pixels to detect the disease-ridden areas for more accurate usage of pesticides. The disease-ridden areas are distinct when edges detection is applied. These edges are then detected by clustering and conveyed using anchor boxes. By applying the chemicals only to disease-ridden areas, the objective of removing the pathogens and curing the disease based on its spread will help in unnecessary application of chemicals to the soil and help in promoting better plant and soil health. This helps in promoting a sustainable method of agriculture and helping farmers in preventing a disease outbreak in large farms by early detection and cure of diseases.

Keywords Leaf disease · Sustainable agriculture · Fuzzy logic · Edge detection · Clustering

V. Shah (✉) · M. Poonawala · A. Shastri
Institute of Artificial Intelligence, MIT World Peace University, Pune, Maharashtra, India
e-mail: vrishanishah20@gmail.com

A. Shastri
e-mail: apoorva.shastri@mitwpu.edu.in

S. Naik
Agribiosis Pvt. Ltd., 004 Daisy Nyati Meadows, Kalyani Nagar, Pune, Maharashtra, India

A. J. Kulkarni and M. Hajiaghaei-Keshteli (eds.), *Advances in Intelligent Systems for Sustainable Agriculture*, Lecture Notes in Networks and Systems 1201,
https://doi.org/10.1007/978-981-97-9839-1_2

1 Introduction

Sustainable agriculture is the production of high-quality crops, resource protection and environment-friendly practices simultaneously being profitable. Conventional agricultural techniques began causing harm to soil productivity and the quality of the crops growing on this soil. This was mainly due to the rush for chemical fertilizers and pesticides. There was a doubt created on the reliability of these techniques. Many alternative methods were sought after which would make the practice more sustainable. Instead of relying on chemicals, more natural methods were considered which would curb the spread of disease and pests while also not causing harm to the resources available for the crops [1].

The industrial model of agriculture treats farms like a factory, considering the plants and animals only as means for production and profit. These methods which may have caused economic growth now to cause major environmental and economic concerns. Sustainable agriculture maintains its productivity and protects the environment both. The reduction of methods which harm the environment, more dependency on natural resources, greater usage of the plant genetics and species, finding patterns in crops and agricultural lands to make conclusions and better resource management are the crux of sustainable agriculture [2].

Plant disease affects the growth and yield of crops. Early and accurate detection helps in maintaining health and curbing the spread of this infection to neighboring leaves and plants. This helps in better yield for the farmer and prevents large economic losses for the farmer. Traditional methods where humans detect these diseases are not feasible and very time consuming, especially in large farms where a large number of crops of different varieties grow [3].

In this paper we identify disease affected areas on plant leaves using Fuzzy Edge Detection and K-Means and Density Based Clustering. Applying Fuzzy Edge detection gives an idea of where the edges are denser which gives information on the areas of the leaf which may have been affected by a disease or pest. The clustering then puts anchor boxes on that area of the plant where the disease occurs. This can help the farmer identify the areas and cure only the affected areas. It helps in the controlled usage of pesticides and chemicals, in turn saving on soil health and plant crops.

2 Literature Survey

In their paper, authors in [4] emphasize the importance of detecting plant and leaf diseases early and the impact it has on agriculture. The growth of disease-free plants is imperative for survival and productivity of plants in large farms and fields. The paper proposes using biosensor methods to detect diseases occurring in fields. Early diagnosis plays an important role in sustainable agriculture, controlling the growth of disease pathogens in the fields and the management of greenhouse systems. It also decreases the emission of harmful pathogen associated diseases and prevents the need for quarantine levels in farms.

As mentioned by Huy-Tan [5], agriculture suffers due to factors like climate change and extreme weather conditions. The importance of the difference made by agriculture in the general economy of the world is very huge and makes a significant difference. There have been many drawbacks in the general approach due to which Artificial Intelligence methods have been introduced to curb the disease spread. These models can be computationally expensive and have around 90% accuracy. The authors have taken a group of data from multiple sources and created a centralized platform. They have also tried to create a lightweight AI model to address this issue.

In the paper, authors in [6] mention the importance of agriculture for survival. It talks about the importance of monitoring plant disease to prevent the spread of the disease and the need for constant detection in large scale farms where the spread can take place very quickly causing harm to the crop yield. To promote a more sustainable and less wasteful method of agriculture, early onset of the disease needs to be detected for measure to be taken. Image based information has proved to be useful in detecting diseases. The authors use CNN to detect early plant diseases. The main objectives of using CNN and detecting varying plant diseases which differ in color, size, shape etc. The model considers the features of the plant and gives an output.

Anupong [7] proposes the use of Machine Learning algorithms to detect and do pest management like detection of bugs, bacteria etc. This helps in improving the crop yield and quality. The contributions of this paper include the extraction of features using deep learning techniques, classification based on RNN and analysis of the results that have been found. The method performs well on crop images and agricultural farm images with a good run speed.

Delnevo et al. [8] use Internet of Things to study and examine plant diseases. They utilize sensors to collect data and use an entirety of data along with other publicly available data to predict plant diseases. They propose an infrastructure for farmers to control spread of disease by using SIoT and crowdsourced images coupled with deep learning for disease prediction. They have used expert knowledge for labeling the dataset and validating the results. The drawback is the methodology's application on very specific datasets making it less versatile.

In this paper, the authors [9] do a comprehensive study of disease detection methods. They talk about the purposes for which disease detection and image analysis is applied. Most of the diseases occur on the leaves or the stems and the visual patterns are complex. A better image pattern understanding technique is required for detecting and understanding disease detection. The multiple methods they state are homogenization, Otsu segmentation, clustering and neural networks, color transformation, texture features, fuzzy c-means etc.

Revathi and Hemalatha [10] use image segmentation by edge detection and image analysis using the proposed HPCCDD algorithm. The work is mainly done for cotton leaves where they identify the disease areas using image analysis. The prediction is also done in three languages and the farmers can work on solving the problem by making the required changes to their yield. They identify the disease areas by applying white lightning. The white lightning denotes the disease affected areas. They image analysis is done and it denotes the disease, symptoms and recommendations for the farmers.

3 Methodology

Edge Detection is the division of the image pixels based on brightness and intensity. Images in the image are the points where the brightness of the edge suddenly changes [11]. Using these edges, the image can then be identified as areas where an anomaly exists and where it does not. The points where there is a large number of edges in comparison to the other parts of the leaf, we can say those areas are majorly affected by some plant disease.

Inclusion of Fuzzy Logic helps in identifying a more accurate representation of leaves using the edge pixels. A greater number of edges identified helps in forming better clusters which can then be used to identify the disease-ridden areas when using clustering. Using edge detection, we can quantify the size and area of the affected part in the leaf [6]. By identifying the particular leaves that are suffering from a particular disease, the chemicals sprayed which are harmful to the environment can be controlled and the particular leaf is cut off the particular plant is removed so as to not spread the disease any further. These chemicals can be very unsustainable and harmful for the environment and curbing their usage can be beneficial to promote a sustainable practice in agriculture [12].

As seen in Fig. 1 (Samir 2019), the disease-ridden parts of the plant are more distinct on applying Fuzzy Edge Detection to the image and it can be easily separated using clustering of the disease pixels of the leaf.

After the edges are detected using Fuzzy Edge Detection, a hybrid clustering algorithm is employed. As mentioned by Mittal et al. [13], there are a variety of clustering algorithms available during the creation of this algorithm concepts form two different algorithms, that is the K Means and density clustering algorithm was used to form clusters. Image segmentation is the partitioning of the image into meaningful smaller regions mainly useful in identification or object detection especially in fields like medical imaging, biometric identification etc. In this case the clustering is helping in identifying the disease affected areas of the leaf by forming clusters around those pixels. This can be done due to the edge pixel clusters that can be extracted through edge detection. Clustering basically is a technique in which the process of partitioning an image into meaningful and visually interpretable regions or segments takes place.

The image is then put through a series of image-processing steps to isolate and create bounding boxes for the disease-infected areas on the leaves. The image is blurred using a 7 by 7 Gaussian blur matrix. The blurring helps in enhancing the features of the image, reducing the intensity of the pixel values and the noise of the image, leaving only the concentrated areas. The core of the methodology lies in the pixel-wise traversal algorithm used to compute bounding box coordinates. The

a) Original plant image for frogeye

b) Edge plant image

c) Original plant image for Beech

d) Edge plant image

Fig. 1 Original and fuzzy edge image of disease-ridden leaves

algorithm iterates through the image pixels, identifying connected components and forming bounding boxes around them. The extent of pixel exploration is controlled via a parameter. Bounding box coordinates are stored in a list, capturing the minimum and maximum pixel indices for both rows and columns. Lastly, to provide a visual representation of the computed bounding boxes, an overlay image is generated. The overlay image highlights the boundaries of each bounding box in a distinctive color. Thicker lines are drawn to enhance the visibility of the bounding box boundaries. All these steps together can help in easy detection of the infected areas and help in localizing it within the confines of the boundary.

As seen in Fig. 2, the disease-ridden areas are highlighted using the box overlay. It is a versatile method which can be changed according to the requirements of the dataset.

In Fig. 3, the flow of the working of the proposed solution is depicted. Initially the input image is preprocessed. The Sobel filter applied gives the gradient value along x-axis and y-axis. These images are taken as the input for the Fuzzy Logic. The membership functions *Ix* and *Iy* are used. Using the rule base for Fuzzy Edge Detection, *Iout* is found to be the output function. On evaluating this and finding

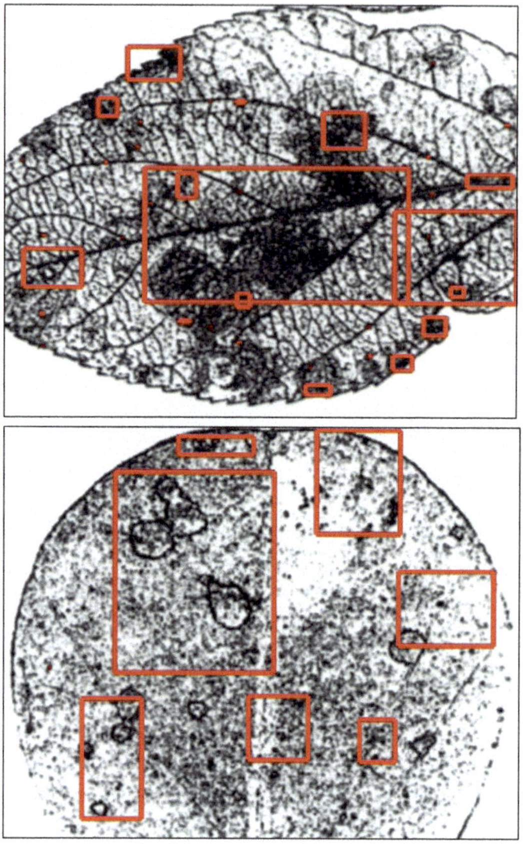

Fig. 2 Box overlay of disease-ridden leaf areas

distinct values, we get the edge image. This is then preprocessed to make it more suitable to apply clustering. Utilizing the K-Means and Density Clustering, the final box overlays are given informing on the disease ridden areas of the leaf.

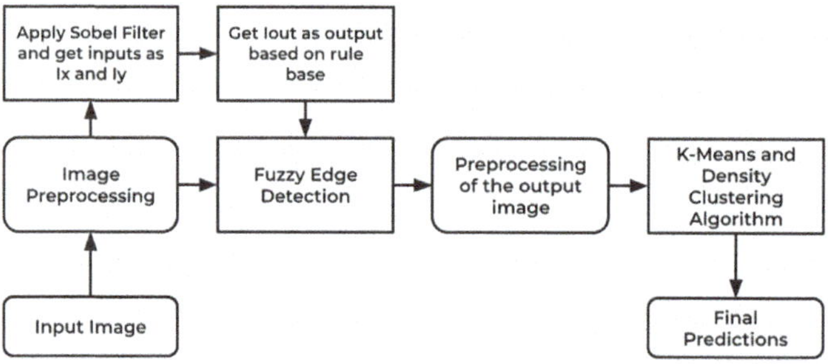

Fig. 3 Edge plant image using canny edge detection-1

4 Results

Figure 4 is the edge detection on the disease-ridden leaf without applying Fuzzy Logic. As seen, the edges are not distinct and we cannot identify the areas where the plant is infected, whereas on applying Fuzzy Logic, the difference is very visibly seen. Hence in detecting leaf disease, the application of Fuzzy Logic is more accurate and beneficial. It helps in finding the disease and thus promoting a more sustainable agricultural practice. Fuzzy Edge Detection is also dynamically self-adaptable, meaning it can adjust the thresholds according to the input image. This in turn reduces the need for human interaction as the algorithm is suitable for images taken in different conditions.

On comparison, using Fuzzy Edge detection proves to be a better fit for the given problem of addressing leaf diseases. The image is more useful when clustering is applied as disease-ridden areas can easily be detected.

To help in accurately locating the diseased spots a clustering algorithm was used. As seen in Fig. 4 a clustering algorithm is further used to improve the disease localization's precision and usefulness. By putting pixels that have similar properties together, clustering helps isolate areas that have similar disease-related traits. The localization of the affected areas makes it easier to apply chemical sprays effectively and precisely. Farmers can maximize resource use, reduce environmental impact, and adopt a more sustainable agricultural practice by precisely identifying the disease-affected areas. This can be seen in Fig. 4 where the infected areas are grouped together in clusters. As can be seen from Figs. 1 and 4 the clustering algorithm is able to detect the trouble region with suitable accuracy and this can be verified by correlating the position of the boundary boxes to the original image. Fuzzy Logic

Fig. 4 Edge plant image using canny edge detection-2

and clustering algorithms work together to increase disease detection accuracy while also facilitating more efficient and ecologically friendly pest and disease control in agriculture.

5 Conclusion and Future Scope

In conclusion, this combination of Fuzzy Edge Detection and K-Means and Density Based Clustering algorithms gives good information about the disease affected areas of the leaf. The anchor boxes give the affected area of the leaf which can be beneficial for the farmer to either cut the diseased leaf or apply pesticide and curb the spread of the disease. This will help in stopping wasteful usage of chemical fertilizers and pesticides and other harmful steps taken to stop the spread of diseases, thus maintaining soil health and promoting sustainable agriculture. In this paper, the detection of the areas has been done, this can be improved by working on the classification of the disease using Deep learning methods. An automated system can also be developed using IoT where the affected areas are automatically sprayed to stop the early onset of disease making it a quick and easy way to curb disease spread in large farms and plants.

References

1. Reganold JP, Papendick RI, Parr JF (1990) Sustainable agriculture. Sci Am 262(6):112–121. http://www.jstor.org/stable/24996835
2. Ikerd JE (1993) The need for a system approach to sustainable agriculture. Agric, Ecosyst Environ 46(1–4):147–160. ISSN 0167-8809. https://doi.org/10.1016/0167-8809(93)90020-P.
3. Sardogan M, Tuncer A, Ozen Y (2018) Plant leaf disease detection and classification based on CNN with LVQ algorithm. In: 2018 3rd International Conference on Computer Science and Engineering (UBMK), Sarajevo, Bosnia and Herzegovina, pp 382–385. https://doi.org/10.1109/UBMK.2018.8566635
4. Ali Q, Ahmar S, Sohail MA et al (2021) Research advances and applications of biosensing technology for the diagnosis of pathogens in sustainable agriculture. Environ Sci Pollut Res 28:9002–9019. https://doi.org/10.1007/s11356-021-12419-6
5. Thai H-T, Le K-H, Nguyen NL-T (2023) Towards sustainable agriculture: A lightweight hybrid model and cloud-based collection of datasets for efficient leaf disease detection. Futur Gener Comput Syst 148:488–500. ISSN 0167-739X, https://doi.org/10.1016/j.future.2023.06.016
6. Parez S, Dilshad N, Alanazi TM, Lee JW (2023) Towards sustainable agricultural systems: a lightweight deep learning model for plant disease detection. Comput Syst Sci Eng 47(1):515–536
7. Wongchai A, Jenjeti DR, Indira Priyadarsini A, Deb N, Bhardwaj A, Tomar P (2022) Farm monitoring and disease prediction by classification based on deep learning architectures in sustainable agriculture. Ecol Model 474:110167. ISSN 0304-3800. https://doi.org/10.1016/j.ecolmodel.2022.110167
8. Delnevo G, Girau R, Ceccarini C, Prandi C (2022) A deep learning and social IoT approach for plants disease prediction toward a sustainable agriculture. IEEE Internet Things J 9(10):7243–7250. https://doi.org/10.1109/JIOT.2021.3097379.

9. Rathod AN, Tanawal B, Shah V (2013) Image processing techniques for detection of leaf disease. Int J Adv Res Comput Sci Softw Eng 3(11)
10. Revathi P, Hemalatha M (2012) Classification of cotton leaf spot diseases using image processing edge detection techniques. In: 2012 International Conference on Emerging Trends in Science, Engineering and Technology (INCOSET). https://doi.org/10.1109/incoset.2012.6513900
11. Yusuf HS, Hagras H (2018) Towards image steganography using type-2 fuzzy logic and edge detection. In: 2018 10th Computer Science and Electronic Engineering (CEEC), Colchester, UK, pp 75–78. https://doi.org/10.1109/CEEC.2018.8674225
12. Gary S, Meng Q, Li B (2022) Leaf disease segmentation and detection in apple orchards for precise smart spraying in sustainable agriculture. Sustainability 14(3):1458. https://doi.org/10.3390/su14031458
13. Mittal H, Pandey AC, Saraswat M et al (2022) A comprehensive survey of image segmentation: Clustering methods, performance parameters, and benchmark datasets. Multimed Tools Appl 81:35001–35026. https://doi.org/10.1007/s11042-021-10594-9

Plant Disease Detection Using Deep Convolutional Neural Network for Corn, Rice, and Banana Crop

Divyam Dholwani, Kaustubh Patil, Vedangi Thokal, Pradnya V. Kulkarni, and Mamta Bhamare

Abstract Among the most widely farmed and eaten crops globally are corn, rice, and bananas. However, a variety of both biological and environmental variables, including weather, pests, soil, and water quality, can influence these crops, leading to illnesses that have a substantial impact on crop quality and output. These illnesses not only cause financial loss but also represent a serious danger to the farming sector. Traditional methods of detecting diseases are insufficient and frequently depend on laborious eye examinations. Consequently, having efficient techniques for identifying and treating these illnesses is crucial. Using a Deep Convolutional Neural Network for the crops of corn, rice, and bananas, this article tackles this problem. A layout in which max-pooling layers are arranged in a deliberate manner after convolutional layers with progressively larger filters to guarantee the identification of certain illness patterns. The findings showed that the accuracy was 99.7% for Bananas, 98.8% for Rice, and 99.8% for Corn. This study aims to empower farmers worldwide and support them in protecting the health and productivity of their crops by showcasing the application of convolutional neural networks in crop disease identification.

Keywords Machine learning · Deep learning · Neural networks · Convolutional neural network

1 Introduction

Some of the major obstacles to worldwide agricultural production and food security include pests and plant diseases Reducing crop losses and promoting sustainable food production need timely identification and resolution of these issues. Nevertheless, conventional disease detection approaches, reliant on visual assessments by human experts, frequently entail labor-intensive processes, consume significant time,

D. Dholwani · K. Patil · V. Thokal · P. V. Kulkarni (✉) · M. Bhamare
School of Computer Science and Engineering, Dr. Vishwanath Karad MIT World Peace University, Pune, Maharashtra, India
e-mail: pradnyav.kulkarni@mitwpu.edu.in

A. J. Kulkarni and M. Hajiaghaei-Keshteli (eds.), *Advances in Intelligent Systems for Sustainable Agriculture*, Lecture Notes in Networks and Systems 1201,
https://doi.org/10.1007/978-981-97-9839-1_3

and are susceptible to errors. Over the past few years, the application of machine learning (ML) and deep-learning (DL) approaches has shown remarkable promise in transforming the field of crop illness detection. These techniques leverage the power of computational algorithms to analyze and interpret large amounts of data, particularly images of diseased plants, enabling the accurate and early identification of various plant diseases. Khatoon et al. [1] provide a comprehensive overview of the latest advancements in deep learning for plant disease detection. It covers various convolutional neural network architectures, data enhancement techniques, and disease classification strategies. Plant health and illness may be recognized with high accuracy using ML and DL models that are trained to collect and analyze information from photos, such as their hue, texture, and shape. Saleem, Potgieter, and Arif's extensive review paper [2], provides a thorough analysis of state-of-the-art deep-learning techniques for plant disease detection and classification.

Convolutional Neural Networks (CNNs) are a unique class of neural networks that process data with a grid topology, making them ideal for image processing. CNNs have shown great potential in revolutionizing the field of plant disease detection, enabling the accurate and prompt recognition of various crop illnesses. By extracting and analyzing features such as color, texture, and shape from images, in a recent study conducted by Andrew J et al. [3], the potential of deep-learning techniques, specifically convolutional neural networks (CNNs), was investigated in terms of their ability to accurately distinguish between healthy and diseased plants. The study found that CNNs demonstrated great precision in this task. The researchers managed to accomplish a remarkable 96% accuracy rate, demonstrating the effectiveness of AI in early disease detection. While Afifi, Alhumamand, and Abdelwahab investigate the use of CNNs for recognizing various diseases on grape leaves [4]. The developed model achieved a remarkable 99.8% accuracy, highlighting the power of AI in differentiating between healthy and diseased plant tissues. Simhadri and Kondaveeti [5] focused on developing a deep-learning model for rice disease detection that performs well even in resource-limited environments with limited computing power and data availability. On the other hand, we used verified datasets related to these crops to train our model. In this research, we propose a mechanism that uses CNNs to detect diseases in two prominent crops, corn and rice. It predicts diseases related to these crops by processing various images and weather datasets. Furthermore, our work presents an architecture that combines photos of plants with weather data. This architecture utilizes simple pooling operations and convolutions and is referred to as CNN augmented with cellular automata rules. This approach aims to effectively identify plant diseases by making use of the combined information from weather data and plant images.

This paper is organized into seven sections, with Sect. 1 providing the introduction to the problem, Sect. 2 providing an extensive literature survey, Sect. 3 providing into the data engineering undergone, Sect. 4 giving the design and methodology used in our work, Sect. 5 providing the results and comparisons to the existing work, Sect. 6 providing the conclusion to our work, and Sect. 7 discussing the future scope of our research. Seeking an in-depth understanding of advanced methods and technologies in the field, the goal is to contribute to the development of more trustworthy and efficient ways for identifying and controlling pests and plant diseases.

2 Literature Review

In the pursuit of global food security, Kutyauripo et al. [6] conducted a comprehensive systematic review, evaluating the application of artificial intelligence (AI) across the entire food production ecosystem. Addressing the threat of plant diseases impacting crop yield and economic losses, Muthukannan et al. [7] propose a solution leveraging smart technology and making use of directly collected images from an agriculture field using a camera. They introduce Feed Forward Neural Network (FFNN), Learning Vector Quantization (LVQ), and Radial Basis Function Networks (RBF) to effectively identify and classify diseased plant leaves, with a focus on beans and bitter gourds.

Reshmi A.M and Prameeja Prasidhan's research [8] stands out for their use of an efficient Convolutional Neural Network (CNN) algorithm. Achieving an impressive 99.5% accuracy. The exact origin of the dataset is not mentioned; however, they state that their dataset consists of over 1000 images of healthy and diseased leaves. The study employs several methods for image processing to detect and classify crop diseases. Examining the landscape of automatic crop disease identification, the study by Justine Boulent et al. (2019) [9] utilizes Convolutional Neural Networks (CNNs) to analyze 19 relevant works with a focus on tomatoes. Most images used in the studies come from public datasets. The most widely used dataset is PlantVillage [10]. In their 2023 research, Pawan Shankar Ghodekar et al. [11] proposed a system, that incorporates algorithms for data pre-processing, CNN model training, and disease classification. The dataset for training and testing was extracted from Kaggle. Notably, the project's integration of OpenCV for feature extraction and CNN for predictive accuracy marks a significant advancement. Abade et al. 2020 [12] reviewed 121 papers published between 2010 and 2019 and found that PlantVillage was the most widely used dataset, while TensorFlow was the most frequently used framework. Bangari et al. [13] conducted a review on disease detection using Convolutional Neural Networks (CNNs), focusing on potato leaf diseases.

Tugrul, Elfatimi, and Eryigit 2022 [14], reviewed the recent surge in applying deep learning, particularly convolutional neural networks (CNNs), to address plant disease detection in agriculture by using images obtained from PlantVillage [10], Kaggle, Food and Agriculture Organization (FOA).In the study conducted by Shelar et al. [15], a remarkable achievement of 95.6% accuracy was attained using early stopping during the model training phase, spanning 50 epochs. They made use of the publicly available PlantVillage dataset [10]. Hassan et al. 2021 [16] showcase the potential of deep convolutional neural network (CNN) models for precise crop disease identification. Trained on a diverse dataset spanning 14 plant species and 38 disease classes obtained from PlantVillage [10], their models, notably EfficientNetB0, excel with accuracy rates reaching 99.56%. Srivastava et al. 2020 [17] present a novel approach to combat plant diseases, employing deep convolutional neural networks (CNNs) for accurate disease detection in crops. The study also mentions that the PlantVillage dataset [10] consists of 53,407 images of healthy and diseased leaves from 14 different plant species. The proposed methodology, emphasizing simplicity and efficiency, marks a significant leap in automating plant disease identification, offering promise for improved precision in agriculture. Pokkuluri and Nedunuri [18] developed a novel approach that integrates Cellular Automata encoding with Convolutional Neural Networks (CNN). They made use of datasets from ECMWF for processing the weather data to predict the crop's condition [19]. The resulting classifiers, CNN-CA-I and CNN-CA-W, demonstrated remarkable accuracies of 92.6% and 90.1%, respectively, marking a significant advancement in the prediction of crop diseases across various crops.

While the aforementioned studies cover a range of models and techniques, there is an over-reliance on common datasets such as the PlantVillage dataset ultimately limiting the diversity in training data. Novel approaches like integrating Cellular Automata with CNNs definitely show innovation; however, there is a gap in transitioning from theoretical or controlled experiments to actual field-level applications that can be adopted by the farmers. Although some studies focus on optimizing models for more efficiency, there still remains a broad gap in addressing computational costs and accessibility for the end-users. The following table provides an overview of the outcome of the literature review and the models used in each study (Table 1).

3 Data Engineering

Any machine learning model's ability to perform well is closely correlated with the caliber and variety of the dataset used for training. The leaves of corn, rice, and banana plants are the main subjects of our research. The creation of the model and its analysis depend heavily on this part. Researchers and practitioners can access all of the utilized datasets for free because they are all operated under public domain licenses.

3.1 Corn Leaf Disease Dataset

It consists of a rich variety of carefully annotated corn leaf images. The images are categorized into 4 classes on the basis of their diseases, common rust, gray leaf spot, blight, and healthy.

Common Rust. A condition resulting from the fungus 'Puccinia sorghi'. Chlorotic specks on the leaf surface are one of the early signs of common rust. These specks

Table 1 Literature review summary: plant disease detection studies

Study	Authors	Model used	Outcome of study	Dataset used
Artificial intelligence applications in the agrifood sectors	Kutyauripo et al	Review of AI in food production ecosystem	Emphasizes AI's potential in enhancing productivity and addressing challenges in food system	Publicly available datasets, exact source not specified [6]
Classification of Diseased Plant Leaves using FFNN and LVQ Algorithm	Muthukannan et al	FFNN, LVQ, RBF for plant disease identification	FFNN outperforms other methods, aiming to enhance crop quality in the Indian economy	Directly collected images from agriculture fields [7]
Leaf Disease Detection Using CNN	Reshmi A.M and Prameeja Prasidhan	CNN algorithm for crop disease detection	Achieves impressive accuracy of 99.5%, providing valuable insights for farmers to mitigate losses	Dataset containing over 1000 plant images, the dataset name is not provided [8]
Convolutional Neural Networks for the Automatic Identification of Plant Diseases	Justine Boulent et al	CNNs for tomato leaf-centric disease detection	Highlights potential of deep-learning techniques in advancing practical tools for sustainable agriculture	PlantVillage [9]
Plant Leaf Disease Detection Using Cnn	Ghodekar et al	CNN for disease classification in Indian agriculture	Integrates OpenCV for feature extraction and CNN for predictive accuracy, contributing to the sustainability of agriculture	Kaggle [11]
Plant Diseases recognition on images using Convolutional Neural Networks: A Systematic Review	Abade et al	Review of datasets and frameworks used in plant disease detection	Identifies PlantVillage as the most widely used dataset and TensorFlow as most frequently used framework	PlantVillage [12]

(continued)

Table 1 (continued)

Study	Authors	Model used	Outcome of study	Dataset used
Potato Leaf Diseases Detection Using Sequential Model	Bangari et al	Review of CNNs in potato leaf disease detection	Concludes CNNs are adept in detection of plant diseases and contribute to high level of disease identification	Publicly available datasets, kaggle [13]
Review on CNNs in Plant Disease Detection	Tugrul, Elfatimit and Eryigit	Survey on CNNs for plant disease detection in agriculture	Emphasizes efficacy of CNNs for prompt disease detection and outlines challenges in plant disease detection methodologies	PlantVillage, Kaggle and Food and Agriculture Organization (FOA) [14]
Plant Disease Detection using Cnn	Shelar et al	CNN for accurate plant disease identification	Achieves remarkable accuracy of 95.6%, showcasing promising potential for detection	PlantVillage [15]
Crop Disease Prediction with CNN Augmented With Cellular Automata	Pokkuluri & Nedunuri	Cellular Automata encoding with CNN for crop disease prediction	CNN-CA-I and CNN-CA-W demonstrate remarkable accuracies, marking significant advancement in crop disease prediction	Publicly available datasets and ECMWF for weather data [18]

grow into tiny, round to elongated, tan patches that might appear in groups. Such spots appear to be scattered around the adaxial and abaxial leaf surfaces (Fig. 1).

Gray Leaf Spot. It is caused by 'Cercospora zeae-maydis'. Two to three weeks before tasseling, lower leaves show symptoms for the first time. The lesions of this disease will appear as yellow or tan rectangular spots but will expand into large cigar-shaped lesions. Leaves appear grayish in color due to the presence of fungal spores (Fig. 2).

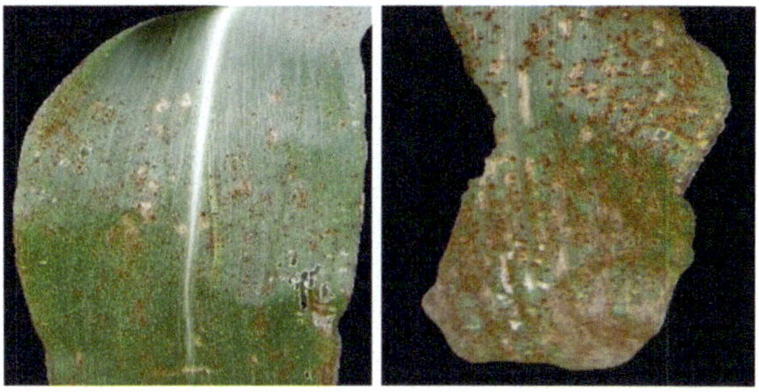

Fig. 1 Common Rust in corn leaves. [20]

Fig. 2 Gray leaf spot [20]

Blight. It is a disease caused by Exserohilum turcicumis that damages corn. It's also known as Turcicum leaf blight (TLB). It can cause significant yield loss in susceptible corn hybrids. Usually, the lowest leaves are the first to show symptoms. The leaf spots are initially greenish-gray and later on turn to pale brown. They are usually long (up to 15 cms). The underside of the leaf under the lesions produces spores, which give the appearance of a dusty green fuzz (Fig. 3).

Healthy. Healthy corn leaves are generally green, but can also be purple or reddish-brown. They are a monocot plant, so they have parallel venation. They can be 0.3–1.5 m long and 5–15 cm wide. The margins of the leaves are entire, meaning they don't have obvious teeth. They can also have narrow longitudinal wrinkles and are sometimes vertically wavy (Fig. 4).

Fig. 3 Corn blight [20]

Fig. 4 Healthy corn leaves [20]

3.2 *Rice Leaf Diseases Dataset*

This dataset consists of a diverse array of diseased rice leaf images. A total of 120 images is present in this dataset. The images are further classified into 3 classes namely, Leaf smut, brown spot, and bacterial leaf blight.

Leaf Smut. Entyloma oryzae is the causing agent of leaf smut. It's a common but mild illness that generates little black spots on rice leaves that are angular and slightly elevated. The spots are linear, rectangular, angular, or elliptical, and measure 0.5–2.0. On a single leaf, there may be many spots, but they all stay unique from one another. When the epidermis becomes moist/wet, it breaks open and releases black spores. The edges of heavily diseased leaves die and become pale, while the leaves themselves turn yellow (Fig. 5).

Brown Spot. It is caused by Cochliobolus miyabeanus. It's also known as Helminthosporium leaf spot, sesame leaf spot, Helminthosporiose, or fungal blight. The appearance of several big spots on the leaves—which may eventually cause the leaf to die entirely—is the most obvious harm caused by this. In cases where the infection affects the seeds, it results in the formation of unfilled grains or seeds that are spotted or discolored. Infected seedlings exhibit small, circular lesions that are yellow-brown or brown in color. These lesions have the potential to encircle the coleoptile (the protective sheath covering the emerging shoot) and cause distortion in both primary and secondary leaves (Fig. 6).

Fig. 5 Rice leaf smut [21]

Fig. 6 Rice brown spot leaf [21]

Fig. 7 Bacterial leaf blight in rice leaf [21]

Bacterial Leaf Blight. It is a highly destructive bacterial disease. It has a significant impact on the rice cultivation worldwide. It affects millions of hectares of rice crops each year, the bacterial agent causing this disease is Xanthomonas oryzae.

Some of the common signs and symptoms are light green to grayish green plants with water-soaked streaks on the leaves. These streaks form larger yellowish lesions with uneven edges. Infected leaves change color to grayish green and curl. As the disease advances, the leaves become yellow or straw-colored and wither (Fig. 7).

3.3 Banana Leaf Diseases Dataset

Numerous illnesses commonly pose a threat to banana agriculture and negatively affect productivity. These illnesses have a negative impact on banana plants, stunting their development, reducing their ability to produce fruit, and sometimes even killing the plants. The dataset includes a wide range of photos that highlight three common banana leaf spot illnesses called cordana, sigatoka, and pestalotiopsis. For a more thorough inspection, pictures of healthy banana leaves are also included. So in total 4 classes are taken.

Cordana. It is a fungal disease called cordana leaf spot (CLS) mostly affects banana trees. Banana diamond leaf spot is the most popular term used to describe this illness. Two fungi species cause CLS, C. johnstonii and C. musae. Even though it is common worldwide, has generally little impact on production (Fig. 8).

Fig. 8 Cordana leaf spot on a banana leaf [22]

Sigotaka. It is a fungal disease that infects the leaves, causing spots that can lead to leaf drying and defoliation. This can affect the quality of the fruit and reduce yield. The fungus *Cercospora musae* is the cause of *sigatoka*, often referred to as Cercospora leaf spot. It is one of the most harmful banana illnesses common in banana plants across the world (Fig. 9).

Pestalotiopsis. Pestalotiopsis palmarum is a fungus that causes a disease in bananas, coconuts, and date palms. The fungus is responsible for the formation of spots on the leaves, blight on the petiole/rachis, and occasionally, bud rot in palms. Pestalotiopsis palmarum can attack every part of the leaf, starting from the base and extending to the tip. Pestalotiopsis disseminata is a species of fungi in the family Pestalotiopsidaceae that causes banana leaf spot (Fig. 10).

Fig. 9 Banana leaf affected by Sigotaka [22]

Fig. 10 Pestalotiopsis in banana leaves [22]

Fig. 11 Healthy banana leaves [22]

Table 2 Summary of our datasets' characteristics

Dataset	Classes	Number of images per class	Total Images
Corn Leaf Dataset	Common Rust, Gray Leaf Spot, Blight, Healthy	1306, 574, 1146, 1162	4188
Rice Leaf Dataset	Leaf smut, Brown spot, Bacterial leaf blight	40 each	120
Banana Leaf Dataset	Sigotaka, Cordana, Pestalotiopsis, Healthy	400 each	1600

Healthy. Healthy banana leaves are large, wide, and slightly rounded with a smooth surface. They are waxy, flexible, and glossy, and can grow up to two meters long and half a meter wide. Healthy leaves are generally dark green in color, flexible, and intact with smooth edges (Fig. 11).

Image characteristics. As a part of pre-processing, images were re-sized, OpenCV library in Python was used to resize our images and retrieve their shape. The desired dimensions for our output images in this case were 180px in height and width.

Data Splitting. Two divisions were made in the dataset, the Training set and the Testing set. The division for the disease detection model was 75–25, implying that 75% of the dataset was used for training, while the remaining 25% was set aside for testing the model. The rationale behind this approach was to ensure sufficient data for training while also having a substantial set for unbiased evaluation. The split was executed by randomly shuffling the dataset and subsequently partitioning it into two disjoint sets. This methodology allowed for a balanced consideration of training effectiveness and the model's capacity to generalize unseen data (Table 2).

4 Proposed Model and Methodology

The methodology for Disease Detection in plants involves leveraging Deep Learning models to successfully identify diseases in plants based on input images. Neural Networks are an essential aspect of Deep Learning models. They consist of several layers of nodes—one convolutional "input layer", other hidden layers, also termed max-pooling layers and an output layer. Every node is connected to another. All nodes have an associated weight and threshold, forming a Neural Net. Among the various types of Neural Nets exist Convolutional Neural Networks (CNN). The main applications of CNNs include image and video detection. Prior to CNNs, object detection relied on manual and extensive feature extraction techniques. Nevertheless, Convolutional Neural Networks (CNNs) implement concepts from linear algebra, specifically involving matrix convolution and multiplication, to detect patterns within images, making them more adaptable. CNNs stand out from other Neural Networks due to their excellent performance with audiovisual inputs. Three different types of layers make up CNNs: the Convolutional Layer, the Pooling Layer, and the Fully connected Layer (also known as the Dense Layer). The network is started with the convolutional layer, which may be followed by other convolutional or pooling layers. The focus of the first layers is on fundamental elements such as colors and edges. As the layers proceed, the CNN becomes more complex, eventually dealing with more significant elements of the object till it successfully performs identification [23] (Fig. 12).

Figure 13 offers a graphical representation of the research methodology's operational flow. The initial stage of the proposed framework focuses on data acquisition from the previously cited sources [20–22]. This meticulous process aims to construct a comprehensive dataset encompassing a diverse range of both healthy and diseased crop images. Following the acquisition, the data undergoes rigorous pre-processing. This crucial step involves essential procedures like normalization, cleaning, and image resizing to ensure consistency within the data, preparing it for optimal utilization by the convolutional neural network (CNN) model. This pre-processed data then forms the foundation for the subsequent model training phase. During training, CNN ingests the meticulously labeled images, facilitating its progressive learning process. Through this process, the model effectively identifies and distinguishes the distinctive patterns associated with various crop diseases. Once the model has achieved adequate training, it undergoes rigorous testing with a separate, unseen dataset. This crucial evaluation stage allows researchers to assess the model's generalizability and

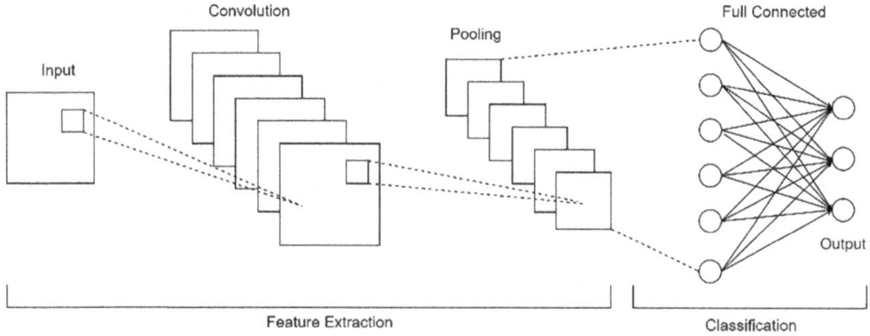

Fig. 12 Convolutional Neural Network Architecture

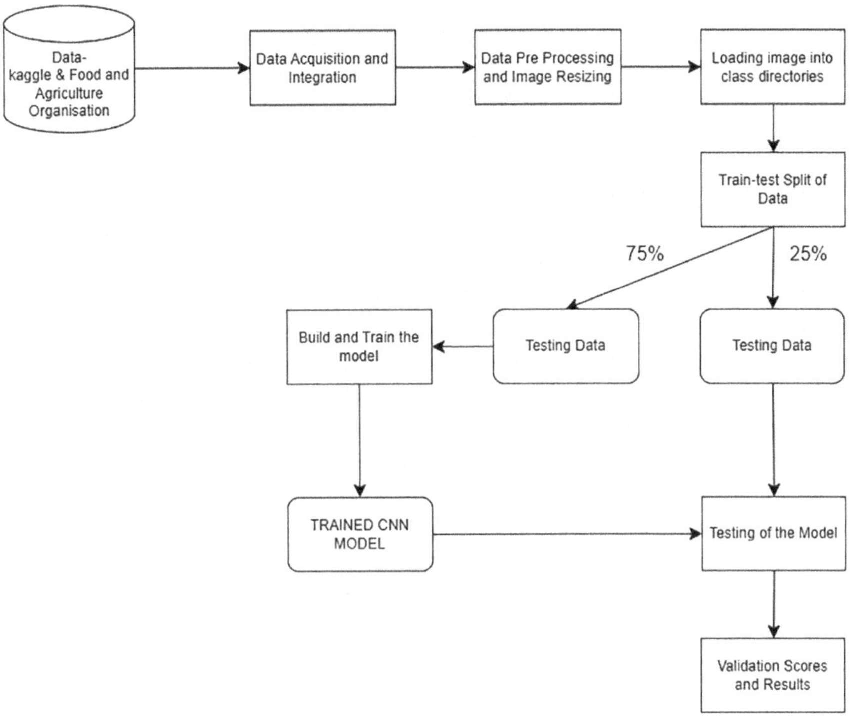

Fig. 13 Flowchart of the methodology followed

its ability to accurately identify diseases in previously unseen crop images. Finally, the obtained results are subjected to a thorough verification process to ensure their accuracy and reliability.

To understand the effective use of CNNs in agriculture applications, a systematic literature review was conducted. The review aimed to gather insights from peer-reviewed research papers available on various platforms. Articles published before the year 2000 were filtered out in order to maintain relevancy. Preparing a Convolution Neural Network organization for pattern detection using the dataset was proposed. The model was tailored sequentially, implying a linear stack of layers. Initially, three convolutional layers are integrated. They apply convolutional operations, aiming to extract features from the images. They all have identical padding and are set up with a 3 × 3 filter size to preserve the input dimensions. The number of filters increases progressively, to enhance detection of crop-specific disease patterns. Max-pooling layers come after every convolutional layer. It downsamples the spatial dimensions while preserving the most important data for further processing. As a result, the computational load is decreased, and the subsequent layers are more focused on essential features, hence streamlining the learning process. This down-sampling technique contributed to our model efficiency by minimizing redundant details and noise. It selects the maximum values within a localized group of pixels hence retaining salient features. Strategic integration of pooling layers between convolutional layers is a key element of the model. Subsequently, a flattened layer is utilized. This layer flattens the multidimensional output obtained from its predecessor layers into a one-dimensional array so that the data is prepared to be input into densely connected layers. Finally, two fully connected or dense layers are utilized. The first layer comprises 128 neurons activated by the rectified linear unit activation function and serves as an intermediate representation. ReLU is one of the most commonly used activation functions. It removes negative values from the training model. ReLU involves simple mathematical operations, which makes it more efficient compared to other activation functions like sigmoid and tanh. This efficiency is particularly important in convolutional neural networks with multiple layers as it helps in speeding up training. Along with this, ReLU zeroes out negative values, which helps the model to focus on relevant information while disregarding noise in the current data. It removes negative values from the training model [24]. The final layer aligns with the number of classes corresponding to the specific crop under consideration. The model is purposefully tailored in a hierarchical fashion. The data then moves through the layers causing the model to gradually evolve its understanding from basic features such as colors and edges to more advanced elements providing a comprehensive analysis of plant health. Its progressive nature captures the patterns that are relevant to the health of crops and offers a thorough method for detecting infections in bananas, rice, and maize early on. This design helped us capture the nuanced patterns specific to crop diseases (Fig. 14).

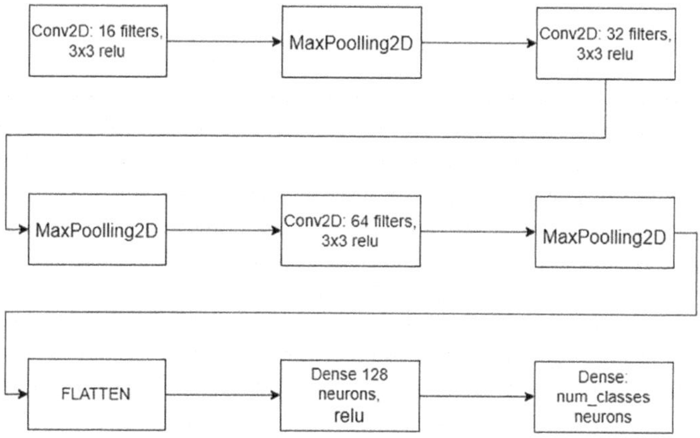

Fig. 14 CNN model representation

5 Results and Discussion

This research work proposes the use of a Convolutional Neural Network for disease detection in corn, rice, and banana plants. The models will be trained on a dataset obtained from the Ministry of Statistics and Programme Implementation of India.

A model has to be tested using fictitious data in order to be evaluated for efficacy. All of the input values in this testing data are utilized to assess the model. The model's performance is further elucidated by comparing its output predictions with the expected results. The testing data encompasses the most recent year, while the remaining data is used for training purposes.

There are two graphs shown for each crop, 'Model Accuracy' and 'Model Loss'. The accuracy of the model was calculated by using the following formula.

$$\text{Accuracy} = \frac{\text{Number of correct predictions}}{\text{Total number of predictions}} \tag{1}$$

and the loss was calculated using the formula.

$$\mathrm{L}_{\mathrm{CE}} = -\sum_{i=0}^{n} t_i \log(p_i), \quad \text{for } n \text{ classes}, \tag{2}$$

where t_i is the truth table and p_i is the Softmax probability for the ith class [25].

The left graph shows the model's accuracy over time, measured as the fraction of crops the model correctly identified as diseased or healthy. The model's loss over time, which is a gauge of how inaccurate the model's predictions are, is displayed on the right graph. The quantity of training epochs is indicated on the x-axis. An epoch denotes the end of an iteration over the training set. The model is exposed to all of the training examples during each epoch, and it modifies its parameters in response to the patterns it discovers and gains knowledge from the data.

5.1 Rice

The accuracy graph starts around 0.38, calculated using Eq. 1, indicating that the model correctly identified about 38% of the rice plants as diseased at the beginning. The accuracy then continues to rise steadily and finally peaking at 30 epochs. The loss in data starts around 3.0, calculated using Eq. 2. Over the course of the 30 epochs, the loss line drops dramatically, reaching a value of 0.007 at epoch 30. This indicates that as the model learned, its prediction errors decreased dramatically (Fig. 15).

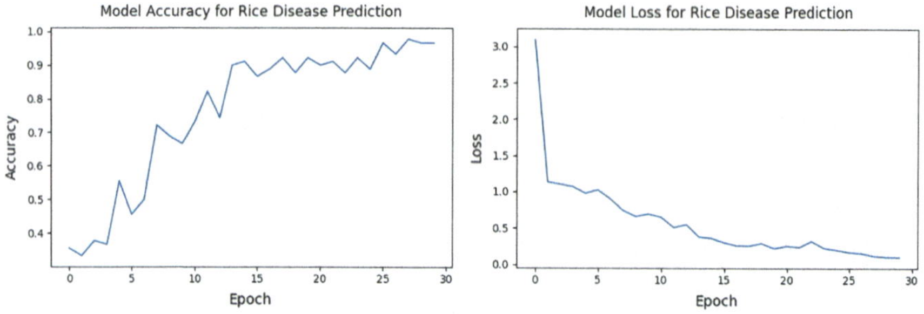

Fig. 15 Model accuracy and loss during training for rice disease detection

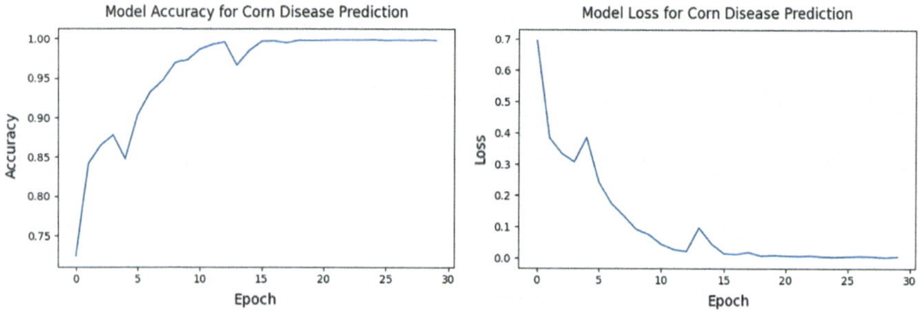

Fig. 16 Model accuracy and loss during training for banana disease detection

5.2 Banana

The accuracy graph starts around 0.5, calculated using Eq. 1, indicating that the model correctly identified about 70% of the banana plants as diseased at the beginning. The accuracy line rises steadily over the 30 epochs, reaching a peak of 0.99 (99%) at epoch 30. The loss graph starts around 1.2, calculated using Eq. 2, and shows how far off the model's initial estimates were from the right answers. The loss line then falls sharply over the 30 epochs, reaching a value close to 0.007 at epoch 30 (Fig. 16).

5.3 Corn

The accuracy graph begins at 0.24, calculated using Eq. (1), meaning that around 24% of the rice plants were initially accurately recognized by the model as unhealthy. After that, the accuracy increases gradually before peaking at 30 epochs. About 0.7 is where the data loss begins. The loss line experiences a significant decline over the course of the 30 epochs, peaking at 0.005 at epoch 30, calculated using Eq. (2) (Fig. 17).

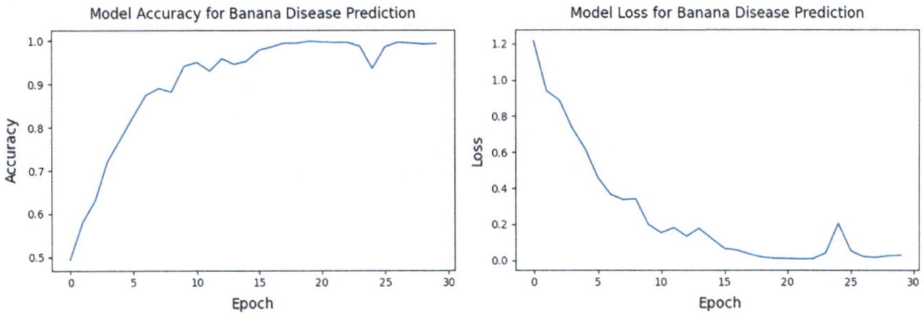

Fig. 17 Model accuracy and loss during training for corn disease detection

Table 3 Model's accuracy and data loss for each crop after 30 epochs

Crop Name	Accuracy	Data Loss
Rice	0.9889	0.0104
Banana	0.9975	0.0052
Corn	0.9986	0.0075

Thus, the results observed are a testimony to the potential application of this CNN model in real-world scenarios to aid in the early and accurate diagnosis of plant diseases (Table 3).

6 Conclusion

The proposed research suggests a convolutional neural network approach for disease detection in corn, rice, and banana. Trained on a dataset obtained from the Ministry of Statistics and Programme Implementation of India, the model exhibits significant accuracy in identifying diseases within these crops. The foundation of neural networks is explained through node layers and provides a fundamental structure to our solution. The CNN was tailored to include convolutional layers with increasing filter size, followed by strategic placement of max-pooling layers that minimize the spatial dimensions of the input. Due to this the learning curve of the model was improved drastically. The hierarchical design resulted in highly dense layers, which proved to be instrumental in capturing nuanced patterns specific to crop diseases. Through meticulous testing and evaluation, the model displayed remarkable efficiency. The learning capabilities of the model are displayed via the accuracy and loss graphs, which indicate the performance peak at 30 epochs. The model showed a staggering accuracy of 98.8% for Rice, 99.7% for Banana, and 99.8% for Corn.

The paper presents a novel framework for agriculture crop disease prediction that focuses on three major crops: banana, rice, and corn. Unlike many existing frameworks that are crop-specific, our approach is designed to be versatile and adaptable to multiple crops. Our framework emphasizes the practical implementation of two essential components, that is, focusing on reducing computational costs and improving accessibility for end-users. Through rigorous testing and validation, our framework has demonstrated superior performance in predicting diseases for banana, rice, and corn, making it a valuable contribution to the field of agriculture crop disease prediction. Overall, our framework offers a versatile and effective solution for predicting crop diseases across multiple crops, setting it apart from existing frameworks that are limited to specific crops.

7 Future Scope

Looking ahead, the future implications of this research are expansive. This includes scaling the model to encompass a wide range of crops will ensure the versatility of the system. The development of real-time monitoring systems with user-friendly interfaces holds the power to empower farmers, allowing them to see instant and actionable insights. Collaborative efforts with agricultural institutions will contribute to the real-world implementation of the model. This trajectory not only promises further technological advancements in the agriculture sector but also a positive impact on global agriculture.

References

1. Khatoon1 S, Maruf Hasan1 Md, Asif1 A, Alshmari1 M, Yap Y-K (2020) Image-based automatic diagnostic system for tomato plants using deep learning
2. Saleem MH, Potgieter J, Arif KM (2019) Plant disease detection and classification by deep learning
3. Andrew J, Eunice J, Popescu DE, Kalpana Chowdary M, Hemanth J (2022) Deep learning-based leaf disease detection in crops using images for agricultural applications
4. Afifi A, Alhumam A, Abdelwahab A (2021) Convolutional neural network for automatic identification of plant diseases with limited data
5. Simhadri CG, Kondaveeti HK (2023) Automatic recognition of rice leaf diseases using transfer learning
6. Kutyauripo I, Rushambwa M, Chiwazi L (2022) Artificial intelligence applications in the agrifood sectors
7. Muthukannan K, Latha P, Pon Selvi R, Nisha P (2015) Classification of diseased plant leaves using neural network algorithms 10(4), ISSN: 1819-6608
8. Reshmi AM, Prasidhan P (2022) Leaf disease detection using CNN. IJCRT 10(3). ISSN: 2320-2882
9. Boulent J, Foucher S, Théau J, St-Charles P-L (2019) Convolutional neural networks for the automatic identification of plant diseases. Front Plant Sci Sec Techn Adv Plant Sci 10
10. PlantVillage Dataset, https://www.kaggle.com/datasets/moazeldsokyx/plantvillage
11. Ghodekar PS, Yermune V, Sable A, Mandhare R, Bhosale M (2023) Plant leaf disease detection using Cnn. Int Res J Modernization Eng Technol Sci 5(4)
12. Abade AS, Ferreira PA, de Barros Vidal F (2020) Plant diseases recognition on images using convolutional neural networks: a systematic review

13. Bangari S, Rachana P, Gupta N, Sudi PS, Baniya KK (2022) A Survey on disease detection of a potato leaf using CNN. 2022 Second international conference on artificial intelligence and smart energy (ICAIS), Coimbatore, India, pp 144–149, https://doi.org/10.1109/ICAIS53314.2022.9742963
14. Tugrul B, Elfatimi E, Eryigit R (2022) Convolutional neural networks in detection of plant leaf diseases: a review. Adv Agric Eng Technol Appl. ISSN: 2077-0472
15. Shelar N, Shinde S, Sawant S, Dhumal S, Fakir K (2022) Plant disease detection using Cnn. ITM Web Conf 44:03049
16. Mahmudul Hassan Sk, Maji AK, Jasiński M, Leonowicz Z, Jasińska E (2021) Identification of plant-leaf diseases using CNN and transfer-learning approach
17. Srivastava P, Mishra K, Awasthi V, Sahu VK, Pal PK (2020) Plant disease detection using convolutional neural network. Int J Adv Res (ISJR) ISSN: 2320-5407
18. Pokkuluri KS, SSSN Usha Devi Nedunuri (2022) Crop disease prediction with convolution neural network (CNN) augmented with cellular automata. Inte Arab J Inform Technol 19(5)
19. ECMWF dataset, https://www.ecmwf.int/en/forecasts/datasets.
20. https://www.kaggle.com/datasets/unknown6874/corn-leaf-disease-dataset, last accessed 2023/12/22
21. https://www.kaggle.com/datasets/vbookshelf/rice-leaf-diseases, last accessed 2023/12/22
22. https://www.kaggle.com/datasets/shifatearman/bananalsd, last accessed 2023/12/22
23. Hasan RI, Yusuf SM, Alzubaidi L (2020) Review of the state of the art of deep learning for plant diseases: a broad analysis and discussion
24. Lin G, Shen W (2018) Research on convolutional neural network based on improved RELU piecewise activation function
25. Boudaa B, Touhami H (2023) A graph attention networks model for session-based recommender systems

Analyzing Various CNN Architectures for Detection of Early Blight and Late Blight in Infected Potato Plants

Varsha Naik, Arnav Gajjar, Inayat Sood, and Shubham Gite

Abstract Diseases in potato plants cause tremendous damage to the yield. CNN constitutes a possible solution for the early detection of early and late blight diseases. This paper does a detailed comparison based on performance and suitability of a new model and Pre Trained models in context of detecting an early blight or a late blight in Potato Crops. This investigation comprises a comparative analysis including their accuracy and loss. Through meticulous testing and training, this study aims to provide cognizance into the efficiency of these architectures for automated identification of blight diseases in potatoes, important for crop disease management. The images after adequate preprocessing and augmentation (by using techniques such as Random Flip and Random Rotation) were feeded to the five pre-trained models-InceptionV3, ResNet152v2, MobileNetV2, DenseNet121 and VGG16 neural network architectures and one custom CNN model GreenNet101 over a total of 10 epochs; out of which MobileNetV2 gives out the best accuracy of over 99.99%.

1 Introduction

The Indian economy's backbone, agriculture contributes 17% to the nation's cumulative GDP from the previous 12.7% in 2012 [1]. It employs an approximate 58% of the nation's population. Its contribution to the economy rose from 17.8% in the year 2019–2020 to 19.9% in the year 2020–2021. Crop failures can have serious detrimental effects on the nation. It can shoot up the unemployment rate or lead to food shortages to name a few. Diseases in plants can be caused by various factors: pathogens, soil health, environmental conditions [2], genetics of the plants, etc. Potatoes are the third most important crop in the world. Potatoes account for a total of 12% of the total world production [3]. Potato plants are prone to two such diseases that can be detected from its effect on the leaves of the plant. They are called early-blight

V. Naik (✉) · A. Gajjar · I. Sood · S. Gite
School of Computer Engineering and Technology, MIT World Peace University, Dr. Vishwanath Karad, Pune, Maharashtra 411038, India
e-mail: varsha.powar@mitwpu.edu.in

A. J. Kulkarni and M. Hajiaghaei-Keshteli (eds.), *Advances in Intelligent Systems for Sustainable Agriculture*, Lecture Notes in Networks and Systems 1201,
https://doi.org/10.1007/978-981-97-9839-1_4

and late-blight infections. Early-blight is primarily caused by the fungus Alternaria solani. Its symptoms include dark lesions with concentric rings on leaves, starting from lower leaves and progressing upward. Early-blight can result in a loss of the crop up to 30% annually and can affect the storability and the crop quality. Oomycete pathogen, Phytophthora infestans, causes Late blight. Its symptoms include water-soaked and dark lesions on leaves that can rapidly expand. Late blight causes a total of $6.7 billion in losses annually. Hence it becomes essentially crucial to detect diseases earlier to mitigate and prevent the prospective losses that can be caused to the crops by the use of fungicides [4], etc. Farmers, therefore, are in a dire need of a new and innovative technology that is not only fast, but extremely accurate. Several methods have been put forth to alleviate this problem. Earlier, these diseases were detected by experts. But now there is a need to minimize human intervention so that faster results are achieved. Several methods are proposed to solve this issue. These include using logistic regression, K-nearest neighbor, support vector machine (SVM) [5], and Deep-learning algorithms [6], particularly Convolutional Neural Networks (CNNs) [7], constitute an extremely popular approach to counter this problem. There are several imaging techniques for plant-disease detection [8]. Deep-learning algorithms find big applications in the field of image processing and object detection. This study's main goal is to identify Neural Network architectures and identify hyper parameters which would be essential to identify diseases in potatoes in the field with both traditional as well as modern embedded automated environments. This paper delves into the application of neural networks for the purpose of detecting potato blight diseases, with a specific focus on evaluating the efficacy of transfer learning models over custom CNN models. Despite custom CNN models offering comparable or even better accuracies in some cases, this paper examines the various popularly used transfer learning [9–12] architectures for the potential advantages they bring to the table. Because pre-trained models have already learned generic features from large datasets, transfer learning has the advantage of requiring less training time and resources. Furthermore, transfer learning can improve model generalization—particularly in situations where there is a dearth of labeled data—helping to create reliable and efficient potato blight detection systems. The study's findings could contribute to the advancement of sustainable farming methods, food security, and a more robust agricultural ecosystem. Furthermore, investigating transfer learning models in this particular setting not only provides useful information for detecting potato blight but also advances the conversation about machine learning's potential applications in agriculture, opening the door for further advancements in crop disease monitoring and management. For this, five (5) Convolutional Neural networks (InceptionV3, ResNet152v2, MobileNetV2, DenseNet121 and VGG16) were subjected to training and testing to choose the most suited hyperparameters impacting the architectures greatly in detection of Early/Late Blight. The parameters examined include, Batch size, number of Epochs and Accuracy. Our contribution is to evaluate the performance metrics of several neural network architectures—InceptionV3, ResNet152v2, MobileNetV2, DenseNet121 and VGG16—with a custom neural network, GreenNet101, for early blight and late blight detection in potatoes.

Key performance metrics such as accuracy, loss, validation loss and validation accuracy at the end of 10 epochs shall be assessed to compare the reliability of these classification models in detecting the disease. A custom neural network was devised to address the intricacies of the problem domain and also measure its performance in comparison to the already available CNN architectures. The study delves into the robustness and generalization capabilities of each model.

2 Related Work

In July 2023, Singh and Yogi [13], concluded that VGG16, InceptionNet, ResNet, NasNet, MobileNet, and VGG19(CNN), appear to be potential for studying effective features for image classification of leaf diseases. In 2021, plant pathology researchers Li et al. [14] employed machine learning and image analysis approaches to diagnose a range of leaf diseases in diverse crops. J. Rashid et al. [15] developed a multi-level deep learning model. It works in two stages. In the first stage, the YOLOv5 image segmentation approach is used to extract potato leaves from the plant photos. In the second stage, a CNN is used to identify early and late blight infections in potato plants based on the leaf images. Using a dataset of potato leaf disease, the suggested deep learning method produced results with 99.75% accuracy. Mohanty, Hughes et al. [16] trained a model to identify 14 crop species and 26 diseases, achieving an accuracy of 99.35% on a held-out test set. Chakraborty et al. [17] trained four CNN models viz. VGG16, VGG19, ResNet50 and MobileNet with the PlantVillage dataset and got the highest accuracy of 92.69% in comparison to other models. They then enhanced the model by tweaking the parameters and got the accuracy of 97.89%. Guanlin et al. [18] used image segmentation on datasets of grape leaves to identify and classify fungal diseases. In this work, K Mean clustering is used to extract color, texture, and form features from the image. The recovered features are then employed by SVM (Support Vector Machine) to identify diseases. In 2019, Geetharamani and Pandian [19] proposed a 9-layer deep CNN for leaf identification. Their proposed model with the average accuracy of 96.46% performed better than transfer learning models.

Chen et al. [20] used DSIFT and a CNN model called LeafNet to extract features from an image. Following that, SVM and MLP were used to categorize the tea leaf images. In June 2023, Feng et al. [21] used ShuffleNetV2*2.They had introduced 3 strategies-reducing depth of network, reducing 1*1 convolutions and introducing an attention model. The improved model of the loss function had reduced to 0.36.Their model had reduced the FLOPs, Number of hyperparameters, model size by 23%, classification accuracy by 0.85% and time to detect a single image was lessened to 3.27 s. Arnaud et al. [22] evaluated six CNN architectures, AlexNet, GoogleNet, SqueezeNet, DenseNet121, EfficientNet b7, and VGG19 and the result came out that DenseNet121 architecture with a SGD optimizer produced the best result, with an accuracy of 98.34% and a 97.37% f1-score. Al-Adhaileh et al. [23] customized a CNN with fewer trainable parameters, less computation time and reduced information loss.

The proposed model outperformed the others with an overall accuracy of 99% In 2022 Akbar et al. [24] introduced a novel LightWeight Convolutional Neural Network (WL- Net) model based on VGG 19 for the detection and classification of Bacteriosis in peach crops. The research encompassed the creation of a dataset comprising 10,000 images (4,500 of Bacteriosis and 5,500 healthy leaves).LWNet demonstrated exceptional accuracy of 99% against established CNN models—LeNet, Alexnet, VGG-16, and a simplified VGG-19—revealed the superiority of the proposed model. (Bedi and Gole, 2021) [25] introduced a novel hybrid model merging Convolutional Autoencoder (CAE) and Convolutional Neural Network (CNN) for automated plant disease detection of Bacterial Spot in peach plants using leaf images from the PlantVillage dataset. The hybrid system achieves exceptional training and testing accuracies of 99.35% and 98.38%, respectively, leveraging a mere 9,914 training parameters. The reduced parameter count accelerates training, expedites disease identification in plants, highlighting its potential for efficient and accurate automated plant disease detection across various plant types. Malvade et al. [26] introduced a deep convolutional neural network (DCNN) framework utilizing pre-trained VGG-16 for automated recognition and classification of diverse stresses in paddy crops during the booting growth stage. Leveraging 30,000 field images encompassing 12 stress categories across 5 paddy crop varieties, the trained models achieved an average accuracy of 92.89% on the validation dataset. Ramesh et al. [27] developed a Deep Neural Network optimized with the Jaya algorithm (DNN JOA) to classify four types of rice plant diseases using 650 leaf images. These images, captured in rural areas of Tamilnadu, underwent clustering to distinguish diseased, normal, and background sections. With a training set of 70% images, validation on 10%, and testing on 20%; the model showcased an accuracy of 97% after a total of a staggering 125 epochs, demonstrating efficacy in disease classification for agricultural applications. Gonzalez-Huitron et al. [28] utilized Transfer Learning models (MobileNetV2, NasNetMobile, Xception, MobileNetV3) to diagnose 10 tomato leaf diseases using 109,290 PlantVillage dataset images. Tests showed that Xception achieved nearly 100% accuracy and for each epoch it takes 2,512 s. Additionally, the model was successfully deployed on a 4 single-board Raspberry Pi computer for practical application. Ji et al. [29] devised a combined CNN model, merging GoogleNet and ResNet50, to detect three grape diseases from 1,619 leaf images sourced from PlantVillage. The UnitedModel they proposed achieved impressive accuracy (98.57%), precision (99.05%), and recall (98.88%), showcasing high performance disease discrimination from healthy leaves. Paymode et al. [9] utilized a VGG16 model to detect multiple crop leaf diseases from 14,421 tomato leaf images collected in Aurangabad, India. They employed diverse image processing techniques and training sessions of 40 epochs for grapes and 30 epochs for tomatoes, achieving impressive accuracies of 98.40% and 95.71%, respectively, in disease classification. In order to close a vacuum in the literature caused by previous studies that frequently concentrated on a small number of diseases, Ahad et al. [30] investigate the efficacy of several convolutional neural network (CNN) architectures in identifying and localizing rice disorders. Using a database of nine common rice diseases in Bangladesh, the study evaluates six CNN-based models: DenseNet121, MobileNetV2, resNext101, Inceptionv3, Seresnext101 and

Resnet152V. The researchers also established an ensemble model called DEX and applied transfer learning to some of these models. When it comes to rice leaf disease detection and localization, the results show that the ensemble framework has the highest accuracy (98%), and that transfer learning improves performance by 17% when compared to the Seresnext101 model. CNN models exhibit a high degree of accuracy when it comes to identifying and classifying rice diseases, indicating their potential for use in plant disease detection. This could have substantial consequences for real-time agricultural systems. With the development of a CNN-based rice leaf detection system, which can help with the timely and early identification of diseases and empower farmers to take prompt action to safeguard their agricultural yields and quality, the research is especially beneficial to farmers in rice-growing regions. The paper delves into the impact of plant diseases on food security, emphasizing the significant crop losses experienced and their implications for global food supply. The paper highlights the limitations of traditional disease classification methods, such as naked-eye observation and laboratory tests, due to their time-consuming and subjective nature. It then explores the potential of machine learning approaches, particularly deep learning using convolutional neural networks (CNNs), in addressing these challenges by automating the process of feature extraction from plant leaf images. The study reviews the latest CNN networks relevant to plant leaf disease classification and identifies the rise of deep learning, particularly CNNs, as efficient and automatic tools for plant disease classification. Additionally, it discusses the future direction of plant disease classification, including the potential for further advancements in automated detection and classification methods, as well as the necessity of high-quality input data for accurate deep learning model results.

3 Methodology

Hence, there's a growing need for smart methods to diagnose plant diseases effectively. Currently, various approaches exist to automatically identify these diseases, reducing human involvement. A thorough review of published papers exploring methods aiming to detect and categorize leaves by their type, color, texture, and size using a substantial sample size was conducted. This system's outcomes hold potential for automating the tedious jobs of identification, which is then followed with the classification of leaf diseases in agriculture.

3.1 Dataset

The dataset that has been taken is from Kaggle called "PlantVillage" [31] by Arjun Tejaswi. It has a total of 15 directories within it that contains images of pepper bell bacterial spots among others that include potato leaves' images and tomato leaves' images. Procurement and extraction of 3 directories, namely, the directories with potato leaves—'Potato Early blight' (1000 files), 'Potato Late blight' (1000 files), 'Potato healthy' (152 files) that are early blight, late blight and healthy potato leaves' images respectively were done. Hence, the models were initially fed a dataset comprising a total of 2152 images.

3.2 *Image Preprocessing*

Preprocessing was applied for a better feature extraction to improve the classification findings' consistency. Label Encoding and Categorical transform and fit was used as well.

3.3 *Data Augmentation*

To create a robust and reliable model that can easily detect, with utmost certainty, early and late blight diseases; sophisticated data augmentation practices were employed to enhance the initial image data available. Two key techniques, namely Random Rotation and Random Flip were used for the data augmentation process. Random Rotation applies random rotations to the images during training and hence helps at introducing variability. Similarly, Random Flip was employed to create mirrored versions of the images by flipping them horizontally and to add diversity to the data. Furthermore, the model is trained on images that were deliberately zoomed, tilted and rotated.

3.4 *CNN Model: GreenNet101*

In this research, a custom Convolutional Neural Network (CNN), GreenNet101, was devised using the Keras Sequential API to analyze images and differentiate between diseased and healthy leaves of the potato plant. A comparison of this new method is made with traditional computer learning techniques. Convolutional and max pooling layers are the first layers in the architecture, and then fully connected layers come next. A layer that uses resizing and rescaling is used for preprocessing the input data. Reversed linear unit (ReLU) activation functions are used in the convolutional layers to introduce nonlinearity. 32 filters with a kernel size of (3,3) make up the first convolutional layer. Max pooling, with a pool size of (2,2), comes next. Similar configurations and 64 filters characterize subsequent convolutional layers. The network architecture is further enriched by max pooling layers after each convolutional layer to downsample the spatial dimensions. The output layer with'n classes' units and softmax activation for multiclass classification, a densely linked layer with 64 units and ReLU activation, and flattening the output are the last layers.

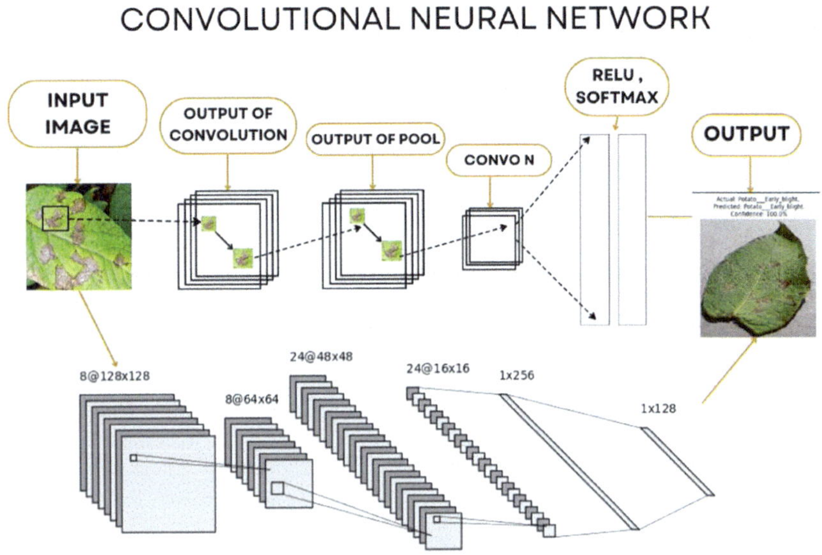

Fig. 1 CNN Model with detailed convolutional and pooling layers

During training, the model is optimized using categorical cross loss, and validation accuracy and loss are monitored to assess generalization performance. The Adam optimizer is employed, and the model undergoes training for multiple epochs, with early stopping to prevent overfitting.

Figure 1 illustrates the proposed convolutional neural community (CNN) structure hired for early blight and past due blight detection in potato plants. The input layer accepts RGB potato plant snap shots of length 128 × 128 pixels. These pics then pass through three convolutional layers, every followed by using a rectified linear unit (ReLU) activation feature and a max pooling layer. The first convolutional layer extracts 8 characteristic maps of length 64 × 64 pixels, whilst the second and third layers extract 24 and 24 function maps of step by step smaller sizes (48 × 48 and 16 × 16 pixels, respectively). These convolutional and pooling layers steadily lessen the spatial dimensions of the function maps at the same time as extracting increasingly more complicated capabilities relevant to blight detection. Finally, a completely related layer with 128 neurons accompanied with the aid of a softmax activation characteristic outputs the anticipated blight magnificence (early blight, overdue blight, or healthy) (Fig. 2).

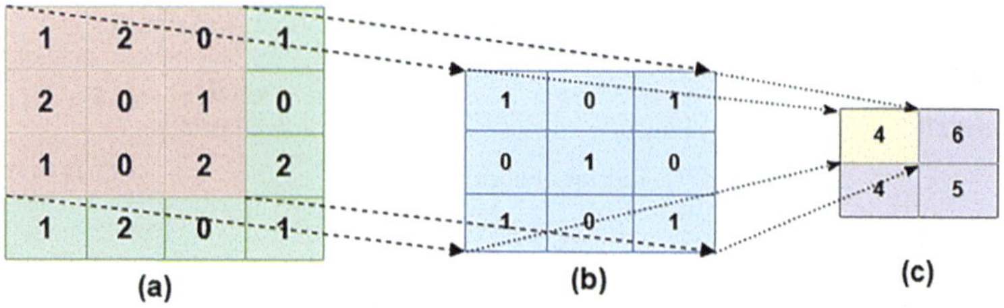

Fig. 2 Example of an operational convolution with a kernel size 3 × 3. **a**. Input matrix representing pixel intensities. **b**. Kernel matrix containing the known weights of the filter. **c**. Output matrix (feature map) resulting from the convolution operation, computed by element-wise multiplication of overlapping regions of (a) with (b), followed by summing the products for each position

3.5 Model Training Configurations

Hyperparameters:

1. Number of epochs: Total iterations over the dataset during training.
2. Early stopping, a practice endorsed to avoid overfitting.

Evaluation Metrics for Performance Assessment:

1. Accuracy: The proportion of correctly classified instances in a potato blight detection image dataset.
2. Loss: A measure of the dissimilarity between predicted and actual classes in the potato blight image dataset.
3. Validation Accuracy: The percentage of correctly classified instances on a separate validation set for potato blight leaves images, indicating model generalization performance.
4. Validation Loss: A measure of dissimilarity between predicted and actual classes on a validation set for potato blight detection images, helping assess model performance during validation.

3.6 Neural Network Architectures

Transfer learning enables the transferring of the knowledge learnt or gained by an already existing pre-trained model to a new one. It has been used for various applications like sentiment analysis, text classification, spam email detection, and many other generic natural language processing and image processing applications. Weiss et al. [32] there are several architectures available. The models chosen to be analyzed for this study were:

1. **InceptionV3:**

InceptionV3 is a convolutional neural network (CNN) structure comprising 48 layers, at the side of convolutional, pooling, and completely associated layers. It incorporates Inception modules offering 1×1, 3×3, and 5×5 convolutions and makes use of global common pooling in advance before the fully connected layers, culminating in approximately 23 million trainable parameters. Known for its depth and multi-scale feature extraction, InceptionV3 aims to balance computational efficiency and accuracy in image classification tasks.

2. **ResNet152v2:**

ResNet152v2 stands out as an advanced form of the ResNet framework, known for its Residual Neural Networks. With a staggering 152 layers, it uses special connections to tackle issues with gradients disappearing during training, making the learning process smoother. By incorporating skip connections and bottle-neck designs, ResNet152v2 achieves top-notch performance in various computer vision tasks, demonstrating its depth while also managing the computational load effectively.

3. **MobileNetV2:**

MobileNetV2 is a neural network architecture that is lightweight and finds its application in mobiles and cell phone devices. It has 53 layers, and it uses special connections to tackle issues with gradients being divided into training and testing. It has about 3.4 million parameters and it behaves as an efficient image classifier and detector.

4. **VGG16:**

VGG16, a CNN architecture, which consists of 16 weight layers, features mostly 3X3 convolutional filters stacked on top of each other which culminates in approx 138 million parameters. It is a simple and uniform structure which has 5 convolutional blocks followed by fully connected layers. It is a high computation model due to its depth and parameter count.

5. **DenseNet121:**

DenseNet121 is a convolutional neural network with densely connected layers. It has 121 layers, which are closely connected and give rise to about 8 million parameters. It reduces redundancy, enhances feature propagation and it maintains efficiency.

After running 10 epochs of each of the above 5 models and the custom neural network, the results have been noted into Table 1 as follows.

Figure 3 is a graph that indicates the training accuracy and validation accuracy achieved via the InceptionV3 version throughout a ten-epoch training run. Training accuracy (red line) regularly increases over a path of epochs, achieving about 98.61% accuracy by means of the stop of epoch 10. The validation accuracy (blue line), which shows the model's relevancy, additionally shows a growing fashion. It peaks at round 91.88% at epoch 5. This hole inside the various 2 traces highlights the importance of validation accuracy for the model's actual basic usual performance.

Table 1 Comparison of various CNN models at the end of 10 epochs

Name	Accuracy (%)	Loss (%)	Validation accuracy (%)	Validation loss (%)
InceptionV3	98.61	3.83	90.95	31.36
ResNet152v2	98.84	3.98	92.11	38.96
MobileNetV2	99.99	0.05	97.22	10.4
VGG16	99.88	0.44	96.29	12.04
DenseNet121	99.88	0.77	96.75	8.79
GreenNet101	98.24	5.61	97.40	5.21

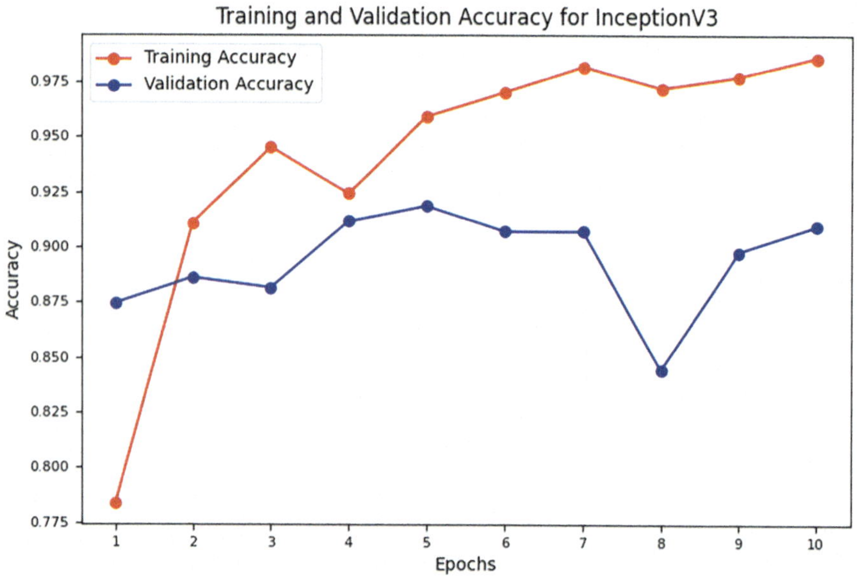

Fig. 3 Plotting InceptionV3's training and validation accuracy during the 10 epochs

Figure 4 is a plot that shows the training accuracy and validation accuracy achieved via the ResNet152v2 version throughout a ten-epoch training run. Training accuracy (red line) regularly increases over a path of epochs, achieving about 99.83% accuracy by means of the stop of epoch 8. The validation accuracy (blue line), which shows the model's relevancy, additionally shows a growing fashion. It peaks at around 98.84% at epoch 9. Interestingly, the validation accuracy upland after epoch 7, while the training accuracy continues to climb slightly.

Figure 5 is a graph that indicates the training accuracy and validation accuracy achieved via the MobileNetV2 version throughout a ten-epoch training run. Training accuracy (red line) regularly increases over a path of epochs, achieving about 99.99% accuracy by means of the stop of epoch 10. The validation accuracy (blue line), which shows the model's relevancy, additionally shows a growing fashion. It peaks at around 97.22% at epoch 10. Interestingly, the validation accuracy upland after epoch 8,

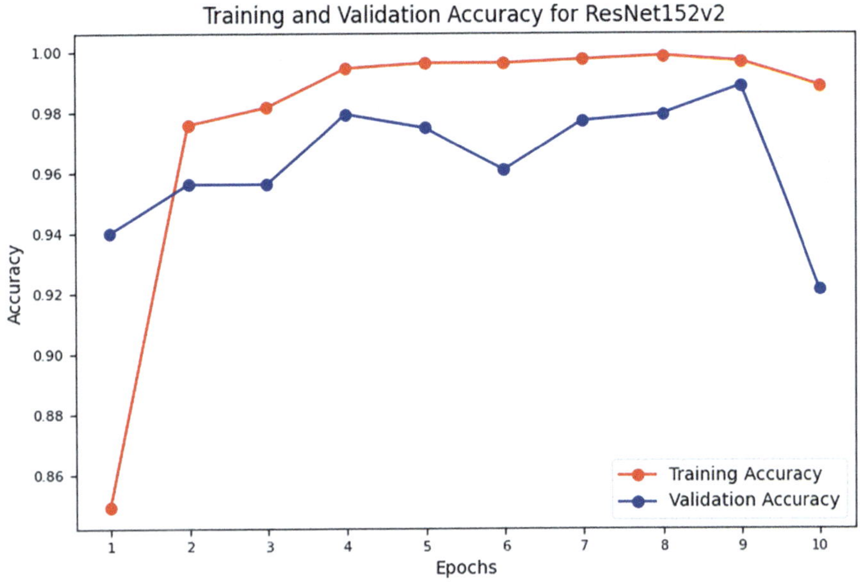

Fig. 4 Graph of training and validation accuracy of ResNet152v2 after 10 epoch

while the training accuracy continues to climb slightly. This observation highlights the importance of careful meta parameters, to optimize the model's relevancy while maintaining high training accuracy.

Figure 6 is a plot that shows the training accuracy and validation accuracy of VGG16 spanning across ten epochs. The training accuracy (red line) almost always increases along a range of epochs, reaching close to 99.88% by the tenth epoch. The validation accuracy (blue line) in addition demonstrates an increasing trend showing the model's relevancy. It reaches about 98.14% during epoch number 8. After the seventh epoch, validation accuracy increases, drops after eight; while training accuracy still increases slightly. This implies that model is over fit with respect to lab data hence requiring further tuning for relevance/validation significance.

Figure 7 is a graph that indicates the training and validation accuracy obtained through the DenseNet121 version over a ten-epoch training run. Training accuracy (red line) frequently goes up over a route of epochs, hitting practically 99.88% precision at the end of epoch 10. The growing fashion is also displayed by the blue line as validation accuracy represents relevancy of model. Reaches about 97.4% approximately at epoch 5. Interestingly, the validation accuracy upland after epoch 3 while training continues to gradually improve slightly.

Figure 8 is a plot that shows the performance of the custom CNN model, Green-Net101 for training and validation accuracy after 10 epochs, on early and late blight detection applied to potato plants. The two models show a gradually increasing trend, which gives evidence to the fact that they possess learning capabilities. Validation accuracy is always lower or near the training one during the whole process, thus

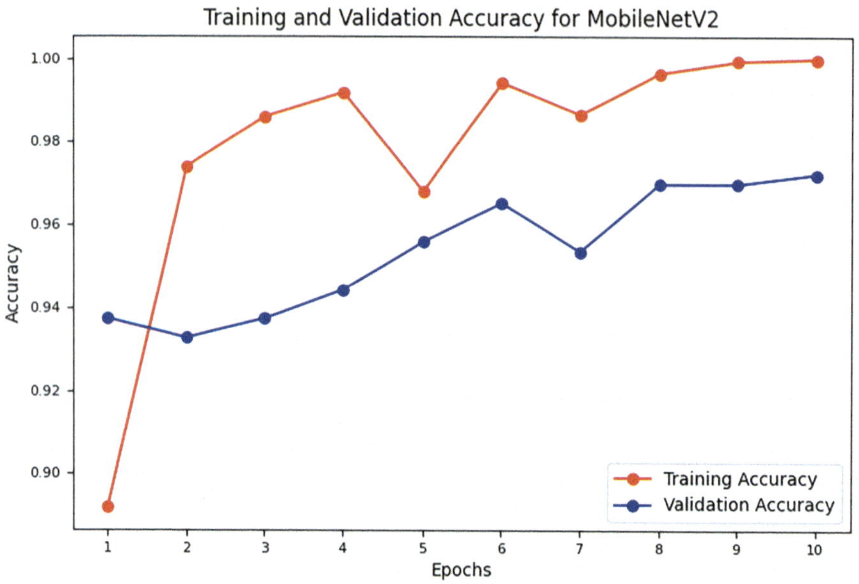

Fig. 5 Plot of the training accuracy and the validation accuracy of the model MobileNetv2 over 10 epochs

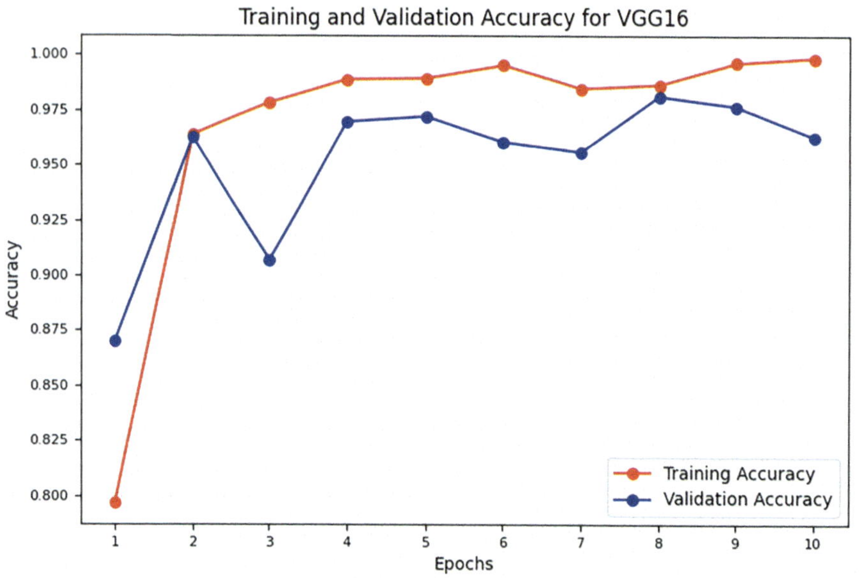

Fig. 6 Graph of VGG16's training accuracy and validation accuracy over the 10 epoch

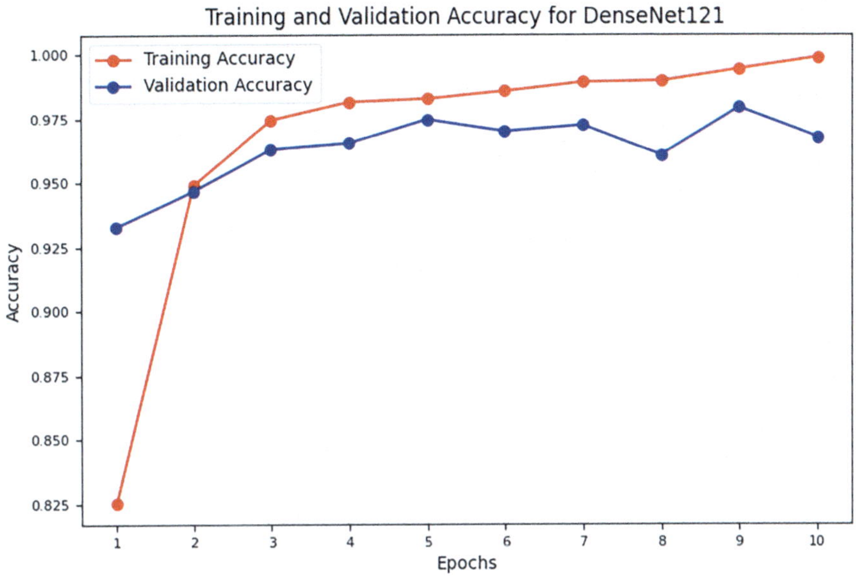

Fig. 7 Graph of training and validation accuracy of DenseNet121 after 10 epochs

indicating overfitting. Nevertheless, the two curves are very close together to show that there is generality in this model besides lowering concerns about overfitting. These findings therefore offer preliminary evidence of the performance efficacy in distinguishing healthy and diseased potato plants by means of using the proposed CNN architecture.

4 Future Scope

Implementing a custom Convolutional Neural Network (CNN), GreenNet101, to identify prevalent diseases in potato plants in Maharashtra would revolutionize farming practices in the region. The custom built model, GreenNet101, has shown an accuracy of 98.24% after 10 epochs. After 50 epochs it showed great improvement and yielded an accuracy of over 99.99%. While our custom-built neural network architecture has demonstrated promising results in the detection of plant diseases, there is still a great deal of room for improvement and research. The research paper's future scope provides opportunities for an in-depth exploration of neural network architectures in the detection of plant diseases by researchers and practitioners. To propel the field forward, we encourage the scientific community to consider fine-tuning our model or explore modifications to address potential limitations and optimize the performance of the model.

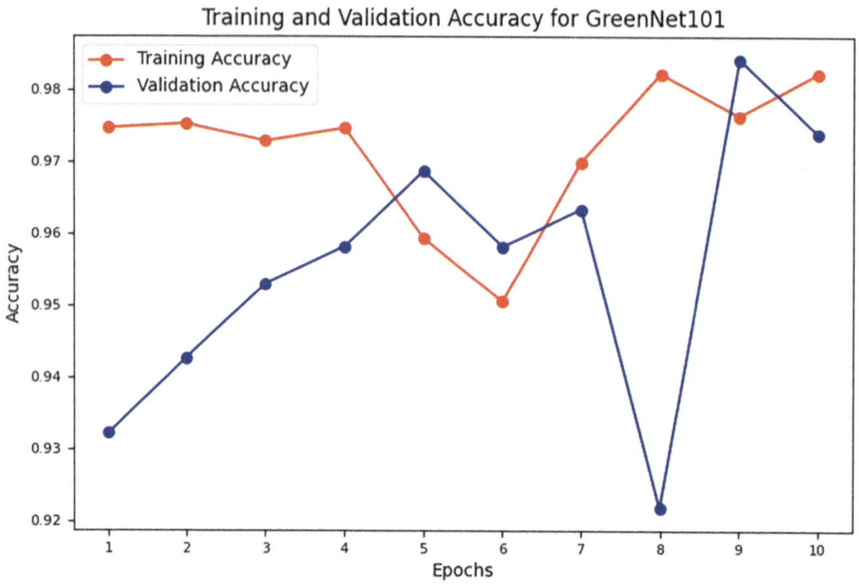

Fig. 8 Graph of training and validation accuracy of custom neural network GreenNet101, after 10 epochs

By customizing disease recognition to suit Maharashtra's specific potato crops, integrating this technology into a user friendly app or video intercom available in Marathi/Hindi/English, conducting farmer-centric training, gathering localized data, offering cost-effective measures, collaborating with local authorities, and providing real-time alerts, Maharashtra's potato farmers can benefit immensely. This approach ensures precise disease diagnosis and targeted solutions, leveraging accessibility, education, data accuracy, affordability, government support, and timely interventions to cater specifically to Maharashtra's potato farming landscape, ultimately bolstering crop yields and securing farmers' livelihoods in the region. This expansion involves integrating advanced learning algorithms to evaluate the severity of diseases in different constituents of plants, that includes their stems, fruits, and leaves. Moreover, the system's utilization of drones for extensive surveying and targeted pesticide application in affected areas showcases an innovative approach to enhancing crop health and management techniques.

5 Result

In the paper, 5 pre-trained models (InceptionV3, ResNet152v2, MobileNetV2, DenseNet121, VGG16) and 1 custom made model (GreenNet101) were taken, trained and tested and then compared based on their results. Each model showed fair and satisfactory results based on their premises and architecture. MobileNetV2 emerged as the best model among the six above-mentioned models.

With the highest accuracy rate of 99.9% among all the 6 models used, MobileNetV2 has outdone all the other models tested in the study. Moreover, it has exhibited an impressively low loss of 0.05%, showcasing its efficiency in minimizing errors during the training process. The combination of good accuracy and very less loss positions MobileNetV2 as the leading choice for detecting diseases in potato plants.

This discovery not only highlights how effective the MobileNetV2 model is but also has important implications for improving the monitoring of crop health and supporting sustainable agricultural practices. In simpler terms, it means that using MobileNetV2 can help in keeping a closer eye on the well-being of crops, ultimately contributing to more sustainable and eco-friendly farming methods.

6 Conclusion

For many years, plant diseases have been a major problem in precision farming. By making optimal decisions based on the outcomes of DL techniques, precision farming has made it possible to detect diseases early and minimize losses. New developments in deep learning yield very accurate results, and quick processing has been made possible by hardware being easily available. The decision-making procedure can be enhanced, though. When evaluated in real-world conditions, the models that are now available do not yield results that are up to the mark. This served as motivation, and following up on the authors' earlier research, a novel method for plant disease detection was put forth in an effort to get beyond the main obstacles to its practical application.

The dataset used in the paper contains photos of leaves in their natural surroundings, in diverse weather conditions and at different angles. As a result, the dataset becomes more extensive, which improves the model's classification accuracy and practical utility. However, an insufficient number of instances is a universal problem since it causes overfitting. To address this issue, numerous augmentation strategies were applied. In this paper, a new custom neural network model was created to address the specific challenges in the given domain. The model was meticulously crafted and trained extensively on the same dataset. It gave an accuracy of 98.24% and loss of 5.61% with a validation accuracy of 97.40% and validation loss of 5.21%. Finally, the MobileNetV2, a two-stage design, was presented for detecting plant diseases efficiently. The trained model attained an accuracy of 99.9% and loss of 0.05% with a validation accuracy of 97.22% and validation loss of 10.4% on the PlantDisease dataset. The main advantage of using this model is that the user can directly apply this model on their dataset. This model doesn't have to be built. It is a pre-existing and a pre-trained model. It is easy to use and yields high accuracy with less loss.

Further research should focus on identifying diseases on different locations of the plant as well as at the different stages of the diseases. MobileNetV2 is recommended to identify diseases easily and accurately.

References

1. Sahoo K, Sethi N (2012) Investigating the impact of agriculture and industrial sector on economic growth of India OIDA. Int J Sustain Dev 5(5):11–21
2. Jones RAC, Barbetti M (2012) Influence of climate change on plant disease infections and epidemics caused by viruses and bacteria. CABI Rev. https://doi.org/10.1079/pavsnnr20127022
3. Mhatre PH, Lekshmanan DK, Palanisamy VE, Bairwa A, Sharma S (2021) Management of the late blight (Phytophthora infestans) disease of potato in the southern hills of India. J Phytopathol. https://doi.org/10.1111/jph.12958
4. Mollah MMI, Hassan N (2023) Efficacy of Trichoderma harzianum, as a biological fungicide against fungal diseases of potato, late blight and early blight. J Nat Pesticide Res. https://doi.org/10.1016/j.napere.2023.100047
5. Sharma S, Anand V, Singh S (2021) Classification of diseased potato leaves using machine learning. In: 2021 10th IEEE International Conference on Communication Systems and Network Technologies (CSNT) (2021). https://doi.org/10.1109/CSNT51715.2021.9509702
6. Barbedo JGA (2018) Factors influencing the use of deep learning for plant disease recognition. Biosyst Eng. https://doi.org/10.1016/j.biosystemseng.2018.05.013
7. Abade A, Ferreira PA, Vidal FDB (2021) Plant diseases recognition on images using convolutional neural networks: A systematic review. Comput Electron Agric. https://doi.org/10.1016/j.compag.2021.106125
8. Singh V, Sharma N, Singh S (2020) A review of imaging techniques for plant disease detection. Artif Intell Agric. https://doi.org/10.1016/j.aiia.2020.10.002
9. Paymode AS, Malode VB (2022) Transfer learning for multi-crop leaf disease image classification using convolutional neural network VGG. Artif Intell Agric. https://doi.org/10.1016/j.aiia.2021.12.002
10. Niu S, Liu Y, Wang J, Song H (2020) A decade survey of transfer learning (2010–2020). IEEE Trans Artif Intell. https://doi.org/10.1109/TAI.2021.3054609
11. Arshad MS, Rehman UA, Fraz MM (2021) Plant Disease Identification Using Transfer Learning. In: 2021 International Conference on Digital Futures and Transformative Technologies (ICoDT2). https://doi.org/10.1109/ICoDT252288.2021.9441512
12. Sinshaw NT, Assefa BG, Mohapatra SK (2021) Transfer learning and data augmentation based cnn model for potato late blight disease detection. In: 2021 International Conference on Information and Communication Technology for Development for Africa (ICT4DA). https://doi.org/10.1109/ICT4DA53266.2021.9672243
13. Singh G, Yogi KK (2023) Comparison of RSNET model with existing models for potato leaf disease detection. Biocatal Agric Biotechnol. https://doi.org/10.1016/j.bcab.2023.102726
14. Li L, Zhang S, Wang B (2021) Plant Disease Detection and Classification by Deep Learning—A Review. IEEE Access. https://doi.org/10.1109/ACCESS.2021.3069646
15. Rashid J, Khan I, Ali G, Almotiri SH, AlGhamdi MA, Masood K (2021) Multi-level deep learning model for potato leaf disease recognition. Electron 2021. https://doi.org/10.3390/electronics10172064
16. Mohanty SP, Hughes DP, Salathe M (2016) Using deep-learning for image-based disease detection. METHODS artic Front Plant Sci 22 Sept 2016 Sec Tech Adv Plant Sci. https://doi.org/10.3389/fpls.2016.01419
17. Chakraborty KK, Mukherjee R, Chakroborty C, Bora K (2022) Automated recognition of optical image based potato leaf blight diseases using deep learning. Physiol Mol Plant Pathol. https://doi.org/10.1016/j.pmpp.2021.101781

18. Li G, Ma Z, Wang H (2012) Image recognition of grape downy mildew and grape powdery mildew based on support vector machine. In: Li D, Chen Y (eds) Computer and Computing Technologies in Agriculture V. CCTA 2011. IFIP Advance International Conference Technology. https://doi.org/10.1007/978-3-642-27275-217
19. Geetharamani G, Pandian JA (2019) Identification of plant leaf diseases using a nine-layer deep convolutional neural network. Comput Electr Eng. https://doi.org/10.1016/j.compeleceng.2019.04.011
20. Chen J, Liu Q, Gao L (2019) Visual tea leaf disease recognition using a convolutional neural network model. Symmetry 2019. https://doi.org/10.3390/sym11030343
21. Feng J, Hou B, Yu C, Yang H, Wang C, Shi X, Hu Y (2023) Research and validation of potato late blight detection method based on deep learning. Agronomy 2023. https://doi.org/10.3390/agronomy13061659
22. Arnaud S, Rehema N, Aoki S, Kananu M (2022) Comparison of deep learning architectures for late blight and early blight disease detection on potatoes. Open J Appl Sci. https://doi.org/10.4236/ojapps.2022.125049
23. Al-Adhaileh MH, Verma A, Aldhyani THH, Koundal D (2023) Potato blight detection using fine-tuned CNN architecture. Mathematics 2023. https://doi.org/10.3390/math11061516
24. Akbar M, Ullah M, Shah B, Khan RU, Hussain T, Ali F, Alenezi F, Syed I, Kwak KS (2022) An effective deep learning approach for the classification of Bacteriosis in peach leave. Front Plant Sci. https://doi.org/10.3389/fpls.2022.1064854
25. Bedi P, Gole P (2021) Plant disease detection using a hybrid model based on convolutional autoencoder and convolutional neural network. Artif Intell Agric 5:90–101. https://doi.org/10.1016/j.aiia.2021.05.002
26. Anami BS, Malvade NN, Palaiah S (2020) Deep learning approach for recognition and classification of yield affecting paddy crop stresses using field images. Artif Intell Agric 4:12–20. https://doi.org/10.1016/j.aiia.2020.03.001
27. Ramesh S, Vydeki D (2020) Recognition and classification of paddy leaf diseases using Optimized Deep Neural network with Jaya algorithm. Inf Process Agric 7(2):249–260. https://doi.org/10.1016/j.inpa.2019.09.002
28. Gonzalez-Huitron V, Leoń-Borges JA, Rodriguez-Mata AE, Amabilis-Sosa LE, Ramírez-Pereda B, Rodriguez H (2021) Disease detection in tomato leaves via CNN with lightweight architectures implemented in Raspberry Pi 4. Comput Electron Agric 181:105951. https://doi.org/10.1016/j.compag.2020.105951
29. Ji M, Zhang L, Wu Q (2020) Automatic grape leaf diseases identification via United Model based on multiple convolutional neural networks. Inf Process Agric 7:418–426. https://doi.org/10.1016/j.inpa.2019.10.003
30. Ahad MT, Li Y, Song B, Bhuiyan T (2023) Comparison of CNN- based deep learning architectures for rice diseases classification. Artif Intell Agric. https://doi.org/10.1016/j.aiia.2023.07.001
31. Kaggle dataset- PlantVillage https://www.kaggle.com/datasets/arjuntejaswi/plant village. Last accessed 30 Dec 2023
32. Weiss K, Khoshgoftaar TM, Wang D (2016) A survey of transfer learning. J Big Data. https://doi.org/10.1186/s40537-016-0043-6

Intelligent Monitoring and Crop Yield Systems-I

GrainLog: An IoT-ML Based Food Grain Monitoring and Shelf-Life Prediction System for the Farmer

Vaidehi Sinha, Chaitanya Kulkarni, Ujjwala Rawandale, Abhishek Hosur, Rishabh Humbare, and Trushita Chaware

Abstract The wastage of foodgrains due to improper storage conditions is a critical issue affecting the Indian agriculture sector. In India 10% of the total foodgrain is wasted every year. This work introduces GrainLog, an advanced solution designed to modernize grain storage by leveraging Internet of Things (IoT) and Machine Learning (ML) technologies. GrainLog enables remote monitoring of granaries through IoT sensors and provides insights via a user-friendly app, helping farmers make informed decisions. The system features a microcontroller-based sensor network for real-time grain condition monitoring, with data processed on a central cloud server. A method is proposed which utilizes ML algorithms trained on extensive datasets, for predicting grain shelf life and enhancing storage management. Its collaborative, crowdsourced data approach amplifies its utility, making it accessible and beneficial for a wide range of users, including those in rural areas. GrainLog aims to reduce food grain wastage, improve supply chain efficiency, and safeguard the economic interests of farmers and intermediaries.

Keywords Machine learning · Internet of things · GrainLog · Agriculture

1 Introduction

Agricultural practices prove to be India's largest source of livelihoods with two thirds of its total population dependent on agricultural produce as their main source of income [1]. The cultivation of food crops takes up the highest percentage in the

V. Sinha (✉) · C. Kulkarni · U. Rawandale · A. Hosur · R. Humbare · T. Chaware
Dr. Vishwanath Karad, MIT World Peace University, Survey No. 124, Paud Road, Kothrud, Pune, Maharashtra 411038, India
e-mail: vaidehisinha30@gmail.com

U. Rawandale
e-mail: ujjwala.rawandale@mitwpu.edu.in

T. Chaware
e-mail: trushita.chaware@mitwpu.edu.in

A. J. Kulkarni and M. Hajiaghaei-Keshteli (eds.), *Advances in Intelligent Systems for Sustainable Agriculture*, Lecture Notes in Networks and Systems 1201,
https://doi.org/10.1007/978-981-97-9839-1_5

total agricultural produce in India. This is due to the fact that India has to supplement its ever-growing population with food in order to keep the growth aspects in a steady state. This leads to cultivation and storage of grains being heavily invested in every budget declaration made by the governments.

The total production of food grains in India is estimated at 329.68 million tons while on the other hand around 74 million tons of food grain wastage happens every year in India [2] which accounts for 22% of foodgrain output. As per the research of Indian Council of Agriculture Research (ICAR), duration and time of storage plays a crucial role in wastage of food along with other parameters like climate conditions, crops and commodities etc.

The majority of this can be attributed to improper storage conditions and inadequate storage facilities for the storage of the food grains. This is the main motivation behind the research and development of GrainLog—to highlight and solve the problem of improper storage of grains in the granaries and warehouses. This is often exacerbated by the 'keep even after the expiration date' practices in the storage warehouses. Letting the grains go to waste when being stored improperly in the warehouse is an injustice to farmer's time, resources, manpower, and irrigation that went behind those crops. Beyond economic losses to the farmers and traders, it also affects food security, supply chain stability and market prices.

The produce in a warehouse can become spoiled due to unfavorable conditions like temperature fluctuations, elevated humidity levels, pest infestations, extended storage and poor-quality control [3]. Usually, stored grains are checked manually by farmers for visual signs of degradation. But this method is not reliable as visual signs develop very late. It also consumes a lot of time, effort and manpower. Further, there is a lack of awareness regarding the actions that need to be taken to control the degradation occurring due to various factors. This poses a need for a technological solution for monitoring of the food grains along with a dynamic system for the prediction of shelf life which can empower the farmers to take crucial and timely decisions with respect to their grains.

Thus, the objectives of the proposed solution, GrainLog are as follows:

- Reduction of the spoilage of the food grains and retaining their quality for a longer period of time.
- Remote monitoring and proactive identification of issues related to the grains allowing farmers to take prompt action for their mitigation.
- Prediction of the quality and shelf life of the grains to maximize the profitability of farmers by allowing them to take in-time sell or hold decisions.

In order to achieve these objectives, the application of Machine Learning (ML) and Internet of Things (IoT) plays a pivotal role as data driven tools are proving to be of great use in solving complex challenges like this in various domains including Agriculture. Considering all above problems, the solution GrainLog is proposed in this work, which aims to revolutionize and optimize primitive grain storage systems.

The proposed solution is a one-of-a-kind data-driven solution which not only tackles the grain storage challenges at the farmer level but also helps to overcome them. Going beyond conventional grain monitoring systems, GrainLog encompasses a range of innovative features like shelf-life prediction, collaborative knowledge sharing, grain quality reports, actionable insights, learning resources, etc. which are less focused on in the existing systems as per the study carried out. The machine learning based shelf-life prediction method proposed in this research is a pioneering approach in grain storage monitoring.

2 Literature Review

In the constantly shifting terrain of agricultural technology, the amalgamation of IoT and ML has brought about a new era of effectiveness and accuracy. This literature review plunges into significant publications that investigate the combination of these technological advances in the fields of agricultural grain monitoring and shelf-life prediction. It strives to give an in-depth assessment of the current landscape and prospective opportunities for IoT and ML use cases in agriculture. The study analyzes noteworthy studies and publications that have contributed significant improvements in their applications in grain quality control, storage optimization, and predictive analytics. This exploration not only highlights the technological advancements in the field but also sheds light on the practical implications and benefits for farmers and the agricultural industry at large. The insights gleaned from this review form the cornerstone for the development of "GrainLog," an innovative IoT-ML based system designed to revolutionize food grain monitoring and extend its shelf-life, thereby enhancing agricultural productivity and sustainability.

2.1 IoT Applications in Agricultural Grain Monitoring

The integration of IoT in agriculture, particularly for grain monitoring, represents a significant shift towards data-driven farming. Reference [4] provides a detailed exploration of how real-time monitoring systems, enabled by IoT technologies, can lead to more efficient farming practices. The work particularly emphasizes the importance of IoT in monitoring environmental conditions, such as soil moisture and weather patterns, which are crucial for grain cultivation and storage. Expanding on the application of IoT in agriculture [5] delves into the transformative effects of IoT in agricultural resource management. It discusses how IoT devices can optimize water usage, fertilizer application and pest control, leading to higher crop yields and better-quality grains. The paper stands out for its focus on the practical applications of IoT in reducing the environmental footprint of farming.

Reference [6] takes a narrower approach, concentrating on the use of IoT for monitoring grain storage conditions. It explores various innovative IoT solutions for tracking parameters like temperature, humidity and grain quality. The study is particularly insightful in discussing how IoT can prevent grain spoilage and reduce post-harvest losses. Deployment of IoT sensors in grain storage facilities is examined [7]. It highlights how these sensors provide continuous monitoring, enabling early detection of spoilage factors like excess moisture or pest infestations. The research provides a compelling case for the adoption of IoT sensors in maintaining the integrity of stored grains. Broader implications of IoT in the agricultural supply chain is investigated in [8]. This study details how IoT can enhance the traceability of grains from farm to table, improve inventory management, and streamline the distribution process, thereby ensuring the freshness and quality of grains upon reaching consumers.

2.2 Machine Learning Techniques in Grain Shelf-Life Prediction

Moving to the realm of ML, [9] serves as a comprehensive review of the application of ML in agricultural settings. The study discusses various ML techniques, such as regression analysis and neural networks, and their effectiveness in predicting crop yields and grain shelf-life. The paper is notable for its discussion on how ML can process vast amounts of agricultural data to make accurate predictions, which are crucial for effective grain storage and management. Role of ML in agricultural data analysis is analyzed in [10]. It explores how ML algorithms can identify patterns in data that are not immediately apparent to humans. The work is particularly relevant in the context of grain shelf-life prediction, as it discusses how ML can be used to analyze data from sensors monitoring grain storage conditions, thereby predicting potential quality degradation.

Reference [11] provides an in-depth analysis of various predictive models used to estimate the shelf-life of stored grains. It evaluates the accuracy of different models, including time-series analysis and machine learning approaches, in predicting shelf-life based on factors like moisture content, temperature, and grain type. It is significant for its direct application in optimizing grain storage strategies. Various ML algorithms in their effectiveness for shelf-life prediction are compared in [12]. It specifically focuses on supervised learning techniques and how they can be trained on historical data to accurately predict the shelf-life of different grain types. It is an essential reading for understanding the practical application of ML in agricultural product management. Reference [13] highlights the use of data analytics, a key component of ML, in the assessment of grain quality during storage and transportation. It discusses how data analytics can be employed to detect early signs of quality deterioration, thereby enabling timely interventions. This study underscores the importance of ML in maintaining the overall quality and safety of grains stored in granaries.

2.3 Comparison with Existing Methodology

Dhande et al. [14] developed a mixed solar dryer that integrates traditional grain drying methods with solar technology to improve postharvest handling. This innovative approach aims to reduce reliance on non-renewable energy sources for drying agricultural produce, offering a sustainable solution that can be particularly beneficial in regions with ample sunlight. The development signifies an important step towards enhancing food security by minimizing postharvest losses, which are a significant concern in general agricultural settings.

Gaur [15] focuses on comparing traditional and modern storage methods in India, emphasizing the enduring relevance of indigenous practices. The study delves into various traditional methods, such as using natural materials and structures to store grains, and evaluates their effectiveness in preserving food quality against pests and environmental factors. The findings suggest that, despite the advent of modern technologies, traditional storage methods still hold significant value, especially in rural areas where advanced facilities may not be readily accessible.

Karthikeyan et al. [16] document the use of traditional tools and agricultural practices in India, with a particular focus on postharvest grain handling. Their work highlights how these indigenous methods, passed down through generations, are not only culturally significant but also practical in their simplicity and sustainability. The study provides insights into the wisdom embedded in traditional agriculture, which can offer valuable lessons for contemporary sustainable farming initiatives.

Kiruba et al. [17] investigates traditional seed storage strategies against insect infestation in Tamil Nadu, India. It explores how ethnic communities utilize local knowledge and readily available materials to create effective storage solutions that protect seeds from pests without relying on chemical treatments. Their findings contribute to understanding the potential of integrating traditional ecological knowledge with modern pest management strategies to enhance food security.

Kumar and Kalita [18] examine the role of traditional storage structures in reducing postharvest losses in developing countries. Their research underscores the importance of these indigenous practices in maintaining grain quality and quantity, thereby strengthening food security in regions where modern storage facilities may be lacking. The study calls for a greater recognition of traditional knowledge systems in crafting comprehensive postharvest loss mitigation strategies.

Mann et al. [19] conduct a SWOT (Strengths, Weaknesses, Opportunities, Threats) analysis of traditional grain storage practices in India. Their work provides a nuanced evaluation of these practices, identifying their inherent advantages in terms of sustainability and adaptability, while also acknowledging the challenges they face in the context of modern agricultural demands. The analysis serves as a basis for developing integrated storage solutions that leverage the strengths of traditional methods.

This comparative study by [20] examines traditional food grain preservation and storage methods in Nigeria and India. It highlights the similarities and differences in how these two diverse cultures utilize indigenous knowledge to enhance food preservation. The research offers insights into the global relevance of traditional practices in ensuring food security and their potential integration into broader agricultural systems.

Reference [21] sheds light on indigenous practices for eco-friendly storage of food grains and seeds. Their study emphasizes the environmental and health benefits of traditional storage methods, which avoid the use of harmful chemicals and synthetic materials. By documenting these practices, the research contributes to the growing interest in sustainable agriculture and the conservation of biodiversity.

Reference [22] reviews ancient farming methods related to seed storage and pest management practices in India. Their work provides a historical perspective on how traditional agricultural practices have evolved to address storage challenges, highlighting the ingenuity and adaptability of ancient farmers. This retrospective analysis underscores the potential of historical knowledge in informing contemporary agricultural innovations.

Reference [23] describes indigenous grain storage structures in South Tamil Nadu, India, detailing their effectiveness and cultural significance. The study explores the architectural and material aspects of these traditional storage solutions, demonstrating how they are well-adapted to local environmental conditions and agricultural needs. The research underlines the importance of preserving these traditional practices as part of the cultural heritage and as viable options for sustainable agriculture.

3 Methodology

The project GrainLog is an innovative solution to the problem of grain spoilage in warehouses that enables smart monitoring and management of granaries by harnessing the usability of the IoT, ML, Sensor Technology and application development. It not only enables the farmers to remotely monitor their granary but also helps them to take necessary actions for maintaining the safety and quality of the grains. It empowers them to take informed decisions and mitigate any risks by utilizing the actionable insights provided through the user-friendly app.

The design of the solution consists of the Hardware part which is used for the on-ground monitoring, the cloud server where the data is processed and the user-friendly smartphone application for interaction with the user. There are 4 basic functions of the app. Firstly, the farmers can remotely monitor the grains. Secondly, notifications or SMS alerts will be sent to them if any abnormal change is detected along with what actions should be taken. For example, if the moisture content of the granary goes beyond the threshold limit farmers will be asked to increase the aeration of the granary. The app will also provide predictions of the shelf life and give suggestions so the farmers can make better decisions. For example, if nearby farmers are reporting insect infestations, other farmers will also be asked to take measures like fumigation. The app also educates the farmer about various protocols existing for the safe storage of the grains (Fig. 1). The following subsections explain the experimentation setup, software and the shelf-life prediction algorithm.

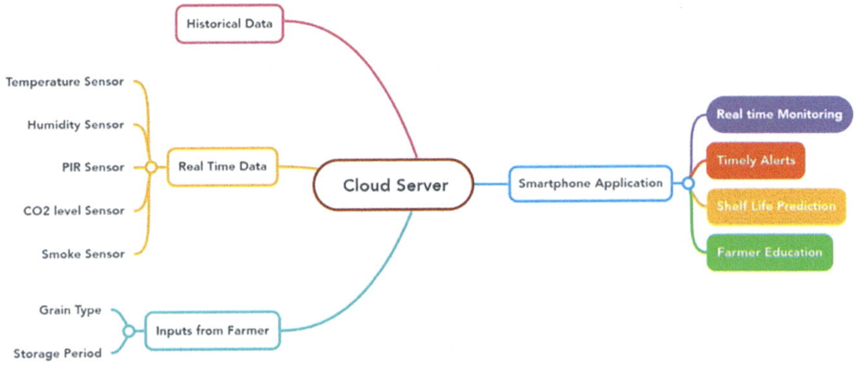

Fig. 1 Workflow of GrainLog

3.1 *Parameter Analysis and Experimentation Setup*

After careful consideration of various parameters affecting the quality of the grains in warehouses the following parameters are chosen for monitoring:

1. Temperature:

Temperature affects the rate of biological and chemical reactions in grains. High temperatures can accelerate the growth of molds, fungi, and insects, leading to spoilage and reduced grain quality while low temperatures can slow down these reactions but may increase the risk of condensation and moisture buildup if not properly managed [24].

2. Humidity:

Humidity levels directly impact grain moisture content. High humidity can promote moisture absorption in grains, leading to caking, mold growth, and mycotoxin production. On the other hand, low humidity levels can cause grain dehydration and loss of weight, affecting both quality and market value.

3. CO_2 Level:

CO_2 is a byproduct of biological activities such as respiration and microbial growth. Elevated CO_2 levels indicate increased microbial activity and metabolic processes within stored grains, which can accelerate spoilage and reduce shelf life. Detecting sudden changes CO_2 levels of more than 400-800ppm helps detect early signs of deteriorating grain quality [25].

4. Smoke:

Smoke detection is critical for fire prevention in grain storage facilities. Fires can result from various causes including electrical faults, mechanical failures, or external factors such as arson.

5. Rodents:

Infestation of rodents can cause contamination, spoilage and spread of diseases in the warehouse hence detecting their presence is crucial for safe keeping the grains.

Figure 2 depicts the experimentation setup consisting of the hardware which consists of various IOT enabled sensors that are connected to an Arduino microcontroller through which they communicate with the cloud using the WIFI module. These sensors include DHT11 Temperature and Humidity Sensor, HC-SR501 PIR Sensor, MQ-135 Smoke sensor and Sensirion SCD40 CO_2 Level Sensor. All the sensors are factory calibrated and their data is validated after testing.

The DHT11 sensor comes with a dedicated NTC that provides accurate readings of humidity and temperature within grain storage facilities within the tolerance level of $\pm 1\%$. Its compact design and cost-effectiveness make it suitable for monitoring environmental conditions crucial for grain preservation. The Sensirion SCD40 sensor uses NDIR technology to precisely measure carbon dioxide (CO_2) levels upto +-50 ppm accuracy. The HC-SR501 Passive Infrared (PIR) sensor enhances GrainLog's features by detecting movement within the storage area. They work by capturing the heat signatures emitted by the rodents as they move. The MQ135 Smoke Sensor monitors air quality by detecting Ammonia (NH3), sulfur (S), Benzene (C6H6), CO_2, and other harmful gasses and smoke. The ideal sensing condition for the MQ135 is 20 °C ± 2 °C at 65% ± 5% humidity.

The electronic circuit is converted to a PCB which is prototyped into a small device measuring 30 cm × 20 cm. The device is powered by a 9v battery and needs Wi-Fi connectivity for its operation. One device covers about 400 sq. ft area.

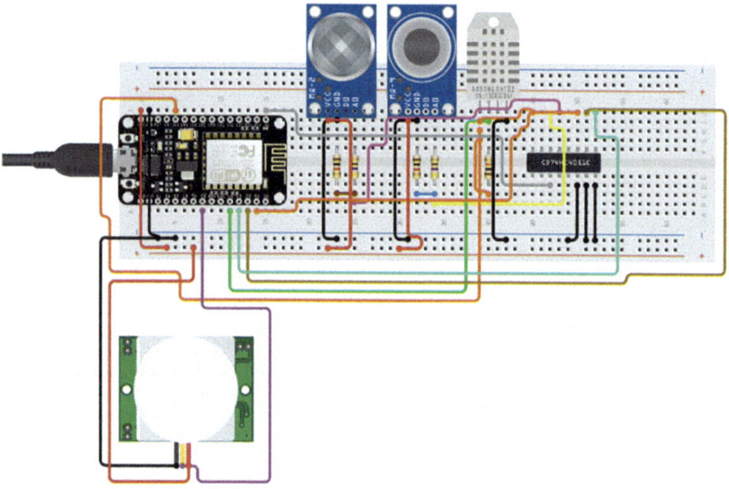

Fig. 2 Experimental setup of GrainLog for sensing different grain parameters

3.2 *Shelf-Life Prediction*

In this paper, we propose a novel approach for predicting the shelf life of grains in storage facilities using advanced machine learning techniques. While the conceptual framework and design have been thoroughly developed, the shelf-life prediction system has not yet been implemented in practice. The primary reason behind this is the unavailability of an existing dataset for training and testing the algorithm. Our methodology integrates a comprehensive data collection process, detailed preprocessing steps, and sophisticated feature engineering. We employ a mathematical formulation that blends time-series analysis with static data modeling, leveraging the strengths of Long Short-Term Memory (LSTM) networks and Gradient Boosting Machines (GBM). The culmination of this approach is an integrated model that combines these two powerful techniques to provide robust and accurate predictions of grain shelf life. This proposed framework is designed to address the complexities of grain storage management and optimize the preservation of grain quality over time. Figure 3 showcases the flowchart of the shelf-life prediction system.

Data Collection

This procedure begins with an extensive data collection process. In the realm of sensor data, we install a network of environmental sensors in the granary to continuously monitor key parameters that influence grain quality. These include temperature, humidity, CO_2 levels, and grain moisture content, recorded at regular intervals, possibly every hour or less. Alongside this real-time monitoring, it's crucial

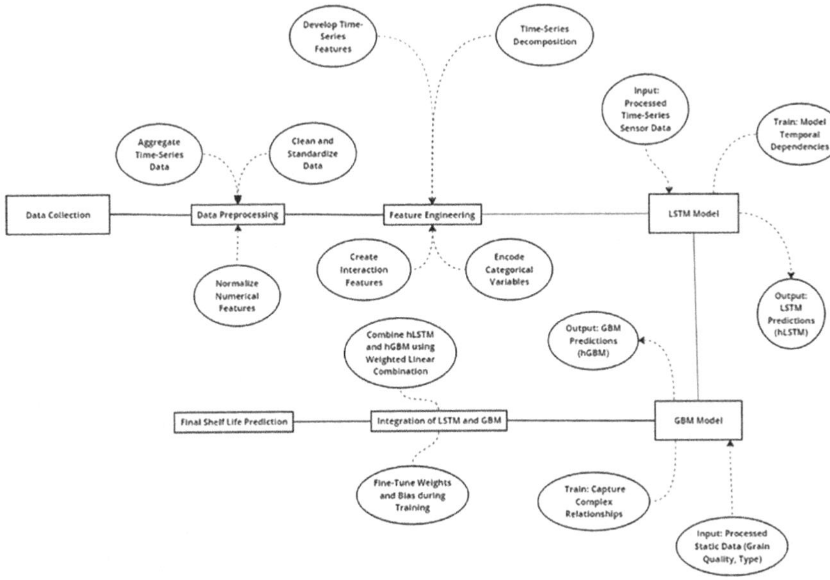

Fig. 3 Flowchart for shelf-life prediction

to collect comprehensive data on the initial quality of the grains being stored. This involves metrics like moisture content, protein content. Additionally, we incorporate a historical perspective by gathering past data on grain storage. This historical dataset includes information on the type of grain stored, conditions under which they were stored, and the actual shelf life observed under those conditions. Along with this the collaborative approach of crowdsourcing the data is used which creates a dynamic system in which the participating farmers also benefit from each other's knowledge and experiences. As the system is deployed and operationalized, it is anticipated that GrainLog will generate its own dataset through real-world monitoring of the storage conditions.

Preprocessing

Once data collection is complete, we proceed to the preprocessing stage, which is pivotal for preparing the data for effective analysis. The initial step involves cleaning the data, which includes addressing missing or erroneous sensor readings through appropriate imputation techniques or outright removal of anomalies. Additionally, the data collected from different sources often need standardization to ensure format consistency. Following cleaning, we normalize the numerical features, especially crucial for the sensor data, to ensure they are on a similar scale. This normalization facilitates the learning process of the machine learning models by mitigating issues that arise from features being in different ranges. For the time-series data from sensors, we aggregate the data into meaningful time windows, like daily averages, to reduce noise and capture more significant trends. Similarly, static data, such as grain type and initial quality metrics, are standardized for consistency across the dataset.

Feature Engineering

Feature engineering is a critical phase where we transform raw data into informative features that can significantly improve the performance of our predictive models. For the time-series sensor data, we develop features like rolling averages, moving averages, and lagged values. These features help capture the temporal dynamics and trends within the data. Furthermore, we apply time-series decomposition techniques to extract underlying trends and seasonal components, which can be highly indicative of environmental influences on grain quality. Another aspect of feature engineering involves creating interaction features. These are designed to model the combined effects of different environmental conditions, such as the interaction between temperature and humidity, which can have a compounded impact on grain shelf life. Lastly, we handle categorical variables, like grain type, through encoding techniques such as one-hot encoding or label encoding, making them suitable for use in machine learning algorithms.

Mathematical Formulation and Machine Learning Model

The core of our methodology lies in the mathematical formulation and development of the machine learning model. Our objective is to create a predictive function that maps our engineered features to the target variable, which in this case is the shelf life of the grains. This function incorporates time-series sensor, static grain quality metrics and grain type as inputs.

f = Predictive Function.

T = Shelf Life of the Grains.

$x_{quality}$ = Static Grain Quality Metrics.

x_{sensor} = Time-Series Sensor Data.

s_{type} = Grain Type.

To effectively model these diverse inputs, a combination of LSTM (Long Short-Term Memory) networks and Gradient Boosting Machine (GBM). LSTM networks are exceptionally suited for modeling the temporal dependencies in the time-series sensor data. These networks are designed to remember long-term dependencies and are particularly effective in handling time-series data under varying conditions. On the other hand, GBM is employed to model the static features, such as grain quality and type. GBM, an ensemble learning method, sequentially builds decision trees, with each tree aiming to correct the errors of its predecessor. This method is adept at capturing complex, non-linear relationships between features.

Integration of LSTM and GBM

The culmination of our methodology is the integration of the LSTM and GBM models. The final prediction model is a sophisticated fusion of these two powerful techniques. We achieve this integration through a weighted linear combination of the outputs from both models. These weights, along with a bias term, are parameters that are fine-tuned during the training process. The formula for this integration is represented as:

$$T = \alpha \times h_{LSTM} + \beta \times h_{GBM} + \gamma,$$

where h_{LSTM} and h_{GBM} are the outputs from the LSTM and GBM models, respectively, and α, β, γ are the learned parameters.

Integrating Long Short-Term Memory (LSTM) networks with Gradient Boosting Machines (GBM) combines the strengths of both models, making it a preferable approach for complex predictive tasks like grain shelf-life prediction. LSTMs are adept at capturing temporal dependencies in time-series data, crucial for monitoring environmental changes over time, while GBMs excel in handling static features and intricate non-linear relationships, essential for assessing grain characteristics and initial quality metrics. This synergy enhances predictive accuracy, ensures adaptability to diverse data types, and improves model generalization, making the integrated approach robust and efficient for real-world applications in agricultural storage management.

3.3 Mobile Application

The mobile application component of the GrainLog project serves as the central interface for users, providing seamless interaction and access to critical information regarding grain storage. It is a multi-lingual and user-friendly smartphone application that empowers farmers with real-time monitoring capabilities, proactive notifications, and insightful predictions to optimize grain management practices. Through its intuitive interface, users can remotely monitor grain conditions, receive timely alerts, and access educational resources aimed at enhancing storage safety and maximizing grain quality.

There are two versions of the homepage based on the shelf life predicted as shown in Fig. 4. Along with this, it displays an array of real-time environmental metrics—temperature, humidity, CO_2 levels, and rodent detection—culminating in a vibrant, color-coded dashboard. Additional features include current weather conditions and agricultural news, offering a holistic view for the user.

The “Device” screen of the application Fig. 5 presents a detailed status overview of the connected monitoring device and its sensors. It shows the device as being connected, with toggles for location, voice control, and alarm settings, and displays the battery level. Below, sensor statuses are graphically represented through trend

Fig. 4 Home Screen of the GrainLog App depicting two scenarios of shelf-life prediction

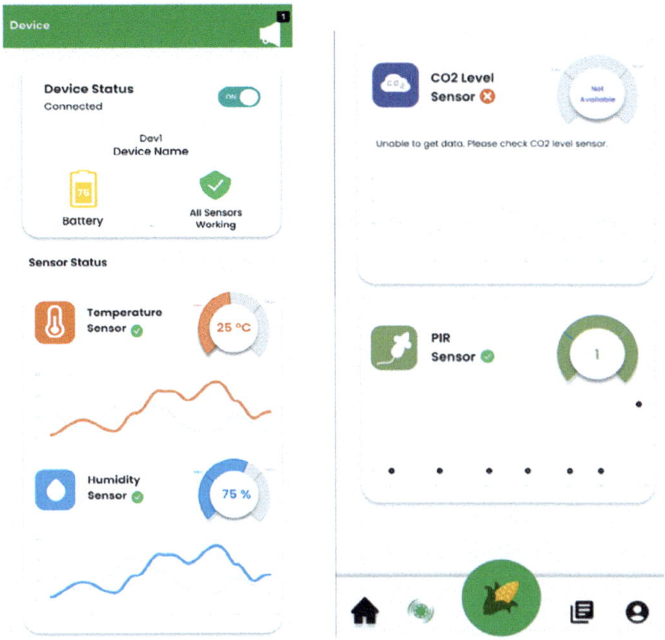

Fig. 5 Device Screen of the GrainLog App depicting the status of the devices connected

lines for temperature, humidity, CO_2 levels, and the PIR sensor, providing the user with a comprehensive real-time and historical data view.

The "Temperature Sensor"- Fig. 6 screen displays a graph with temperature trends over time, including a seven-day average, minimum, and maximum temperature readings. It indicates the sensor's operational status and provides an informative note on the importance of temperature for crop growth and development. Additionally, the screen offers a section for historical data, presenting a table with specific date ranges alongside their corresponding minimum, maximum, and average temperature readings, offering a detailed historical perspective on the temperature conditions within the storage environment. Similar screens are curated for all the sensors.

The "Grain" screen Fig. 7 provides a comprehensive overview of a specific batch of grain, starting with basic details such as grain type, quantity, and storage date. It progresses to a "Grain Report" section that includes a "Grain Quality Rating" and detailed textual analysis under "Grain Condition," which assesses the current state of the. The "Shelf Life" subsection predicts the remaining shelf life of the grain, with insights into environmental factors that may affect it. "Recommendation" provides actionable advice based on the current humidity levels within the warehouse. At the bottom, the "Ideal Parameters for Wheat Storage" table lists the optimal ranges for key storage conditions to maintain grain quality.

Figure 8 shows a user profile page, summarizing personal details and displaying grain information such as grain type, quantity, in-date, and remaining shelf life.

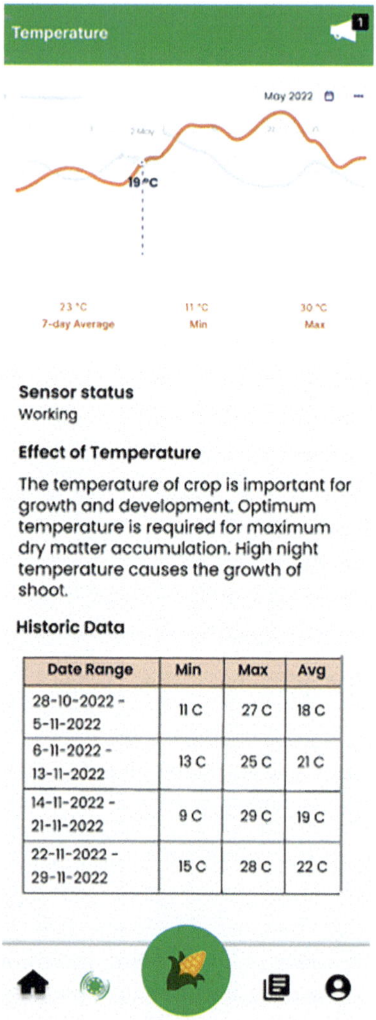

Fig. 6 Sensor-wise screen of the GrainLog App—DHT11 Sensor

Along with this it shows Additional Resources Section comprising of 'Education'—a selection of articles with a videos tab that provide educational content, 'Communities' - categories like 'Regional' and 'Issues' for users to explore and 'Articles'—users can search for and read different articles focusing on topics under 'Agriculture'.

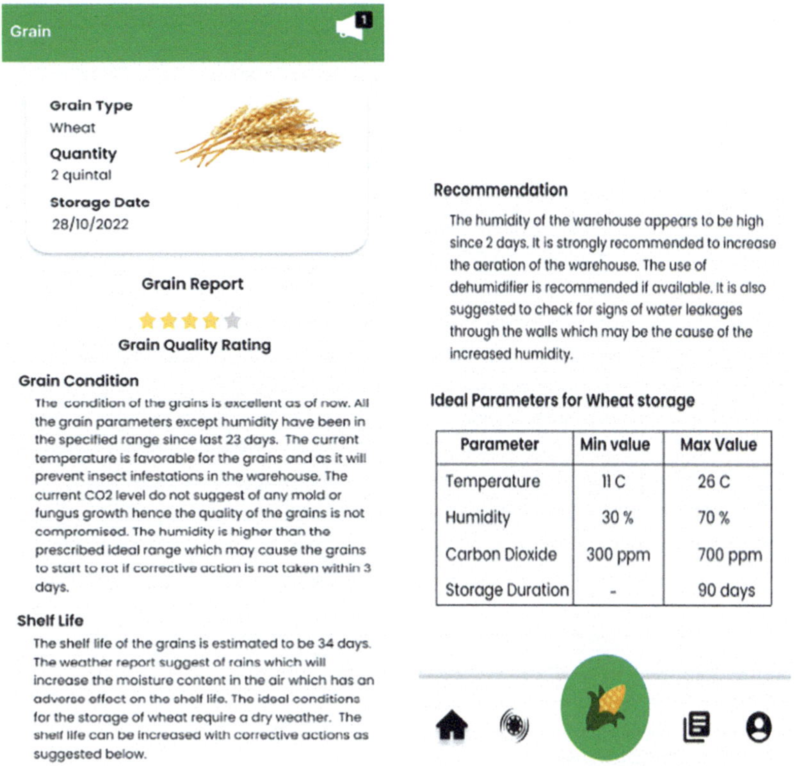

Fig. 7 Grain Screen of the GrainLog App depicting the status and optimal condition for every grain

Fig. 8 User Profile Page and Additional Resources (Articles, Videos, and Communities)

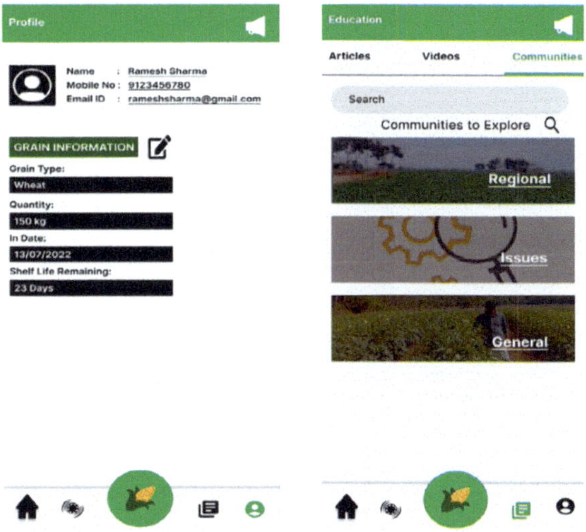

4 Results and Analysis

The implementation of GrainLog yielded promising results across various key performance indicators, demonstrating its efficacy in addressing the challenges associated with food grain storage. The technological robustness of GrainLog along with its user-friendliness, scalability, cost-effectiveness and multilingual usage make it widely accessible including in rural areas catering to a diverse range of farmers and intermediaries, from small to large scale.

Table 1 compares the features of GrainLog with those found in various research papers. GrainLog combines Internet of Things (IoT) and machine learning to offer a complete solution for keeping track of grains, controlling pests and moisture, and ensuring the quality of food. GrainLog stands out because it aims to be sustainable, affordable, and easy to use on a large scale.

The IoT enabled sensors showed commendable performance, accuracy and agility in capturing the real-time data. The values of the parameters reported by the sensors were validated with rigorous testing ensuring reliability in the reported values. The system was also able to generate alerts whenever there was an abnormal change in the parameters. The remote monitoring was also successful allowing access regardless of the physical location.

Compared to the already existing systems, the incorporation of the shelf-life prediction model sets it apart. Once a sufficient amount of data is collected, the proposed machine learning algorithms can be implemented and evaluated for their effectiveness in predicting grain storage outcomes and supporting decision-making processes. Its accuracy is bound to increase as more data is generated and the model

Table 1 Comparison with state-of-the-art infrastructure

Nature	Proposed system GrainLog	Reference [4]	Reference [7]	Reference [10]	Reference [12]	Reference [14]
IoT Integration	✓	✓	✓			
Machine Learning/ Analytics	✓			✓		✓
Grain Monitoring	✓	✓	✓			✓
Pest Control	✓					
Moisture Control	✓		✓			
Scalability	✓	✓	✓			
Sustainability	✓					
Cost-Effectiveness	✓					
Food Quality Monitoring	✓				✓	✓
Shelf-Life Estimation	✓				✓	

is refined on it. The accompanying smartphone application elevates the user experience by presenting a comprehensive report, surpassing the capabilities of traditional systems that merely display raw sensor values. Additionally, the application serves as an educational tool, addressing the problem of farmer unawareness regarding warehouse management protocols, offering quick access to vital information.

The community-driven aspect of GrainLog, emphasizing the involvement of the agricultural community, fosters a collaborative environment where farmers can share and learn from each other's experiences. This communal knowledge exchange is crucial for the continuous improvement and adaptation of the system to diverse farming practices and local nuances. Thus, GrainLog demonstrates a potential for a notable economic and socio-economic impact, minimizing supply chain disruptions, stabilizing market prices, and protecting the financial interests of farmers and intermediaries.

GrainLog, while offering significant benefits to the farmers, also exhibits some limitations. Dependency on consistent internet connectivity may hinder its effectiveness in remote areas with poor network coverage. The maintenance of the hardware also can potentially add to operational overheads. The availability of the dataset is the biggest challenge for the implementation of the proposed ML algorithm. Hence, until the generation of sufficient data by the system itself the performance of the ML algorithm might not be satisfactory.

5 Conclusion

In the present scenario, there is more wastage than usage of food grains due to the extended 'keep even after expiration date' of the grains in many storage warehouses along with other issues such as lack of cold storages, poor support, etc. The GrainLog system emerges as a pivotal innovation in the realm of agricultural technology, adeptly tackling the pressing challenge of food grain wastage with its advanced monitoring capabilities and access to real-time data. This solution exemplifies the critical role of smart technology in bolstering food security and minimizing post-harvest losses, marking a significant leap towards sustainable farming. As GrainLog continues to evolve, the potential for enhancing its functionality through cutting-edge sensor technology, sophisticated machine learning algorithms, and comprehensive data analytics is immense. Furthermore, broadening the mobile application's user base and fostering integration with existing agricultural management systems could significantly refine grain monitoring and management, setting a new standard for efficiency and sustainability in agricultural practices.

References

1. https://www.fao.org/india/fao-in-india/india-at-a-glance/en/
2. https://pib.gov.in/PressReleaseIframePage.aspx?PRID=1968931
3. Kumar D, Kalita P (2017) Reducing postharvest losses during storage of grain crops to strengthen food security in developing countries. Foods 6(1):8. https://doi.org/10.3390/foods6010008.PMID:28231087;PMCID:PMC5296677
4. Ray PP (2017) Internet of things for smart agriculture: technologies, practices and future direction. J Ambient Intell Smart Environ 9:395–420.https://doi.org/10.3233/AIS-170440
5. Chai K, Azizi MMF, Romeli S, Lau HY (2023) Deployment of IoT sensors in the revolution of agriculture. RA J Appl Res 9. https://doi.org/10.47191/rajar/v9i6.01
6. Charlotte L, Vimalraj S, Sherwin L, Monisha R, Murugan R (2022) Automated food grain monitoring system for warehouse using IOT. Measur Sensors 24:100472. https://doi.org/10.1016/j.measen.2022.100472
7. Kodali R, John J, Boppana L (2020) IoT monitoring system for grain storage, 1–6. https://doi.org/10.1109/CONECCT50063.2020.9198549
8. Taj S, Imran A, Kastrati Z, Daudpota S, Memon R, Ahmed J (2023) IoT-based supply chain management: a systematic literature review. Internet of Things 24:100982. https://doi.org/10.1016/j.iot.2023.100982
9. Liakos, Konstantinos G, Busato P, Moshou D, Pearson S, Bochtis D (2018) Machine learning in agriculture: a review. Sensors 18(8):2674. https://doi.org/10.3390/s18082674
10. Benos L, Tagarakis AC, Dolias G, Berruto R, Kateris D, Bochtis D (2021) Machine learning in agriculture: a comprehensive updated review. Sensors 21(11):3758
11. Cui F, Zheng S, Wang D, Tan X, Li Q, Li J, Li T (2023) Recent advances in shelf life prediction models for monitoring food quality. Compreh Rev Food Sci Food Safety 22:1257–1284. https://doi.org/10.1111/1541-4337.13110
12. Srinivasagan R, Mohammed M, Alzahrani A (2023) TinyML-sensor for shelf life estimation of fresh date fruits. Sensors 23(16):7081. https://doi.org/10.3390/s23167081
13. Walker CK, Assadzadeh S, Wallace AJ, Delahunty AJ, Clancy AB, McDonald LS, Fitzgerald GJ, Nuttall JG, Panozzo JF (2023) Technologies and data analytics to manage grain quality on-farm—a review. Agronomy 13(4):1129. https://doi.org/10.3390/agronomy13041129
14. Dhande HK, Shelare SD, Khope PB (2020) Developing a mixed solar drier for improved postharvest handling of food grains. Agric Eng Int CIGR J 22(4):166–173
15. Gaur M (2016) Traditional V/S modern methods of storage in India. PARIPEX—Indian J Res 5(3):406–407
16. Karthikeyan C, Veeraragavathatham D, Karpagam D, Firdouse SA (2009) Traditional tools in agricultural practices. Indian J Traditional Knowl 8(2):212–217
17. Kiruba S, Sam Manohar Das S, Papadopoulou S (2006) Prospects of traditional seed storage strategies against insect infestation adopted by two ethnic communities of Tamil Nadu, southern peninsular India. Bull Insectology 59(2):129
18. Kumar D, Kalita P (2017) Reducing postharvest losses during storage of grain crops to strengthen food security in developing countries. Foods 6(1):8. https://doi.org/10.3390/foods6010008
19. Mann S, Dixit AK, Tushir S, Bashir AA (2016) Traditional grain storage practices in India: SWOT analysis. In: Navarro S, Jayas D, Alagusundaram K (Eds), Proceedings of the 10th international conference on controlled atmosphere and fumigation in stored products (CAF2016), CAF Permanent Committee Secretariat, Winnipeg, Canada, pp 500–503
20. Mobolade AJ, Bunindro N, Sahoo D, Rajashekar Y (2019) Traditional methods of food grains preservation and storage in Nigeria and India. Ann Agric Sci 64(2):196–205. https://doi.org/10.1016/j.aoas.2019.12.003
21. Prakash BG, Raghavendra KV, Gowthami R, Shashank R (2016) Indigenous practices for eco-friendly storage of food grains and seeds. Adv Plants Agric Res 3(4):00101. https://doi.org/10.15406/apar.2016.03.00101

22. Shaila M, Begum N (2021) Ancient farming methods of seed storage and pest management practices in India—a review. Plant Arch 21(Supplement–1):499–509. https://doi.org/10.51470/PLANTARCHIVES.2021.v21.S1.077
23. Sundaramari M, Ganesh S, Kannan GS, Seethalakshmi M, Gopalsamy K (2011) Indigenous grain storage structures of South Tamil Nadu. Indian J Traditional Knowl 10(2):380–383
24. Coradi PC, Maldaner V, Lutz É et al (2020) Influences of drying temperature and storage conditions for preserving the quality of maize postharvest on laboratory and field scales. Sci Rep 10:22006. https://doi.org/10.1038/s41598-020-78914-x
25. Ramachandran RP (2022) Integrated approach on stored grain quality management with CO_2 monitoring-a review. J Stored Products Res 96:101950, ISSN 0022*474X, https://doi.org/10.1016j.jspr.2022.101950

Intelligent Systems for Digitalization and Automation of Agriculture Equipment

Omm Sadul, Smruti Phirake, and Puskaraj D. Sonawwanay

Abstract The growing need for sustainable and effective farming methods has led to increased demand and integration of Intelligent Systems in the digitalization and automation of agricultural equipment. This review paper explores the transformative impact of the Internet of Things (IoT), Artificial Intelligence (AI), Machine Learning (ML), and Data Analytics in the realm of smart agriculture. The synergy of these technologies not only enhances precision farming but also facilitates data-driven decision-making for optimizing resource utilization and crop management. The paper delves into both current and emerging applications of these Intelligent Systems in agriculture, emphasizing their capacity to transform conventional farming methodologies. Additionally, the review includes a comprehensive case study comparing the effectiveness of Drone-mounted pesticide spraying with traditional manual pesticide spraying methods. This comparative analysis sheds light on the operational efficiency, cost-effectiveness, and environmental impact of each approach. By evaluating the real-world implications of these technologies, the paper aims to provide valuable insights into the practical implementation and benefits of Intelligent Systems in the agricultural sector, laying the groundwork for a future in farming practices that is both technologically advanced and sustainable.

Keywords Smart agriculture · Internet of things · Artificial intelligence · Drone technology · Pesticides

1 Introduction

Individuals worldwide eagerly anticipate the introduction of innovative devices and technologies designed to simplify their daily lives. While the procedure is limitless and encompasses both recognized and unidentified elements, professionals and

O. Sadul · S. Phirake · P. D. Sonawwanay (✉)
School of Mechanical Engineering, Dr. Vishwanath Karad MIT World Peace University, Pune, India
e-mail: puskaraj.sonawwanay@mitwpu.edu.in

A. J. Kulkarni and M. Hajiaghaei-Keshteli (eds.), *Advances in Intelligent Systems for Sustainable Agriculture*, Lecture Notes in Networks and Systems 1201,
https://doi.org/10.1007/978-981-97-9839-1_6

innovators across diverse fields consistently endeavor to introduce cutting-edge innovations to benefit society to the fullest extent [1]. Agriculture holds a central position in the economies of developing nations, contributing substantially to the Gross Domestic Product (GDP) and serving as a vital source of sustenance essential for human survival [2, 3].

Acknowledged as a crucial catalyst for global societies and economies, agriculture confronts various challenges that necessitate creative and inventive solutions. Concerns like ensuring food security, promoting the sustainable utilization of natural resources, addressing climate change, managing the increasing demand for food, controlling environmental footprint, and preventing biodiversity loss require effective attention and resolution. In reaction to this, the agricultural industry is adopting cutting-edge technologies, employing a range of sensors for tracking variables such as temperature, air humidity, soil moisture, and UV light. These sensors operate on microprocessor-based systems [4]. The potential advantages of these technological advancements encompass accelerated production, minimized material wastage and production costs, enhanced safety measures, and novel opportunities for testing and prototyping. AI has the capability to gather and process vast quantities of data in real time, leading to substantial enhancements in the dynamic assessment of decision-making and food shelf life. This, in turn, offers notable advantages in minimizing food loss and waste. Robots have the capacity to replace humans in dangerous environmental situations that produce both mental and physical stress, such as in activities like fish processing and slaughtering [5].

Integrating methods that leverage data, intelligent decision support, and computational modeling holds promise for enhancing efficiency, bolstering productivity and profitability, and concurrently addressing the imperative for heightened environmental responsibility in agricultural production [6]. Contemporary farming practices rely on accurate and current data regarding agricultural activities for successful production. Farms are increasingly dependent on digital technologies like sensing and monitoring devices, smart equipment, and advanced analytics.

The transformation of agricultural production into intelligent farming systems is rapidly occurring due to the swift advancements in technology such as cloud computing, ML, robotics, IoT, augmented reality, and big data. Smart Farming represents the evolution of Precision Agriculture, illustrated in Fig. 1. This advancement involves managerial activities that rely not only on accurate location data but also on event triggers, contextual information, and situational awareness [7].

A smart farming system can be conceptualized as a cyber-physical control cycle that effectively combines monitoring and sensing, intelligent analyses and planning, and efficient control of farm operations across all pertinent agricultural processes, emphasizing a comprehensive approach to farm management. In the realm of intelligent agricultural systems, farmers have the capability to remotely oversee and manage operations using digital information in (near) real time, eliminating the need for on-site direct observation and manual tasks. As a result, farmers receive automatic notifications in the event of an issue or any anticipated problems. From their desk or smartphone, individuals can assess the condition of the field or stable by examining

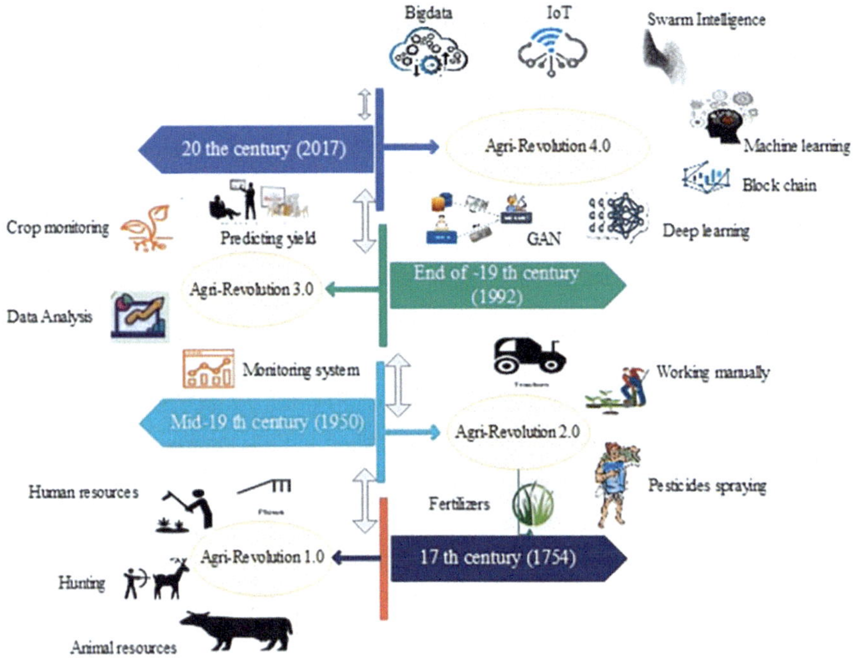

Fig. 1 Agricultural evolution from version 1.0 to 4.0: an overview [7]

a detailed digital representation of the relevant machine, animal, or plant. Simultaneously, ML algorithms enhance the digital perspective through specialized analyses and recommendations tailored to specific objects. Agricultural practitioners have the ability to model corrective and preventative measures, assessing their influence on digital representation. Ultimately, the selected interference has the capability to be implemented from a distance, allowing the farmer to once again utilize the digital interface to confirm the resolution of the anticipated issue. Anticipated is the gradual evolution of autonomous functionality in the smart farm management cycle, eliminating the need for direct involvement by the farmer [8].

2 Smart Agriculture

Agriculture revolves around the creation, management, and optimization of processes to achieve optimal output. The primary hurdle lies in automating tasks through the utilization of machines and robots. This necessitates the development of novel technological innovations by researchers to establish a seamless agricultural industry. Essentially, the goal is to implement an automated and systematic approach. An intelligent farming system, illustrated in Fig. 2, consistently observes the agricultural surroundings, encompassing factors like climate, water, humidity, livestock, and crop yield. It autonomously makes informed decisions or offers suggestions and insights for human consideration [9, 10].

Smart agriculture entails an innovative approach to farming, incorporating cutting-edge technologies to enhance both the quality and quantity of agricultural products [11]. Representing the future prospects of food cultivation and the implementation of novel processes for worldwide food security, smart agriculture empowers farmers to optimize resources while enhancing yields. This optimization encompasses a reduction in the use of fertilizers, manpower, seeds, and water. In developed nations, large-scale smart farming is underway, leveraging technologies like the IoT, AI, and the analysis of agricultural data. The integration of agricultural data analytics offers valuable insights to farmers, contributing to the advancement of smart agriculture, and leading to increased yield security and substantial crop productivity [12]. The evaluation of crop performance across diverse regions, considering specific field conditions and the economic repercussions of natural disasters on yield production has been conducted utilizing agricultural data analytics facilitated by ML [7, 10]. To achieve this, comprehensive and multi-scale AI models have been employed to forecast crop behavior in diverse scenarios and specific field conditions. Presently, mobile technologies have permeated even the most remote regions in developing nations. The extensive embrace of intelligent mobile technology in these remote regions provides a distinctive capacity to link rural producers with urban consumers and forge connections with global investors, facilitating investment and knowledge transfer. Through the appropriate platform, there exists an opportunity to create sustainable value, enhance food safety, expand financial inclusion, and ultimately reinforce underprivileged farmers to optimize the usage of available agricultural resource [12, 13].

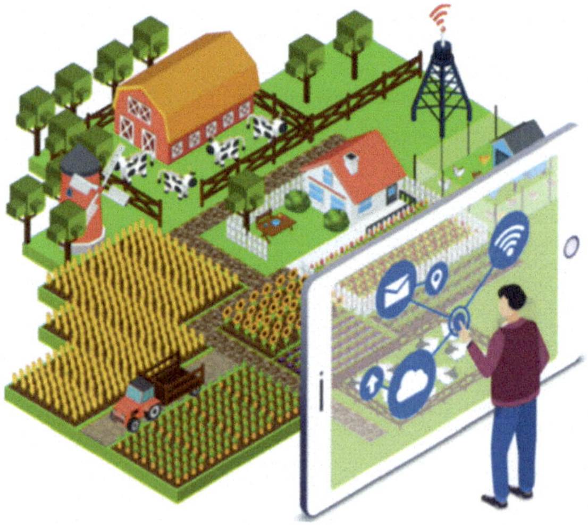

Fig. 2 A prototype of smart farming [10]

2.1 Smart Agriculture Using IoT

The well-integrated, expansive, all-encompassing, and open attributes of the IoT render it highly suitable for smart agriculture. Within the IoT smart agriculture platform, there is a seamless incorporation of automation equipment sourced from different institutions. These devices are readily adaptable to the farm's intelligent system, promoting the exchange of data among diverse elements and enabling automation through standard digital conduct. Some examples of such systems encompass real-time water conservation, farm monitoring, and weather prediction. Similar to other IoT applications depicted in Fig. 3, these use cases data processing, and data collection and necessitate sensors, categorizing them as instances of "Smart Agriculture (SA)" [14–16].

The subsequent points outline several key advantages of employing IoT for the implementation of smart agriculture:

i. Agility: Enhanced operational agility is a notable benefit derived from the integration of IoT in agriculture. This allows farmers to swiftly respond to significant variations in air quality, climate, humidity, or the condition of soil and crops.
ii. Clean Process: Precision farming through the integration of IoT-based technologies not only conserves energy and water, contributing to an eco-friendlier approach to agriculture but also significantly diminishes the usage of pesticides and fertilizers.

Fig. 3 Diverse implementations of IoT in agricultural settings [16]

iii. Implementation of automation in sowing, irrigation, and harvesting processes can result in resource conservation, reduction of human error, and overall cost reduction.
iv. Precise monitoring of yield rates per field over time facilitates the precise prediction of a farm's forthcoming crop yield and economic worth.
v. The utilization of the IoT is poised to advance community farming, especially in remote regions. IoT applications can promote services facilitating the exchange of data and knowledge among communities, fostering increased communication between landowners and agricultural experts.
vi. There have been various accusations of food theft, encompassing adulteration, counterfeiting, and artificial manipulation. Such fraudulent activities pose health risks and can have adverse economic consequences. The application of IoT technologies offers potential solutions to multiple aspects of food theft, addressing issues such as procedural integrity, data integrity, individual transparency, and product validity. The utilization of the IoT can also play an impressive role in guaranteeing operational and analytical regulatory adherence in the food industry.
vii. An asserted advantage of IoT is the ability to remotely control equipment. Leveraging IoT to ascertain optimal timing and methods for applying fertilizers or pesticides holds the potential to minimize costs and reduce wastage.
viii. Embracing IoT presents the prospect of unlocking novel business opportunities, providing independent farmers the means to break free from the influence of intermediaries. This shift enables them to forge direct relationships with customers, leading to enhanced profitability.
ix. IoT has the capability to facilitate real-time monitoring of farm properties and equipment, serving to deter fraudulent activities, accelerate the substitute of components, and ensure the timely execution of routine maintenance tasks.
x. Analytics-Driven Decision Making: IoT integration in smart agriculture enables data collection and analysis, empowering farmers to make insightful decisions based on real-time information. This data-driven technique enhances overall farm management, optimizing resource allocation, and improving productivity while reducing uncertainties in decision-making processes. [17, 18].

Farmers have the ability to closely observe their crops on a micro-scale, employing sensors and field mapping to conserve input resources and minimize environmental impact. This strategic utilization allows farmers to enhance cost-effectiveness by judiciously applying limited resources when necessary. Various types of sensors, including electrochemical sensors, location sensors, optical sensors, etc., depicted in Fig. 4, furnish essential data such as longitudes and latitudinal positions via GPS satellites. Additionally, they offer soil-related information like organic matter and soil moisture content. Electrochemical sensors play a pivotal role in precision agriculture, delivering crucial information such as pH and soil nutrient levels. The application of sensor technology for intelligent farm monitoring brings numerous benefits to farmers in this contemporary era [16, 19].

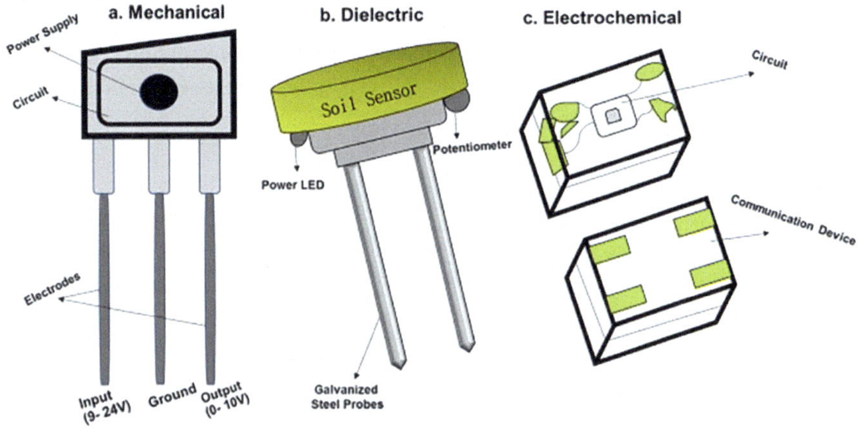

Fig. 4 Images depicting soil sensors employed in outdoor farming [16]

2.2 *Smart Agriculture Using AI*

In envisaging the future of agriculture, there is a forecast that farming will undergo a revolutionary transformation with the incorporation of AI and Robotics. The implementation of AI may not always be overtly evident to end-users, particularly farmers in this context. When envisioning the future of agriculture, predictions point to a groundbreaking transformation facilitated by the incorporation of AI and Robotics. The incorporation of AI may not always manifest explicitly for end-users, specifically farmers in this context. In the future, farmers will likely be required to optimize resource utilization for equivalent food production, leading to reduced water consumption and decreased chemical usage. The agriculture industry is actively seeking novel products, practices, and technologies to address evolving requirements, with precision agriculture emerging as a valuable approach for meeting these diverse needs [20].

Numerous emerging technologies, particularly those related to web technologies, generate extensive datasets that are shareable and accessible by a broad spectrum of users. In the context of smart agriculture, a diverse array of data sources is essential. This includes a variety of formats for sensors, images, strings, numbers, and other types of information. The term 'smart agriculture' primarily pertains to the planning and monitoring of soil and water usage, as well as the management of crop health with a focus on minimizing pollutants, including pesticides and herbicides [10].

The agricultural sector and sustainable farming practices are increasingly incorporating contemporary applications of Machine intelligence systems and resolutions. Scholars over the past several decades demonstrated a fervent interest in the multidimensional realm of sustainability across various disciplines. Furthermore, AI stands out as the technology widely employed by numerous organizations for diverse purposes. The utilization of AI not only enhances the optimization of available resources but also facilitates the collection of pertinent information concerning crops, including disease detection and corrective plant growth assessments [21].

2.3 Smart Agriculture Using ML

As technology continues to advance, ML has emerged as a pivotal asset within the agricultural sector. It functions as a primary decision support system for the integration of smart agriculture. The goal of ML is to formulate high-performance models capable of addressing diverse challenges encountered in smart farming. This depends on several elements including environmental conditions, demography, data variety, and soil quality [7, 22].

As agriculture becomes more intricate, the utilization of smart farming has emerged as a solution to tackle contemporary challenges. ML has played a substantial role in advancing smart farming, finding applications in automating and optimizing various processes. These include but are not limited to reducing food waste, adapting to automatic disease detection, managing water resources, planning missions, climate changes, and optimizing pesticide use. The majority of efforts have been concentrated on data acquisition and management across different stages to facilitate model development [7, 23].

3 Smart Farming Applications

3.1 Current Applications

Throughout history, technologies have often emerged and been adopted for specific applications, giving rise to analogous uses in entirely distinct fields. Industrial robots have become firmly entrenched in the manufacturing sector, prompting a growing exploration of their applications, along with robotics in broader terms, in various domains beyond manufacturing. This is particularly evident in roles where human involvement is risky, unfeasible, unavailable, or deemed inappropriate. The knowledge gained from the manufacturing sector serves as the foundation for these novel applications, collectively referred to as "Agriculture 4.0." [24, 25].

Facilitated by the core technologies outlined in the preceding section, a surge in the creation of applications was initiated in the mid-2000s and gathered considerable traction in the early 2010s, fundamentally transforming the way the world produces and acquires food. Various software, services, and methodologies were developed with the objective of enhancing data utilization and operational efficacy within the industry.

The efficiency of farms is enhanced through technology-enabled farming applications, enabling an increase in food production while utilizing fewer resources. Maintaining the productivity and profitability of a farm stands as the foremost priority for farmers. Similar to any enterprise, the farm's enduring viability relies on its profitability. To enhance profitability, farms need to operate with productivity and efficiency. This involves the effective management of land and water resources, cultivation of crops, and care for animals, all while optimizing the use of energy and chemicals and addressing capable risks. Web-based software applications employing ML algorithms and predictive analytics are commonly designed to aid in making decisions and managing farm-related tasks. Organizations like Granular or Farm Logs offer a range of web applications to agricultural businesses, utilizing data gathering and analysis to generate forecasts and recommendations. These instruments empower farmers to make choices that enable them to monitor and manage expenses, strategize production, and make well-informed judgments [26].

3.2 *Emerging Applications*

Implementing automation in agriculture will necessitate an incremental approach commencing with human operators and advancing toward reduced human involvement. The shift has been occurring as various applications have emerged to tackle a range of technical and social challenges. A Tel-Aviv-founded company named Prospera, utilizing AI and computer vision, represents the subsequent advancement in transitioning toward automation. It assists farmers in analyzing data collected from their fields. Prospera has created an independent system for cultivation practices that utilizes its pre-existing algorithms to offer guidance to farmers and concurrently operate center pivots. Collaborating with Valley Irrigation, a prominent player in the center pivot irrigation sector, Prospera can gather data from varied sources such as sensors, soil probes, drones, and satellite imagery. Subsequently, they employ their algorithms to analyze the gathered information. Following the analysis, Prospera has the capability to transmit operational instructions directly to the center pivot for robotic functioning, simultaneously alerting the grower to initiate the operation [26, 27].

4 Data Analytics

Assessing and tracking weather-related factors is a crucial undertaking in agriculture, as sunlight and rainwater play essential roles in the process of photosynthesis and the hydration of plants. Predicting weather conditions can be achieved through the utilization of past weather data. Consequently, farmers can proactively equip themselves for potential challenges, including situations like pesticide exposure or irrigation needs right before rainfall [28]. Improving both the quality and quantity of crops necessitates ongoing gathering of in-field data and routine surveillance of the plant's condition. Consequently, the cumulative volume of data, encompassing diverse types, significantly expands. The data categorized as big data typically necessitates specialized hardware and methodologies for processing. Processing large datasets benefits from employing strategies such as clustering, classification methods, and rule-based learning techniques. Owing to the extensive volume of data, it is advisable to conduct this processing on remote cloud-based systems equipped with appropriate hardware capable of managing substantial agricultural data. The discussion of edge computing is pertinent at this juncture, as its primary objective is to minimize the overall volume of data transmitted to the cloud [14, 29, 30]. The process involves preprocessing agricultural data gathered on local edge devices prior to its transmission to remote cloud computers, as illustrated in Fig. 5.

Leveraging big data analytics in the domain of production planning contributes to the enhancement of both demand forecasting and production planning processes. The proliferation of sensors and smart devices on farms is leading to a surge in generated data, driving a transformation toward highly data-centric and data-empowered agricultural processes. Currently, the incorporation of data science into specific agricultural approaches enables rapid data processing to facilitate real-time decision-making in precision farming. Nevertheless, further investigation is necessary to explore the manipulation and transformation of big data into smaller datasets tailored to specific challenges or precise agricultural domains. Numerous information systems have been developed for the purpose of agricultural management, catering to the organization

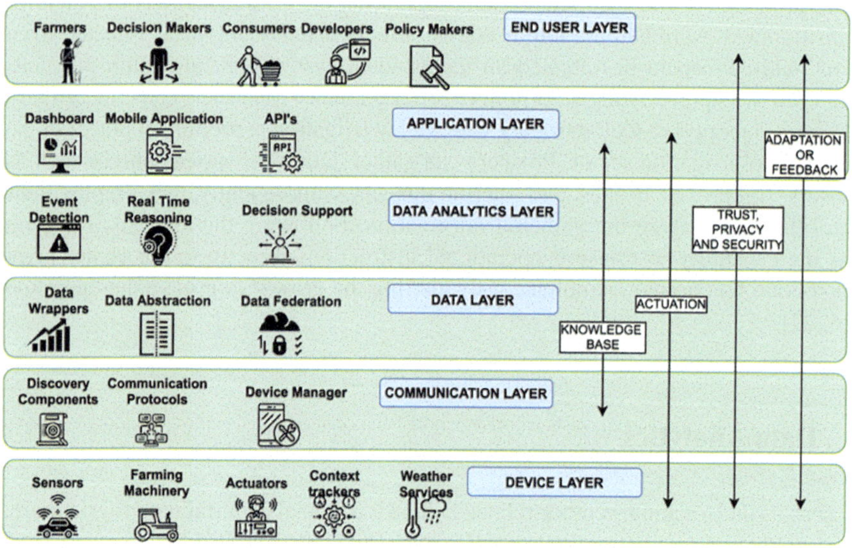

Fig. 5 The structured arrangement of the agricultural Internet Of Things (Agri IoT) ecosystem [14]

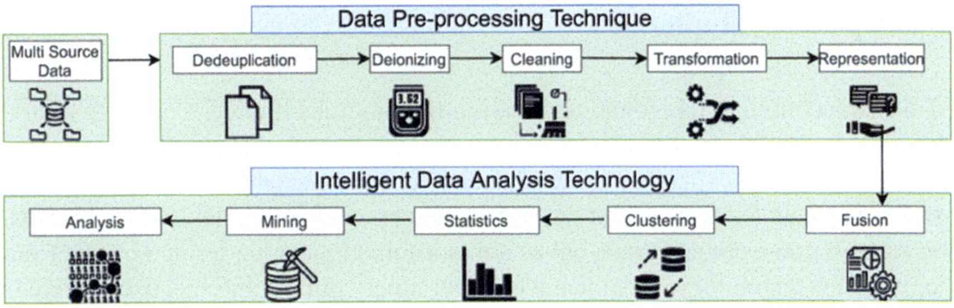

Fig. 6 Diagram illustrating various processing and analytical methods for agricultural Big Data [14]

and upkeep of various types of data [14, 20]. These systems facilitate operations related to data management, data storage, and data analytics, as illustrated in Fig. 6.

Accurate data analysis plays a vital role in agriculture, contributing to increased operational dependability and productivity. In accordance with the demands of IoT applications, data analytics can be classified into the subsequent types:

i. Memory-based analysis
ii. Extensive analytics
iii. Business Intelligence analytics
iv. Offline data analytics
v. Real-time analytics

Memory-based data analysis involves the static analysis of information stored in storage memory. Memory analysis enhances the overall efficiency of a business intelligence system, providing BI users with faster responses compared to traditional disk-based BI, especially for queries that involve prolonged execution times in a large database.

Extensive analytics is synonymous with big data analytics. The intricate process involves the examination of extensive data sets to derive valuable insights, including market trends, consumer preferences, and correlations, which can aid companies in making well-informed business decisions. Both data analytics and business intelligence involve leveraging data to reveal insights beneficial for the company; however, it is crucial to highlight a distinct difference between the two. In simple terms, business intelligence focuses on the current state, while data analytics is oriented toward future insights. The approach of real-time analytics encompasses the application of logical reasoning and statistical analysis to data, aiming to generate insights for expediting and enhancing informed decision-making. In specific situations, real-time analytics denotes the immediate execution of analyses upon the arrival of newly generated data. Image analysis is extensively employed in agriculture for various purposes, such as detecting diseases in fruits, leaves, and stems, evaluating fruit quality, identifying weeds, and managing irrigation. In recent times, the agricultural sector has employed image processing and IoT technologies to enhance crop manufacturing efficiency [17, 31].

5 Scope and Challenges

IoT-based technology has found application in numerous limited-scale agricultural fields, yet its widespread deployment on a large scale remains pending. A significant obstacle in this expansion is the financial burden associated with deploying and installing IoT-tagged sensors and accessories across extensive agricultural lands [11]. The issue of connectivity stands out as a major impediment hindering farmers from adopting new technology. Despite having over ninety million Internet users, there is a lack of precise reports in organizational research regarding the specific usage of the Internet or smartphones by farmers for agricultural objectives [1]. Insufficient knowledge and literacy about the possibilities, functioning, and application of IoT-based technology in agriculture can lead to the underutilization of intelligent systems. The execution of IoT-connected accessories and smart sensors in rural farm areas, where farmers may be less acquainted with advanced technologies, presents additional challenges [11]. The expenses associated with the development and upkeep of these innovations and technologies would be considerable. The impact of physical wear, wind, humidity, precipitation, sand, and extreme temperatures and tear raises the overall cost of hardware and software employed in agriculture [1]. Consequently, a significant challenge associated with IoT devices is their outdoor installation, exposing them to adverse environmental conditions such as heavy rainfall wind, dust, and temperature fluctuations. These harsh environmental factors may lead to unforeseen mechanical failures in sophisticated devices [11].

The environmental traits present challenges not just for the hardware but also for the network layer. Wireless communication is prevalent in agricultural implementations, primarily because it eliminates the need for wiring costs. The environment significantly influences wireless link quality, with multipath propagation effects and contributions to background noise being identified as major factors leading to low link quality. Hence, the transfer of data must employ resilient and dependable technologies that align with the demands and challenges posed by the rural environment [32, 33].

6 Case Study

6.1 Background

Each day, over 200 thousand individuals join the global population, and projections anticipate that the total world populace will reach 9.6 billion by the year 2050. This will lead to an increased need for food, requiring higher crop yields to fulfill the demand. Hence, there is an urgent requirement for the advancement of the agricultural sector. The limited crop production stems from various constraints, but employing drone technology in agriculture has the capability to overcome these challenges. The objective of this case study is to offer a comprehensive comparison of employing modern drone technology for automated pesticide spraying with the conventional approach of manual pesticide application practiced by farmers. The investigation assesses crucial elements including efficiency, cost efficiency, environmental consequences, and overall efficacy in managing pests.

The results contribute to the continuous discussion on enhancing agricultural methods by incorporating state-of-the-art technologies [34].

6.2 Methodology

6.2.1 Manual Spraying

The conventional method for applying pesticides typically involves the use of a manual mechanical sprayer, which is widely recognized as the predominant tool for this purpose. The application of pesticides through manual spraying has repercussions on human health, potentially causing ailments such as cancer, hypersensitivity, asthma, and various other disorders. Moreover, conventional approaches exhibit various drawbacks, including the utilization of additional chemicals, environmental contamination, narrow coverage of areas, scarcity of agricultural labor, and reduced spray uniformity. Traditional approaches result in increased expenses for pesticide application and exhibit diminished efficacy in pest and disease management. In order to address these limitations, the utilization of a drone-mounted sprayer is being implemented [34].

6.2.2 Drone-Based Pesticide Spraying

To date, traditional approaches to pesticide application remain predominant in various regions across the globe. To address the limitations of manual mechanical sprayers, the utilization of drone-mounted sprayers has been introduced. Introducing drone-mounted sprayers in the field has enhanced coverage capabilities improved the effectiveness of chemicals, and streamlined the spraying process, making it more efficient and expeditious. In contemporary times, drones exhibit the capability to transport pesticide tanks of up to 40 L and autonomously follow pre-mapped routes for crop spraying as needed. Drones demonstrate substantial commitment in effectively covering fields that are difficult for tractors and aircraft to reach [34]. Visual representations of drone-mounted sprayers are provided in Fig. 7.

The integration of AI systems has brought about a revolutionary shift in the application of semi-controlled drones for farm surveillance. The choice of employing a semi-controlled drone was solely determined by the sensor output, depicted in Fig. 8.

Drones have the capability to conduct aid in irrigation, apply fertilizers, soil health scans, and monitor crop health. Additionally, they offer valuable data analysis to approximate farming yield. In recent years, significant transformations have been

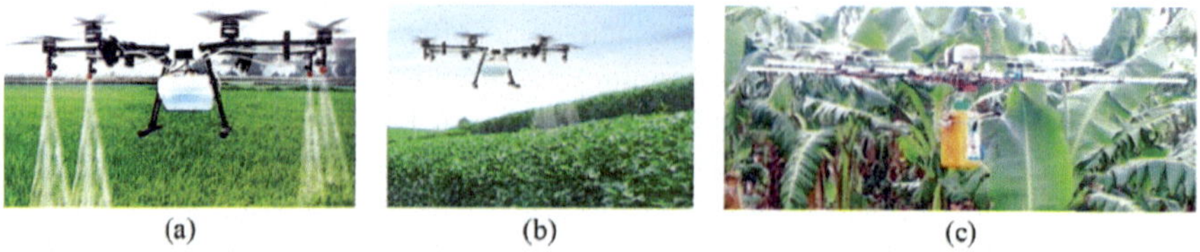

Fig. 7 The spraying system is employed in rice fields, **b** tea plantations, and **c** banana orchards [34]

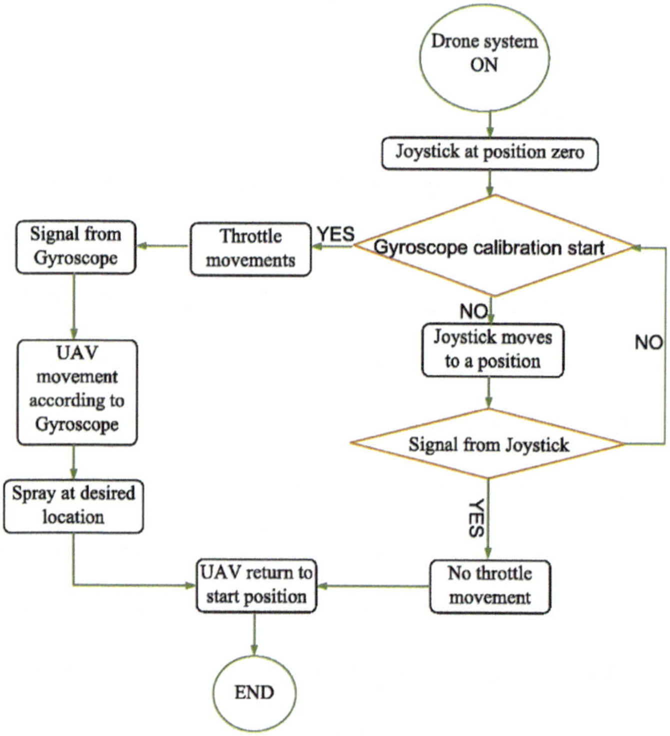

Fig. 8 Workflow diagram of sprayer drone [34]

witnessed in both spraying systems and drone flight controllers. The spraying system has progressed from a semi-controlled device to an AI-based fully automated system [34]. The elements employed in a fully automatic pesticide spraying system are depicted in Fig. 9.

In 2020, Karan Kumar Shaw et al. developed an octocopter with a lightweight spraying system. The payload determination considered factors such as tank storage size (6 L), fluid density, nozzle specifications for a fine spray, and the pump. To meet payload requirements, the system design included 8 Brushless Direct Current (BLDC) motors, a 12 V pump, an LI-PO battery, an FPV camera, an Electronic

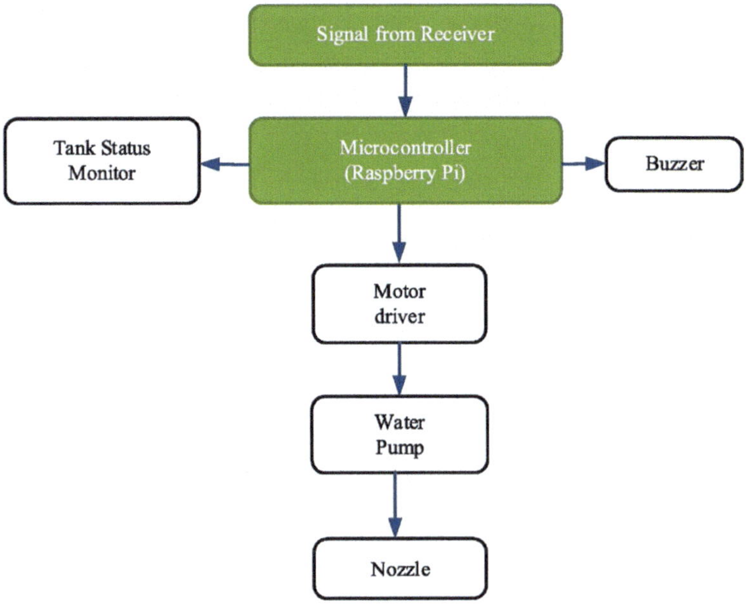

Fig. 9 Automated pesticide spraying system units [34]

Speed Controller, a video transmitter, and a propeller. Although the octocopter design proved effective for farm monitoring, there was a crucial shift from a manually controlled system to an AI-based autonomous system to enhance overall performance.

Unmanned helicopters are primarily employed for the application of pesticides or fertilizer spraying, leveraging their substantial payload capacity. Nevertheless, the use of multi-copters in pesticide spraying is steadily on the rise. Multi-copters offer enhanced stability during spot spraying, providing a beneficial advantage in this application. Notably, drone cameras have undergone substantial progress in terms of weight, size, and resolution. The development of drone controllers has advanced from fundamental microcontrollers to incorporate AI-enabled platforms like Raspberry Pi and Arduino Uno. Ongoing advancements in data transmission, data analysis, and embedded systems have led to a continuous shift in drone technologies from partially regulated to fully mechanized systems. The incorporation of ML in drones has paved the way for the development of agro-friendly systems. Nevertheless, hurdles persist in the adoption of drone technology in the agriculture sector, demanding solutions to boost the acceptance rate of drones [34].

Table 1 Comparison of manual pesticide spraying and drone-based pesticide spraying

Sr no	Parameters	Drone-based pesticide spraying	Manual pesticide spraying
1	Time	Demanding less time and exhibiting greater effectiveness Faster application, covering large fields in a shorter time frame	Require more time and are less effective Slower application, especially in large fields
2	Cost Effectiveness	Initial investment in drones can be high, but operational costs may be lower in the long run	Lower initial investment, but higher labor costs and maintenance for manual spraying
3	Safety	Reduced human exposure to pesticides, enhancing safety for operators	Higher risk of exposure to pesticides for manual operators
4	Spray Efficiency (Precision)	More efficient, covering larger areas in less time	Lower precision, higher likelihood of overspray, and potential environmental impact
5	One-time Capacity	Up to 40-L pesticide tank	Ranges from 12 to 20 L
6	Coverage	Enhanced coverage ability	Less area coverage
7	Environmental Effect	Potential reduction in environmental impact due to precise application	Extra chemicals usage increasing risks of hazards
8	Weather Dependency	Can operate in various weather conditions, depending on the drone's specifications	Limited by weather conditions, such as wind speed and rain, for manual spraying
9	Terrain Accessibility	Great potential in covering the fields with difficult access	Limited to smaller and accessible areas

6.3 Comparative Analysis

The comparison between manual pesticide spraying and drone-based pesticide spraying is given in Table 1.

7 Analysis and Conclusions

In this paper, an in-depth exploration of intelligent systems in agricultural equipment has been presented, highlighting their transformative impact on modern farming practices. The integration of smart technologies in agriculture has paved the way for increased efficiency, precision, and sustainability. One significant case study that exemplifies the application of intelligent systems is the comparison between drone-based pesticide spraying and traditional manual pesticide spraying. The adoption of intelligent systems, particularly in drone-based pesticide spraying, showcases a paradigm shift in agricultural practices.

The results of our analysis reveal that drone-based spraying exhibits superior efficiency and precision compared to manual methods. Drones equipped with advanced GPS technology can cover expansive fields in a fraction of the time it takes for manual applications. This not only enhances productivity but also allows for the timely treatment of crops, mitigating the risks associated with pest infestations. Furthermore, the precision offered by drone-based spraying is a key factor in minimizing environmental impact. The ability to precisely target specific areas for pesticide application reduces the risk of overspray, thus lowering the overall amount of chemicals released into the environment. This aligns with the growing global emphasis on sustainable farming practices and environmental stewardship. While the initial investment in drone technology may be higher, our discussion also considers the long-term cost-effectiveness. Operational costs for drone-based spraying, including maintenance and manpower, may be more sustainable in the long run compared to the ongoing expenses associated with manual labor, especially in larger agricultural settings. The safety aspect is another critical consideration. Drones significantly reduce human exposure to potentially harmful pesticides, enhancing the overall safety of agricultural operations. In contrast, manual spraying poses inherent risks to the health and safety of workers who are directly exposed to the chemicals during application.

Our study also underscores the flexibility and adaptability of drone-based systems in navigating challenging terrains and varying field conditions. This adaptability positions intelligent systems as versatile tools capable of addressing the diverse and dynamic needs of modern agriculture.

From the above analysis, the following conclusions can be obtained:

1. The incorporation of advanced technologies like AI, robotics, and data analytics is revolutionizing agriculture. Illustrated through smart farming systems, this amalgamation provides farmers with instant data and intelligent decision support, resulting in enhanced efficiency, productivity, and profitability.
2. Smart farming improves agricultural operations, tackling key issues like food security, sustainability, and environmental impact. Through constant monitoring and analysis of the farm environment, smart farm systems support informed decision-making, optimizing resource efficiency and boosting productivity.
3. The incorporation of ML in agricultural data analysis offers predictive insights into crop behavior, enhancing yield security. This data-centric strategy, coupled with precision farming, holds the potential to transform agriculture and ensure sustainable food production.
4. IoT applications empower farmers with timely responses to dynamic conditions, optimizing resource utilization and minimizing environmental impact. Furthermore, IoT enhances transparency and accountability in the agricultural supply chain, tackling issues like food theft, fraud, and regulatory compliance.
5. The imminent transformation of agriculture through AI and ML signals a revolutionary shift. Integration into smart farming not only boosts efficiency and cuts resource use but also tackles challenges, marking AI as a pivotal decision-support system for sustainable practices.

6. The amalgamation of AI, computer vision, and data analytics in precision farming, along with progress in robotics and automation, promises a sustainable and productive future for agriculture. Ongoing technological advancements at the intersection of these fields are pivotal in transforming farming practices, overcoming challenges, and fostering a resilient agricultural ecosystem.

References

1. Anand S, Sharma A (2022) Comprehensive analysis of services towards enhancing security in IoT-based agriculture. Measurement: Sensors, 24. https://doi.org/10.1016/j.measen.2022.100599
2. Wanyama J, Soddo P, Nakawuka P, Tumutegyereize P, Bwambale E, Oluk I, Mutumba W, Komakech AJ (2023) Development of a solar powered smart irrigation control system Kit. Smart Agric Technol, 5.https://doi.org/10.1016/j.atech.2023.100273
3. Anastasiou E, Fountas S, Voulgaraki M, Psiroukis V, Koutsiaras M, Kriezi O, Lazarou E, Vatsanidou A, Fu L, Bartolo F Di, Barreiro-Hurle J, Gómez-Barbero M (2023) Precision farming technologies for crop protection: a meta-analysis. Smart Agric Technol, 5.https://doi.org/10.1016/j.atech.2023.100323
4. Routis G, Roussaki I (2023) Low power IoT electronics in precision irrigation. Smart Agric Technol, 5. https://doi.org/10.1016/j.atech.2023.100310
5. Derossi A, Di Palma E, Moses, JA, Santhoshkumar P, Caporizzi R, Severini C (2023) Avenues for non-conventional robotics technology applications in the food industry. Food Res Int 173:113265. https://doi.org/10.1016/j.foodres.2023.113265
6. Thomas RJ, O'Hare G, Coyle D (2023) Understanding technology acceptance in smart agriculture: a systematic review of empirical research in crop production. Technol Forecasting Soc Change, 189.https://doi.org/10.1016/j.techfore.2023.122374
7. Sharma V, Tripathi AK, Mittal H (2022) Technological revolutions in smart farming: current trends, challenges & future directions. In: computers and electronics in agriculture (Vol 201). Elsevier B.V. https://doi.org/10.1016/j.compag.2022.107217
8. Verdouw C, Tekinerdogan B, Beulens A, Wolfert S (2021) Digital twins in smart farming. Agric Syst, 189.https://doi.org/10.1016/j.agsy.2020.103046
9. Jararweh Y, Fatima S, Jarrah M, AlZu'bi S (2023) Smart and sustainable agriculture: fundamentals, enabling technologies, and future directions. Comput Electr Eng, 110.https://doi.org/10.1016/j.compeleceng.2023.108799
10. Ayoub Shaikh T, Rasool T, Rasheed Lone F (2022) Towards leveraging the role of machine learning and artificial intelligence in precision agriculture and smart farming. In: Computers and electronics in agriculture (Vol 198). Elsevier B.V. https://doi.org/10.1016/j.compag.2022.107119
11. Rajak P, Ganguly A, Adhikary S, Bhattacharya S (2023) Internet of things and smart sensors in agriculture: scopes and challenges. J Agric Food Res 14:100776. https://doi.org/10.1016/j.jafr.2023.100776
12. Goel RK, Yadav CS, Vishnoi S, Rastogi R (2021) Smart agriculture—Urgent need of the day in developing countries. Sustain Comput Inform Syst, 30.https://doi.org/10.1016/j.suscom.2021.100512
13. Bacco M, Barsocchi P, Ferro E, Gotta A, Ruggeri M (2019) The digitisation of agriculture: a survey of research activities on smart farming. Array, 3–4. https://doi.org/10.1016/j.array.2019.100009
14. Raj M, Gupta S, Chamola V, Elhence A, Garg T, Atiquzzaman M, Niyato D (2021) A survey on the role of internet of things for adopting and promoting agriculture 4.0. J Netw Comput Appl (Vol 187). Academic Press. https://doi.org/10.1016/j.jnca.2021.103107

15. Idoje G, Dagiuklas T, Iqbal M (2021) Survey for smart farming technologies: challenges and issues. Comput Electr Eng, 92.https://doi.org/10.1016/j.compeleceng.2021.107104
16. Pathmudi VR, Khatri N, Kumar S, Abdul-Qawy ASH, Vyas AK (2023) A systematic review of IoT technologies and their constituents for smart and sustainable agriculture applications. Scientific African, 19.https://doi.org/10.1016/j.sciaf.2023.e01577
17. Sinha BB, Dhanalakshmi R (2022) Recent advancements and challenges of Internet of Things in smart agriculture: a survey. In: Future generation computer systems (Vol 126, pp. 169–184). Elsevier B.V. https://doi.org/10.1016/j.future.2021.08.006
18. Prakash C, Singh LP, Gupta A, Lohan SK (2023) Advancements in smart farming: a comprehensive review of IoT, wireless communication, sensors, and hardware for agricultural automation. Sens Actuators A 362:114605. https://doi.org/10.1016/j.sna.2023.114605
19. Paul K, Chatterjee SS, Pai P, Varshney A, Juikar S, Prasad V, Bhadra B, Dasgupta S (2022) Viable smart sensors and their application in data driven agriculture. Comput Electron Agric, 198.https://doi.org/10.1016/j.compag.2022.107096
20. Sharma R, Kamble SS, Gunasekaran A, Kumar V, Kumar A (2020a) A systematic literature review on machine learning applications for sustainable agriculture supply chain performance. Comput Oper Res, 119.https://doi.org/10.1016/j.cor.2020.104926
21. Obaideen K, Yousef BAA, AlMallahi MN, Tan YC, Mahmoud M, Jaber H, Ramadan M (2022) An overview of smart irrigation systems using IoT. Energy Nexus, 7. https://doi.org/10.1016/j.nexus.2022.100124
22. Mathi S, Akshaya R, Sreejith K (2022) An internet of things-based efficient solution for smart farming. Procedia Comput Sci, 218:2806–2819. https://doi.org/10.1016/j.procs.2023.01.252
23. Taneja M, Byabazaire J, Jalodia N, Davy A, Olariu C, Malone P (2020) Machine learning based fog computing assisted data-driven approach for early lameness detection in dairy cattle. Comput Electron Agric, 171.https://doi.org/10.1016/j.compag.2020.105286
24. Javaid M, Haleem A, Singh RP, Suman R (2022) Enhancing smart farming through the applications of agriculture 4.0 technologies. Int J Intell Netw 3:150–164. https://doi.org/10.1016/j.ijin.2022.09.004
25. Coulibaly S, Kamsu-Foguem B, Kamissoko D, Traore D (2022) Deep learning for precision agriculture: a bibliometric analysis. In: Intelligent systems with applications (Vol 16). Elsevier B.V. https://doi.org/10.1016/j.iswa.2022.200102
26. Charania I, Li X (2020) Smart farming: Agriculture's shift from a labor intensive to technology native industry. In: Internet of Things (Netherlands) (Vol 9). Elsevier B.V. https://doi.org/10.1016/j.iot.2019.100142
27. Debauche O, Mahmoudi S, Manneback P, Lebeau F (2022) Cloud and distributed architectures for data management in agriculture 4.0: Review and future trends. J King Saud Univ Comput Inform Sci 34(9): 7494–7514. King Saud bin Abdulaziz University. https://doi.org/10.1016/j.jksuci.2021.09.015
28. Violino S, Figorilli S, Ferrigno M, Manganiello V, Pallottino F, Costa C, Menesatti P (2023) A data-driven bibliometric review on precision irrigation. Smart Agric Technol 5:100320. https://doi.org/10.1016/j.atech.2023.100320
29. Avşar E, Mowla MN (2022) Wireless communication protocols in smart agriculture: a review on applications, challenges and future trends. In: Ad Hoc Networks (Vol 136). Elsevier B.V. https://doi.org/10.1016/j.adhoc.2022.102982
30. Alwis S De, Hou Z, Zhang Y, Na MH, Ofoghi B, Sajjanhar A (2022) A survey on smart farming data, applications and techniques. Comput Industry 138. Elsevier B.V. https://doi.org/10.1016/j.compind.2022.103624
31. Wolfert S, Ge, L, Verdouw C, Bogaardt MJ (2017) Big data in smart farming—a review. Agric Syst 153:69–80. Elsevier Ltd. https://doi.org/10.1016/j.agsy.2017.01.023

32. Tzounis A, Katsoulas N, Bartzanas T, Kittas C (2017) Internet of Things in agriculture, recent advances and future challenges. Biosyst Eng 164: 31–48. Academic Press. https://doi.org/10.1016/j.biosystemseng.2017.09.007
33. Pachayappan M, Ganeshkumar C, Sugundan N (2020) Technological implication and its impact in agricultural sector: an IoT based collaboration framework. Procedia Comput Sc 171:1166–1173. https://doi.org/10.1016/j.procs.2020.04.125
34. Hafeez A, Husain MA, Singh SP, Chauhan A, Khan MT, Kumar N, Chauhan A, Soni SK (2023) Implementation of drone technology for farm monitoring & pesticide spraying: a review. Inform Process Agric 10(2): 192–203. China Agricultural University. https://doi.org/10.1016/j.inpa.2022.02.002

Machine Learning Methods for Crop Yield Prediction Across Diverse Agricultural Environments

Atiya Khan, C. H. Patil, Amol D. Vibhute, and Shankar Mali

Abstract Machine learning technologies have transformed decision-making and data-processing methods, resulting in global economic development. This study investigates the possible benefits of incorporating machine learning techniques into modern agricultural systems. This paper provides a summary of the latest machine learning utilization in agriculture today, outlining both major challenges and opportunities. Forecasting crop yields is dependent on a wide range of interdependent elements, such as crop sowing and harvesting season, soil, temperature, humidity, rainfall, and management practices. The current research examines the performance of machine learning approaches for predicting crop output using a huge dataset of weather conditions, soil, season, area, and crop production parameters collected from 33 Indian states between 1966 and 2017. From the result analysis, it is inferred that the CatBoost regressor combined with SHAP and CatBoost Regressor models' performance was slightly poor for training dataset. According to the results of the other models, the Random Forest Regressor, out of the six machine learning models, had the highest accuracy during the training phase. During the validation stage, models such as CatBoost Regressor with an average R2 values varied between 0.967 and 0.866, while CatBoost Regressor + SHAP, where averaged R2 value ranged from 0.934 to 0.9171, produced the most accurate findings. While SVR performed better in the training phase, RF yielded more accurate results in other combinations. The RMSE is reduced to average yield with optimal crop yield and production data.

Keywords Machine learning · Crop yield prediction · Accuracy · CatBoost · Random forest regressor · SHAP

A. Khan · C. H. Patil (✉) · S. Mali
School of Computer Science, Dr. Vishwanath Karad MIT World Peace University, Pune, Maharashtra, India
e-mail: chpatil.mca@gmail.com

A. D. Vibhute
Symbiosis Institute of Computer Studies and Research (SICSR), Symbiosis International (Deemed University), Pune 411016, Maharashtra, India

A. J. Kulkarni and M. Hajiaghaei-Keshteli (eds.), *Advances in Intelligent Systems for Sustainable Agriculture*, Lecture Notes in Networks and Systems 1201,
https://doi.org/10.1007/978-981-97-9839-1_7

1 Introduction

India has prospered traditionally owing to its primary and fundamental reliance on agriculture. Crop cultivation made it possible for people to meet their needs on their own. Any country's foundation is its agriculture, which is essential to determining the GDP (gross domestic product). Strong and abundant harvests are necessary for the government to provide a sufficient supply of food for its people. However, a variety of geographies, erratic weather patterns, regional variances, disease prevention strategies, and the aftermath of natural disasters make it difficult to monitor crops efficiently. Traditional techniques for classifying agricultural products and estimating yields frequently fall short of meeting these challenges. Predicting crop yields is a significant but challenging issue that is essential to both sustainable intensification and effective use of natural resources. The various research focuses on particular case studies have used machine learning to estimate crop productivity. It's possible that the information and techniques they employed won't apply to different crops and areas.

In agriculture, crop yield prediction is very important. All farmers try to estimate how much crop production they will receive. In the past, farmers used their prior knowledge of a particular crop to predict future yields. The meteorological conditions, pest activity, and harvest planning efficiency are the main elements that affect agricultural productivity. Making sensible choices in agricultural risk management involves an accurate understanding of past crop yield data [1].

In an era of changing climate, limited resources, and evolving environmental circumstances, modern agriculture must find a way to feed a growing world population. Researchers are focusing on machine learning in particular to improve our capacity to forecast agricultural yields with accuracy by addressing these complications. The application of machine learning techniques to a variety of situations is the main emphasis of this study, which explores how these techniques can be integrated into the agricultural sector.

As agriculture expands into new geographic areas with diverse temperatures, soil types, and agronomic approaches, the demand for tailored crop yield forecasting systems grows. In conditions with this level of variance, traditional forecasting approaches usually fail to deliver reliable forecasts. To address this gap, the project at hand will investigate and implement machine learning algorithms that can adapt to the unique characteristics of varied agricultural environments. Farmers and other stakeholders will benefit from more precise and valuable forecasts that take into account the specific difficulties caused by different climates and farming approaches.

This study aims to increase resource use and boost food security by addressing the unique needs of different agricultural environments. In this regard, machine learning has been employed as a cutting-edge technology to revolutionize crop production forecasts while also improving agricultural robustness and efficiency.

2 Related Literature Work

Crop yield forecasting has become essential for managing the world's food crises and promoting sustainable agricultural practices in an era of rapidly growing populations and climatic uncertainty. In agricultural yield prediction, theory-driven process-oriented models for crops and data-driven machine learning algorithms are often considered as diametrically opposed. These models can be difficult to calibrate because of the complications of cropping systems, the diversity of space and time, the scarcity of available data, and the sensitivity of the parameters. As a result, the author's research may help machine learning become more widely used in agriculture by pointing to the possibility of meta-modeling for precise crop yield prediction [2]. As a result, farmers can create strategic plans for raising global agricultural productivity through the monitoring of a variety of factors. The integration of machine learning has enhanced the ability to predict agricultural productivity, and statistical models have aided in the extrapolation of past data. Farmers and researchers can increase agricultural yield prediction with the assistance of advanced computational techniques [3].

Deep learning is an area of machine learning (ML) that involves learning about representations in artificial neural networks (ANNs). Deep learning has been effectively used in a variety of industries, including computer vision, chatbots, and medicine. More recently, it has been applied in agriculture, mostly for crop classification, yield prediction, and estimation [4]. In some research, crop yields were predicted by combining CNN and RNN while taking climate into account. Using the Gaussian process component in CNN, this result was aided by the utilization of the LSTM algorithm with a Caffe-based deep learning regression model to anticipate national agricultural yield [5]. The author [6] uses spatial, spectral, and temporal information from multiple remotely sensed images at the same time to analyze all of those distinct features using SSTNN (spatial-spectral-temporal neural network), an innovative deep learning framework for agricultural production prediction.

The approach integrates a temporal dependency capturing module with a spatial-spectral component for learning into one convolutional network [6]. Two different machine learning techniques, Support Vector Machine (SVM) and Random Forest (RF), were used in conjunction with thorough feature selection to ascertain the main drivers of the temporal and spatial variation in crop yield caused by meteorological and soil hydrological factors, as well as improvements in breeding and management [7].

The most popular applications of machine learning (ML) techniques include automation, situation recognition, prediction, and recommendation analysis. The stages of data analysis are usually followed by the crop analysis technique. The addition of several classifiers, which are adjusted and assessed to determine which ones are best suited for the incoming data, is a major enhancement. Additionally, our methodology includes feature augmentation and reduction strategies that are necessary to highlight relevant aspects that improve crop detection and let farmers choose the best features for precise predictions [8]. Several studies have made use of deep learning techniques. The widespread use of CNN [27] for agricultural production prediction produced this outcome since it outperformed other deep learning algorithms like faster R-CNN, LSTM, DNN, and hybrid models. In the study, the most often used evaluation performance indicator was the RMSE, which was followed by MAE, R2, MAPE, and MSE [9].

The objectives of the study are

- To evaluate and compare different approaches for forecasting seasonal cropping.
- To evaluate the accuracy of the proposed prediction techniques.
- To forecast crop yield through machine learning utilizing historical agricultural data.
- To explore agricultural yield testing through the integration of optimization techniques.

This paper is organized as follows: The introduction is the first portion that provides a summary and an in-depth understanding of the study work. Section 2 presents an appropriate literature overview. Section 3 includes the techniques and information on the dataset and the machine learning algorithm used for the experiment. The proposed methodology is explained in Sect. 4. The results of the experiment are examined in Sect. 5. Finally, Sect. 6 includes a conclusion and recommendations for further work.

3 Methods and Material

The process of gathering and quantifying data by the programmer using variables related to the study is known as the data acquiring technique. A number of sources, including the Rajasthan government's official website, provided the data that was used. The most harvested crops in the state, including wheat, bajra, jowar, rapeseed bajra, jowar, mustard, barley, maize, and onion, have been documented and collected from each of the 33 districts in the state. The datasets were obtained from various official government websites. The official Government website provided the data from 1997 to 2017. A portion of the data was acquired via the Kaggle database, this is an Internet website where researchers can share their study data.

This dataset discussed in Table 1 includes agricultural statistics from 1997 to 2020 for several crops grown in different Indian states. Important information about crop yield prediction is provided by the dataset, which includes crop years, cropping seasons, states, crop types, cultivated areas, quantity, production, annual rainfall, fertilizer and pesticide usage, and estimated yields.

A correlation matrix for each of the study's variables for the variables you mentioned: year, average rainfall, pesticides, fertilizer, yield, and production are displayed in Fig. 1. The correlation matrix states a strong relationship among cultivation area, yield, and production of crops. The crop yield prediction has the strongest link with crop year, and crop production of that year.

Table 1 Sources for dataset used for yield estimation and prediction

Sr. no.	Data	Variable(s)	No. of samples	Attributes	Duration	Dataset source
1	State-wise monthly rainfall data	State, district, months, and annual rainfall	14,528	18	1966 to 2015	[10]
2	District-wise crop types' yield and production data	State, district-wise yearly area, yield, and production	16,147	83	1966 to 2017	[11]
3	State-wise soil-type data	State, district-wise, soil-type percentage	314	6	NA	[12]
4	Agriculture crop yield in the Indian states' dataset	State, district-wise, crop types, years, seasons, areas, pesticide usage, annual rainfall, fertilizer usage, production, and calculated yields	19,690	10	1997 to 2020	[13]

3.1 *Pre-processing*

A sizable dataset is required for the data mining program. Information obtained from many sources tends to remain unfiltered. It might contain contradictory, out-of-date, or incomplete information. Thus, in this procedure, such redundant data needs to be filtered. It is necessary to normalize the information. Numerous "NA" values in the provided data collection are filtered using Python. Robust scaling and normalization are connected; however, since the dataset comprises numerical data, the interquartile range is utilized in place of standardizing the data. The data size is reduced by a factor of 0 to 1 through normalization [14].

3.2 *Machine Learning Models*

3.2.1 Random Forest Regressor

Random Forest (RF) is a prominent machine learning technique built on the supervised learning paradigm. It can be applied to machine learning tasks like classification

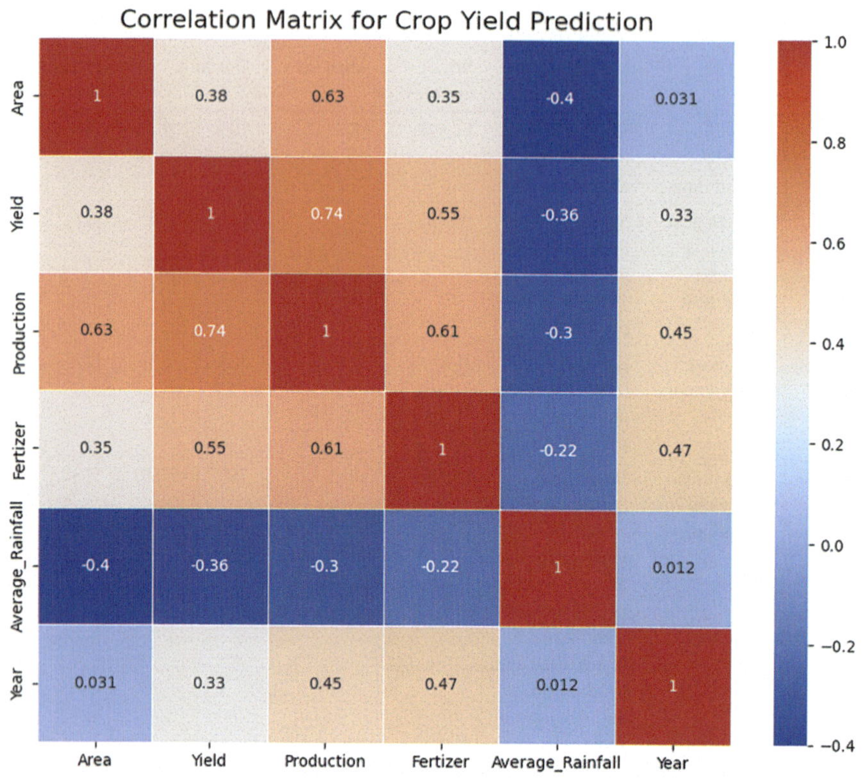

Fig. 1 Correlation matrix of dataset used in study

and regression. It is based on the concept of ensemble learning, which is described as the combination of numerous classifiers to solve a challenging issue and enhance model performance. Combines numerous trees with "bagging" to create a single model. Using boot strapping, K samples are randomly chosen from the original dataset. The RF model is developed by using these training data to create decision trees [15].

The Random Forest regressor is capable of handling sparse data and non-linear relationships between input features and output variables [16]. It can also handle high-dimensional data with many input features and is resistant to over-fitting, a common issue with machine learning models. This may offer an attribute significance measure, highlighting which input attributes are important to the prediction.

3.2.2 Support Vector Regressor (SVR)

As a regression algorithm, Support Vector Regression (SVR) can be used to work with continuous variables rather than classification kernels. It is possible to predict continuous values using the SVR, including integers and floating numbers. With a few minor variations, the support vector regression employs the same ideas as the support vector machines for classification. The support vector regression approach aims to reduce the difference between anticipated and real data by selecting an optimal line within a predetermined or threshold error value. In addition to providing a decent prediction model, SVR acknowledges the presence of nonlinearity in the data. The variation between the actual value and predicted value in the SVR model error prediction is determined to evaluate the model [16].

3.2.3 Linear Regression

Regression analysis can be used to analyze the statistical association among a dependent variable with several independent variables. Simple linear regression [17] and multiple linear regression (polynomial) are two methods for performing regression analysis. Only in cases where there are linear relationships between the data can one use the simple linear regression approach.

Simple linear regression employs the model in Eq. 1, where ε represents the mean zero of an undiscovered random error conditionally on a scalar variable x. In this model, the conditioned prediction of y rises by β1 units for each unit of rise in x.

$$y = \beta 0 + \beta 1\mathrm{x} + \varepsilon \tag{1}$$

Assuming, however, that our data are non-linear. Due to its inability to provide a best-fit line, linear regression is ineffective in this kind of situation. Polynomial regression, which illustrates a curvilinear connection between independent and dependent variables, can therefore be utilized to address this problem. Where there is no linear association between the dependent and independent data regression, a machine learning model is utilized to predict the dependent variables. Equation 2 is a derived polynomialfrom linear regression.

$$y = \beta 0 + \beta 1 \times 1 + \beta 2X_1^2 + \beta 3X_1^2 + \cdots + \beta nX_1^n \tag{2}$$

In this equation, y is the variable that is dependent, β is the value of the regression coefficient, and x is the variable that is independent. Equation 2 depicts the polynomial regression.

3.2.4 CatBoost Regressor

CatBoost is a combination of several category features. To merge all of the dataset's category features with their permutations in the current tree, CatBoost employs a greedy algorithm. CatBoost employs a novel technique called ordered boosting to modify the gradient estimation approach used in the traditional algorithm. With the help of this technique, gradient bias-induced prediction shift may be eliminated, further improving the model's capacity for generalization [18].

3.2.5 Gradient Boosting

One of the most popular optimization techniques for training machine learning models is gradient descent, which reduces the discrepancies between actual and predicted outcomes. A potent machine learning technique for predicting agricultural productivity is gradient boosting. In particular, gradient boosting algorithms, such as CatBoost, XGBoost, and LightGBM, are frequently chosen because of their effectiveness and efficiency. Performance can be optimized by adjusting a number of hyperparameters in gradient boosting methods. Common hyperparameters include the learning rate, number of trees (iterations), tree depth, and regularization terms. To identify the combination that best matches your dataset, use hyperparameter tuning.

3.2.6 SHAP

The SHAP (SHapley Additive ExPlanations) approach is based on the principles of Shapley from cooperative game theory. The prediction job is the "game" in machine learning, where features are the "players." A logical approach to assigning the model's prediction to each feature while accounting for feature interactions is provided by SHAP values. Predicting an output (such as crop yield) using input features (such as rainfall, pesticides, and fertilizer) is the "game" in a machine learning environment.

A gradient boosting library is called CatBoost. Its ability to handle non-numerical features gives it a significant advantage over contending libraries. It is possible to use categorical features without changing them. This indicates that a CatBoost model's SHAP values are simple to understand. Every category feature will have a single SHAP value.

4 Methodology

A comprehensive overview of the technique used in the current study for data processing and analysis is given in this section. Figure 2 illustrates the process of sequential nature through a series of activities. The collection includes harvest and climatic metrics from many sources. It includes agricultural statistics for several crops grown in different Indian states between 1997 and 2020. Various pre-preprocessing steps were carried out, including correlation analysis, null value elimination, and model construction for cross-validation and hyperparameter tuning, to guarantee data integrity. Following that, a thorough evaluation of the data models was carried out, which resulted in the display of predicted outcomes.

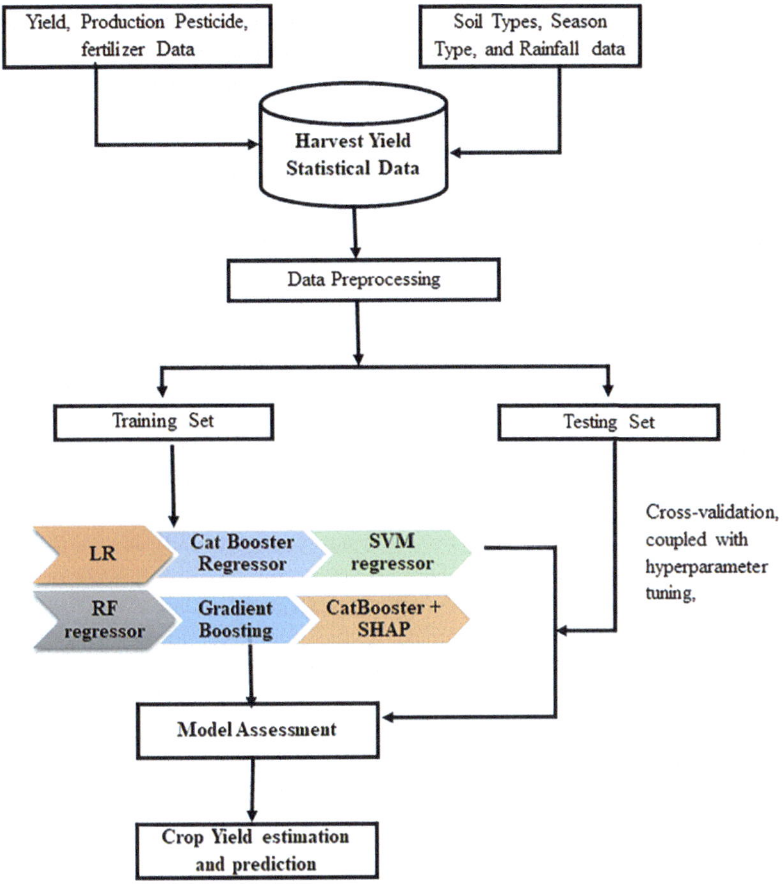

Fig. 2 Methodology of proposed system

Figure 2 describes the architecture of the proposed model. Google Colab was used to conduct the experiment described in this paper. The dataset is collected from the Official Government website and Kaggle website. The datasets are preprocessed, and null values are dropped to guarantee accuracy. It is necessary to import specific libraries in order to use preprocessing tools and machine learning techniques. Building and predicting the models would be performed swiftly and efficiently with the help of these libraries [19]. NumPy, pandas, matplotlib, seaborn, train_test_split, StandardScaler, and cross_validate were among the libraries that were imported. Machine learning methods [23, 24] are used for classifying and forecasting datasets. It selects a subset of features from the attributes of the dataset and divides the attributes into training and testing datasets. Crop yield prediction systems [28] facilitate enhanced planning and decision-making for higher yields. The suggested system includes a prediction module that forecasts the yield of main crops using a number of machine learning classification algorithms [29]. Models including

Gradient Boosting, SVM Regressor, Bagging, Linear Regression, Random Forest Regressor, and CatBoostRegressor and CatBoostRegressor + Sharp were imported.

5 Results and Findings

5.1 Result Assessment

The experiment starts with a pre-processed dataset on crop yield and production. Following that, the necessary packages and libraries are uploaded, while data preparation is completed. The dataset is then divided into two parts: training and testing. Following that, several machine learning-based techniques are utilized to increase yield estimation and prediction accuracy. Figure 3 depicts the average yield for Maharashtra across some years.

The dataset includes variable parameters like rainfall, production quantity, crop yield, cultivation area, year, district level, and state level data [30]. A total of 16,000 records has 20 crop labels like rice, wheat, maize, barley, sunflower, soybean, sugarcane, cotton, vegetables, onion, kharif sorghum, pearl millet, finger millet, groundnut, sesamum, castor, and oilseeds which make up the crop prediction dataset.

Figure 4 displays the extensive use of fertilizer for the crops. The use of fertilizer has increased with the year for a better crop production. Figure 5 shows the average rainfall recorded for the Maharashtra state over the years.

The selected features contribute to almost 74% of the variation in crop production, according to the linear regression model's R-squared value of 0.74. The model's lower RMSE of 2.83 indicates that, on average, predicted and actual yields differ by 2.83

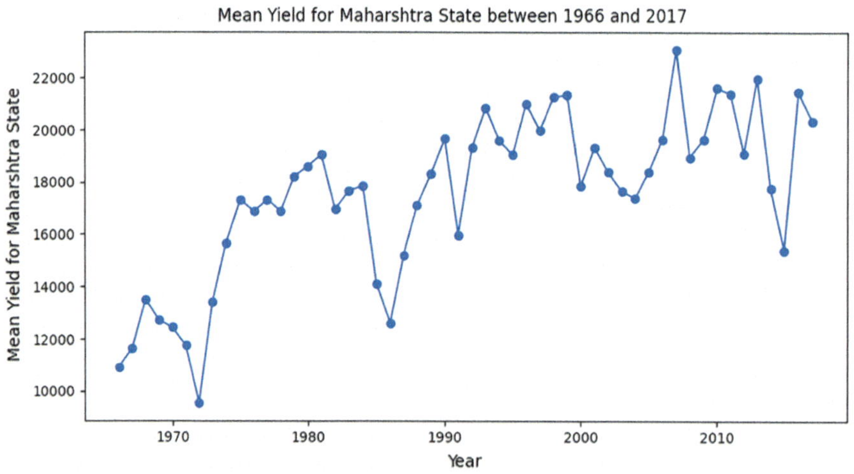

Fig. 3 Mean yield for Maharashtra state

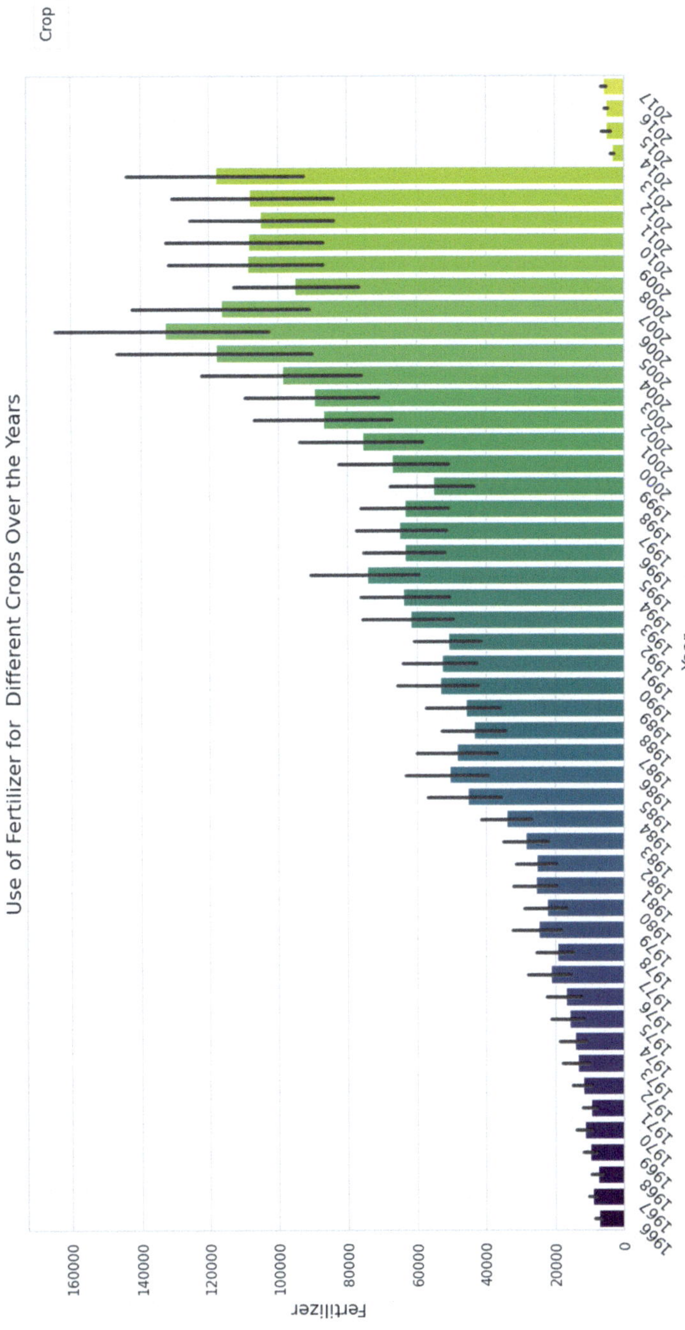

Fig. 4 Use of fertilizer over the years

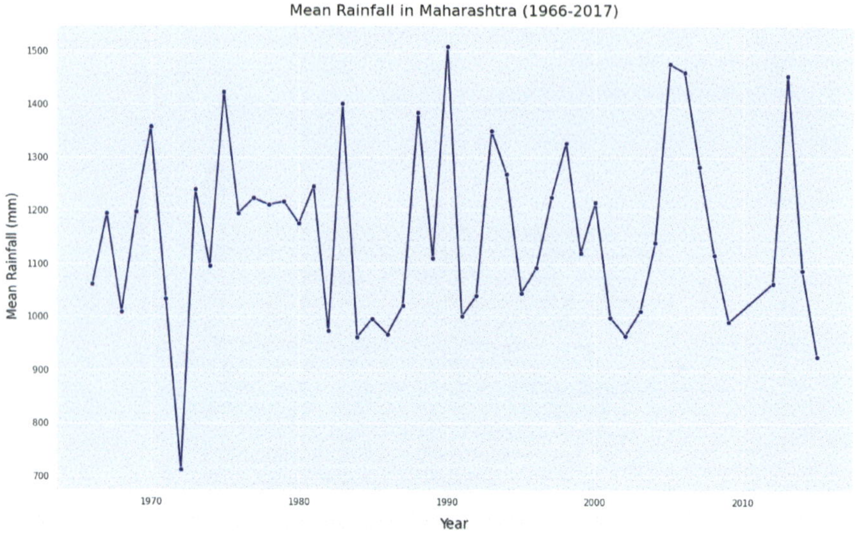

Fig. 5 Average rainfall in Maharashtra over the years

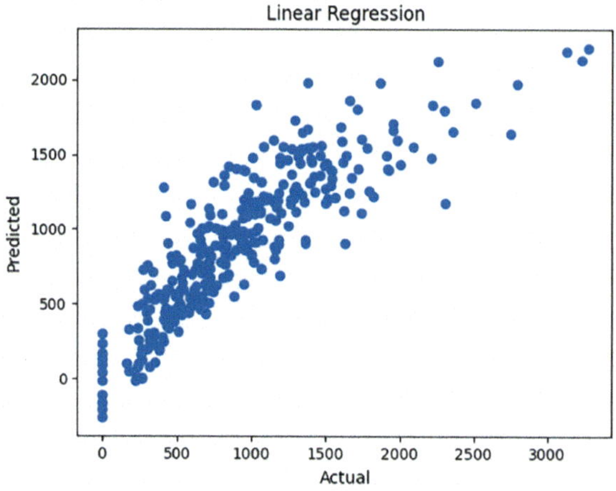

Fig. 6 Linear regression output shows yield estimation predicted versus actual values

units. Figure 6 shows the result of linear regression model depicting relationship between actual and predicted values.

5.2 *Performance Assessment*

A variety of machine learning techniques are needed to predict agricultural productivity. Though there are numerous machine learning approaches available, the choice of an approach is determined by both the application's needs and the algorithm's predicted accuracy. To verify the prediction accuracy of classification algorithms, measures are applied, such as Mean Absolute Error (MAE) and Root Mean Squared Error (RMSE) [20].

5.2.1 Root Mean Square Error

The root mean square error, or RMSE, measures the difference between the actual and projected values [21]. Lower RMSE values demonstrate better fit, while higher RMSE values indicate poorer fit. The RMSE number is not negative. Positive and negative mistakes should be indicated by larger values. If the model's primary purpose is prediction, RMSE is an accurate measure of how well it predicts response and constitutes the most important fit criterion to consider when interpreting it. The average magnitude of the error is also measured by the RMSE quadratic scoring rule.

5.2.2 Mean Squared Error

The average of the squares of errors is measured as mean squared error (MSE) or mean squared deviation (MSD). The mean squared is the difference between the expected and actual numbers.

5.2.3 R-Squared

R-squared (R^2) is a measure of statistical significance which indicates the proportion of variability in the dependent variable (the output) that is predictable from the independent variables (features) in a regression model. R-squared is a useful measure for determining how well a model takes into account crop production variability according to certain parameters for the context of yield estimation.

5.3 Results Prediction

Using newly supplied data, the trained model forecasts the result. For the purpose of making the model predictive for new input, we have saved the trained model in a file. The trained predictive model in this system is stored in a binary stream. The model is then evaluated using a testing dataset. The several machine learning algorithms are part of this prediction model, use training data to learn features, and then apply those properties to forecasts.

Figure 7a compares actual and predicted outcomes using a linear regression model. This illustration shows the linear regression model's ability to capture linear relationships between factors while generating accurate predictions. The graphical depiction in Fig. 7b, c shows the degree of concordance or discrepancy between the actual data points and predicted values using random forest regressor and support vector regressor respectively. Figure 7d shows the result using a gradient boosting regressor, the actual data versus projected data evaluation involves investigating the alignment between the actual observed values in the dataset. The graphical representations in Fig. 7e, f show the degree of consistency or variance between observed data points and expected outcomes produced by the CatBoost Regressor and CatBoost Regressor + SHAP. In Fig. 7g, h, the comparison graphs typically show two lines: one reflecting the anticipated yield values produced by the machine learning models [26], and another line depicting actual yield values observed in the field. The study of these graphs sheds light on the efficiency of every approach in identifying the fundamental patterns and trends in the data [25]. If the predicted values closely match the actual values at different time points or areas, it means that the machine learning model was trained and generalized from the training data. This indicates a high degree of confidence in the capacity of the model to anticipate future crop yield.

In this work, we make use of widely used and reliable machine learning techniques, such as Random Forest Regressor, Support Vector Regressor, Linear Regression, Gradient Booting, CatBoost Regressor, and SHAP. These methods are used to identify the main physical and atmospheric characteristics affecting both temporal and geographical changes in the yield of the crops. A comprehensive feature selection procedure is used in the study to separate the most important predictors from a group of variables that are closely related to one another.

5.4 Comparative Analysis with Competitive Models

In addition to the cultivated areas, the geographical distribution of mean rainfall may assist in explaining the geographic spread of the projected scores for each technique. The rainy season is clearly the primary cause of crop production variability when the total amount of precipitation falls below around 800 mm in a rainy season. A positive linear relationship indicates that varying the amount of rainfall can result in varying crop yields [22]. Table 2 compares the six models' performance on the training and validation datasets in terms of RMSE, MAE, MSE, adjusted R-squared, and variance score [31].

Table 2 compares the ranking scores of statistical indicators used to evaluate the efficiency of machine learning model during training and testing. The CatBoost Regressor and CatBoost + SHAP models performed a bit poorly. The Random Forest Regressor was the most accurate of the six machine learning models during the training stage, which was in line with the results of the other models.

The recommended approach significantly outperforms the other baseline methods in terms of average R2, RMSE, and MAPE, as shown in Table 2. The suggested

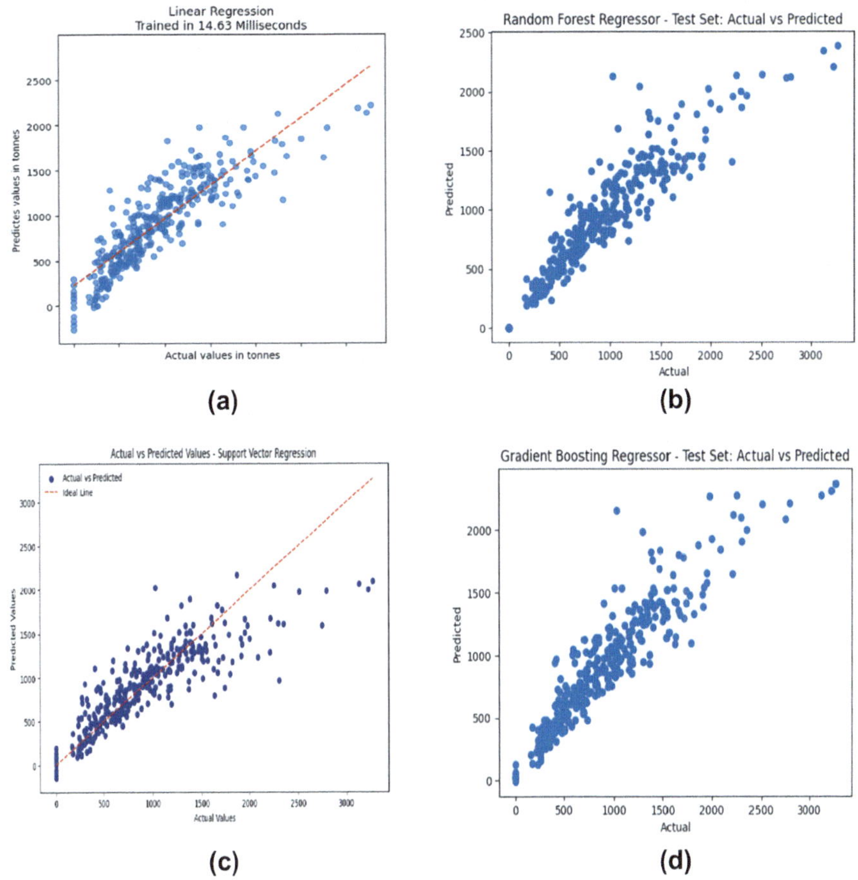

Fig. 7 Output of machine learning model displaying yield estimation predicted versus actual values. **a** Predicted and actual values using Linear Regression for crop between 1966 and 2017 , **b** Predicted and actual values using Random Forest Regressor for crop between 1966 and 2017 , **c** Predicted and actual values using Support Vector Regressor, **d** Predicted and actual values using Gradient Boosting Regressor, **e** Predicted and actual values using CatBoost Regressor, **f** Predicted and actual values using CatBoost Regressor +SHAP, **g** Comparison between Predicted and actual values using Random Forest Regressor, **h** Comparison between Predicted and actual values using Support Vector Regressor

approach provides the best performance in crop prediction across the years. More specifically, the CatBoost Regressor and CatBoost + SHAP models reduce the average RMSE in the testing stage compared with the LT, RF regressor, and SVR, respectively.

The results show that during the evaluation stage, the CatBoost Regressor has remarkable averaged R2 values that vary from 0.967 to 0.866, but the CatBoost Regressor + SHAP has an average R2 value ranging from 0.934 to 0.9171. These

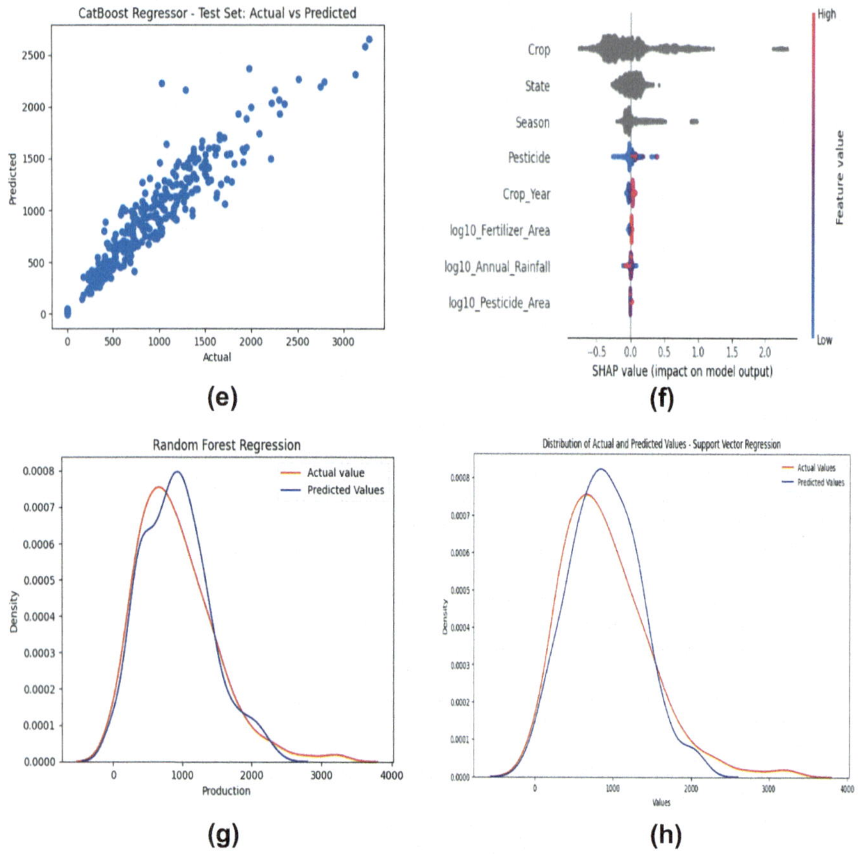

Fig. 7 (continued)

findings demonstrate the greater accuracy of these models in predicting agricultural yield in Maharashtra. Although random forest beats support vector regression during training, its interactions with other models, especially the CatBoost regressor, resulted in more accurate predictions.

The R2 values of the conventional techniques, LR and SVR, are less than those of other models, which is inappropriate for evaluating crop yield prediction. RF and CatBoost outperform the other baseline techniques, as evidenced by their R2 scores of 0.97 and 0.96, respectively.

5.5 *Discussion*

A variety of factors like some under our control and others far from it affect crop production. Factors like crop or seed variety, cultivation techniques, and fertilizer

Table 2 Statistical validation of various models for agricultural yield prediction

Crop										
Algorithm	Train					Test				
	RMSE	R2	MAE	MSE	Variance score	RMSE	R2	MAE	MSE	Variance score
Linear regression	263	0.769	191	696	0.7691	283	0.7495	206	802	0.7498
Random support regressor	81.94	0.977	52	671	**0.9777**	223	0.843	148	501	0.843
Support vector regressor	298	0.726	198	892	0.7164	321	0.678	206	103	0.6857
Gradient boosting	155	0.919	111	242	**0.9196**	219	0.849	146	482	0.8496
CatBoost regressor	99	0.967	73.90	982	**0.9674**	206	0.866	140	428	0.866
CatBoost + SHAP	167	0.93464	114	507	**0.9347**	184	0.9171	123	339	**0.9172**

treatment are all factors under our control. Beyond human control, non-controllable factors include meteorological conditions like precipitation, air and soil temperatures, atmospheric humidity, moisture in the soil, and solar radiation. Any variations from the ideal ranges for these characteristics may have an adverse impact on plant growth and productivity. All of these weather factors must be taken into account in order to make accurate crop output estimates, although data availability may provide some challenges.

In the proposed research work we have focused on utilizing temperature, rainfall, pesticide usage, fertilizer used, and production data to forecast yields, for the crops all over India. Several machine learning algorithms were used in this work to evaluate crops according to their basic properties, producing valuable findings. The study demonstrated that crop yield prediction is difficult to accurately forecast. The results highlighted the significance of carefully choosing features in machine learning algorithms to maximize accuracy when analyzing agricultural data for crop yield prediction.

6 Conclusion

Machine learning manages the increasing volume of data while also assisting in understanding the complexity of modern scientific research. Machine learning techniques emerged as potent resources to handle huge datasets and comprehend the nuances of agricultural systems. This study tackles the requirement for reliable crop production prediction, a vital part of modern agriculture, by leveraging state-of-the-art machine learning methods.

Predicting crop yield precisely is important for optimizing agricultural production, decreasing waste, and facilitating global food security. Conventional processes frequently fail to address the complexities of current agricultural systems. The integration of machine learning algorithms intends to deal with this gap, granting more precise as well as accurate forecasts for improved decision-making in agriculture. Our study uses Gradient Boosting, Random Forest, CatBoost Regressor, and SHAP models to forecast crop yield in Maharashtra. The accuracy of every model is thoroughly evaluated and compared throughout the validation step. The approaches comprise the training and validation phases, analyzing the models according to their R2 values. The machine learning algorithms like Random Forest Regressor, CatBoost Regressor, and CatBoost Regressor + SHAP are evaluated for their ability to predict crop yield and productivity.

The paper analyzes the performance of the gradient boosting and RF models with the CatBoost regressor and SHAP models in predicting crop yield across the state of Maharashtra. The findings indicated that, among the models taken into consideration during the validation stage, CatBoost Regressor, with an average R2 value ranging from 0.967 to 0.866, and CatBoost, Regressor + SHAP, with an average R2 value ranging from 0.934 to 0.9171, produced the most accurate results. RF outperformed SVR in the training phase, but in other combinations, RF produced findings that were more accurate. For prediction of enhanced crop output and productivity, all three of the machine learning algorithms, like Random Forest Regressor, CatBoost Regressor, and SHAP, are strongly recommended.

In conclusion, a combination of machine learning models, especially the CatBoost Regressor and SHAP, is highly effective in predicting agricultural yield in the Maharashtra area. It is recommended that more advancements in the agriculture sector, using machine learning, address the challenges of increasing data volumes and scientific complexities in agriculture.

References

1. Paudel D, Boogaard H, de Wit A, Janssen S, Osinga S, Pylianidis C, Athanasiadis IN (2021) Machine learning for large-scale crop yield forecasting. Agric Syst 187(June 2020):103016. https://doi.org/10.1016/j.agsy.2020.103016
2. Kallenberg MGJ, Maestrini B, van Bree R, Ravensbergen P, Pylianidis C, van Evert F, Athanasiadis IN (2023) Integrating processed-based models and machine learning for crop yield prediction. arXiv:2307.13466, https://doi.org/10.48550/arXiv.2307.13466
3. Rao Kolipaka VR et al. (2023) Crop yield prediction using machine learning and deep learning techniques. Int J Recent Innov Trends Comput Commun 11(10):582–589. https://doi.org/10.17762/ijritcc.v11i10.8542
4. Khaki S, Wang L (2019) Crop yield prediction using deep neural networks. Front Plant Sci 10:1–10. https://doi.org/10.3389/fpls.2019.00621

5. Bhimavarapu U, Battineni G, Chintalapudi N (2023) Improved optimization algorithm in LSTM to predict crop yield. Computers 12(1). https://doi.org/10.3390/computers12010010
6. Qiao M, He X, Cheng X, Li P, Luo H, Zhang L, Tian Z (2021) Crop yield prediction from multi-spectral, multi-temporal remotely sensed imagery using recurrent 3D convolutional neural networks. Int J Appl Earth Obs Geoinf 102:102436. https://doi.org/10.1016/j.jag.2021.102436
7. Lischeid G, Webber H, Sommer M, Nendel C, Ewert F (2022) Machine learning in crop yield modelling: a powerful tool, but no surrogate for science. Agric For Meteorol 312:108698. https://doi.org/10.1016/j.agrformet.2021.108698
8. Elbasi E, Zaki C, Topcu AE, Abdelbaki W, Zreikat AI, Cina E, Shdefat A, Saker L (2023) Crop prediction model using machine learning algorithms. Appl Sci 13(16). https://doi.org/10.3390/app13169288
9. Oikonomidis A, Catal C, Kassahun A (2023) Deep learning for crop yield prediction: a systematic literature review. New Zeal J Crop Hortic Sci 51(1):1–26. https://doi.org/10.1080/01140671.2022.2032213
10. https://data.gov.in/catalog/rainfall
11. http://data.icrisat.org/dld/src/crops.html
12. https://indiawris.gov.in/wris/
13. https://www.kaggle.com/datasets/akshatgupta7/crop-yield-in-indian-states-dataset/data-details for pesticides, fertilizer used for crop production
14. Kamath P, Patil P, S S, Sushma, S S (2021) Crop yield forecasting using data mining. Glob Trans Proc 2(2):402–407.https://doi.org/10.1016/j.gltp.2021.08.008
15. Srivastava AK, Safaei N, Khaki S, Lopez G, Zeng W, Ewert F, Gaiser T, Rahimi J (2022) Winter wheat yield prediction using convolutional neural networks from environmental and phenological data. Sci Rep 12(1):1–14. https://doi.org/10.1038/s41598-022-06249-w
16. Talaat FM (2023) Crop yield prediction algorithm (CYPA) in precision agriculture based on IoT techniques and climate changes. Neural Comput Appl 35(23):17281–17292. https://doi.org/10.1007/s00521-023-08619-5
17. Duddigan S, Shaw LJ, Sizmur T, Gogu D, Hussain Z, Jirra K, Kaliki H, Sanka R, Sohail M, Soma R, Thallam V, Vattikuti H, Collins CD (2023) Natural farming improves crop yield in SE India when compared to conventional or organic systems by enhancing soil quality. Agron Sustain Dev 43(2). https://doi.org/10.1007/s13593-023-00884-x
18. Zhang Y, Zhao Z, Zheng J (2020) CatBoost: a new approach for estimating daily reference crop evapotranspiration in arid and semi-arid regions of Northern China. J Hydrol 588:125087. https://doi.org/10.1016/j.jhydrol.2020.125087
19. Durai SKS, Shamili MD (2022) Smart farming using machine learning and deep learning techniques. Decis Anal J 3:100041. https://doi.org/10.1016/j.dajour.2022.100041
20. Palanivel K, Surianarayanan C (2019) an Approach for prediction of crop yield using machine learning and big data techniques. Int J Comput Eng Technol 10(3):110–118. https://doi.org/10.34218/ijcet.10.3.2019.013
21. Sivanandhini P, Prakash J (2020) Comparative analysis of machine learning techniques for crop yield prediction 9(6):289–293. https://doi.org/10.17148/IJARCCE.2020.9647
22. Sarr AB, Sultan B (2023) Predicting crop yields in Senegal using machine learning methods. Int J Climatol 43(4):1817–1838. https://doi.org/10.1002/joc.7947
23. Apolo-Apolo OE, Pérez-Ruiz M, Martínez-Guanter J, Valente J (2020) A cloud-based environment for generating yield estimation maps from apple orchards using UAV imagery and a deep learning technique. Front Plant Sci 11:1–15. https://doi.org/10.3389/fpls.2020.01086
24. Ghazaryan G, Skakun S, König S, Rezaei EE, Siebert S (2020) Crop yield estimation using multi-source satellite image series and deep learning center for remote sensing of land surfaces (ZFL), University of Bonn , 53113 Bonn , Germany Remote Sensing Research Group (RSRG), University of Bonn , 53115 Bonn , Germa, 5163–5166
25. Anjana KAK, Sana A, Bhat BA, Kumar S, Bhat N (2021) An efficient algorithm for predicting crop using historical data and pattern matching technique. Glob Transitions Proc 2(2):294–298. https://doi.org/10.1016/j.gltp.2021.08.060

26. Van Klompenburg T, Kassahun A, Catal C (2020) Crop yield prediction using machine learning: a systematic literature review. Comput Electron Agric 177:105709. https://doi.org/10.1016/j.compag.2020.105709
27. Khaki S, Wang L, Archontoulis SV (2020) A CNN-RNN framework for crop yield prediction. Front Plant Sci 10:1–14. https://doi.org/10.3389/fpls.2019.01750
28. Agarwal S, Tarar S (2021) A hybrid approach for crop yield prediction using machine learning and deep learning algorithms. J Phys Conf Ser 1714(1). https://doi.org/10.1088/1742-6596/1714/1/012012
29. Morales A, Villalobos FJ (2023) Using machine learning for crop yield prediction in the past or the future. Front Plant Sci 14:1–13. https://doi.org/10.3389/fpls.2023.1128388
30. https://github.com/Drinkler/Yield-Prediction/
31. https://github.com/ankitaS11/Crop-Yield-Prediction-in-India-using-ML

Soil Health Analysis and Automatic Prediction of Macro and Micronutrient Using Machine Learning Algorithms

Varsha Atul Shukre, Surabhi Thatte, Aishwarya Sameer Karandikar, Chaitali Pradeep Kannurkar, Amey Madhukar Patil, and Bhavesh Sudhir Jagtap

Abstract This paper explores the analysis and automatic prediction of macro- and micronutrients using machine learning approaches to comprehend soil health. Through this research we are exploring the analytical capabilities of Machine learning algorithm for imputing missing values specifically for macro and micronutrients of soil. The various types of machine learning algorithms used are Decision Tree, Liner Regression and Ensemble techniques like Random Forest, XGBoost and ADABoost. A total of 25 villages (with varying chemical and texture properties) from Ambegaon Taluka, Pune District, Maharashtra were sampled at a depth of 0–30 cm for laboratory analysis. It is composed of the macronutrients potassium (K), phosphorus (P), and nitrogen (N). However, other micronutrients such as Zinc (Zn), Iron (Fe), Copper (Cu), Manganese (Mn), Sulphur (S), Boron (B), Organic Carbon (OC), Electrical Conductivity (EC) and pH are also considered. R^2 values using Random Forest (RF) for pH(0.9221), OC (0.9992) N (0.99936), P (0.50291), K (0.9561) and Zn (0.9694); using Decision Tree EC (0.9486), K (0.9651) and B (0.9999) and using XGBoost N (0.9993), S (0.995), Fe (0.9662), Cu (0.9988) and Mn (0.9907). It can be concluded

V. A. Shukre
Department of Computer Science and Applications , Dr. Vishwanath Karad MIT World Peace University, Pune, India
e-mail: varsha.sontakke@mitwpu.edu.in

S. Thatte (✉)
Symbiosis Center for Medical Image Analysis, Symbiosis International University, Pune, India
e-mail: surabhi.thatte@gmail.com

A. S. Karandikar
Dragonfly Financial Technologies, Pune, India

C. P. Kannurkar
Bizmetric India Pvt. Ltd., Pune, India

A. M. Patil
Celebal Technologies, Jaipur, India

B. S. Jagtap
CRG Solutions Pvt. Ltd., Bengaluru, India

A. J. Kulkarni and M. Hajiaghaei-Keshteli (eds.), *Advances in Intelligent Systems for Sustainable Agriculture*, Lecture Notes in Networks and Systems 1201,
https://doi.org/10.1007/978-981-97-9839-1_8

with this research that ML algorithms can be utilized for imputing missing values of soil nutrients. They can be deployed as secondary measures for reducing the cost and efforts of repeating primary tests for soil nutrient testing in case of unacceptable values calculation through these methods. Also, a longitudinal study considering crop cultivation and other environmental aspects can help in formalizing a standard for this domain.

Keywords Machine learning algorithms · Macro and micronutrient prediction · MAE · R^2 score · RMSE · Soil health analysis

1 Introduction

To ensure better yield and health of crops, many farmers use the knowledge about soil composition. This can be helpful in deciding upon the farming parameters like crop selection, fertilizer selection, water requirement and other decisions to support good plant health and better yield each season. This aspect is very important to provide sustainable agriculture management.

The most important tests that are performed on the soil are to understand its macro and micronutrients composition. Where N, P, and K are macro nutrients and are the most studied parameter. To get this information there are three widely used techniques which are laboratory test, use of sensor devices and satellite imaging data. All these techniques have been used by many researchers across the globe focusing on a wide variety of problems like understanding a single crop yield or just the soil composition for better choice of fertilizer etc. [1].

There are standard scientific laboratory tests which are applied to collect this information; however, the procedure takes time. Similar difficulties arise when working with sensors and satellite images. Also, all these techniques have been extensively used only for understanding the N.P and K values. A major gap exists for their use in understanding macronutrients.

Macronutrients are nutrients that plants require in large quantities i.e. N, P, K, S, magnesium (Mg), and calcium (Ca) and micronutrients are elements that plants require in little amounts (micro) such as Fe, Zn, Cu, B, chlorine (Cl), manganese (Mn), and molybdenum (Mo) Plant nutrients are essential and cannot be substituted by any other components. A shortage of nutrients can cause poor plant growth and lower yield. If there is a deficiency in the plant for these vital nutrient elements, it will develop abnormally. If the three primary nutrients—N, P, and K—are given to the plant, it will grow healthily. As a result, if we know the amount of N, P and K nutrient content in the soil that will be planted, we can estimate the amount of NPK fertilizer that must be applied to the soil. However, it is found that there is lack of research with respect to micronutrients.

The methods for analyzing the NPK content of soil includes Chemical analysis in which the soil sample is prepared and analyzed in laboratory using various chemical tests. The most common chemical tests for NPK analysis are the Kjeldahl method for N, the Bray-Kurtz method for P, and the flame photometer method for K. In Spectroscopy: the reflectance or absorption of light in different spectral regions is measured. Near-infrared spectroscopy (NIRS) can detect near-infrared light by measuring the reflectance or absorption of near-infrared light to determine the levels of NPK in a sample. X-ray fluorescence (XRF) method uses X-rays to excite the atoms in a soil sample which then emit typical X-rays that can be used to identify the NPK elements present in the sample. However, these tests can be time-consuming and cumbersome. They may often lead to inaccurate or missing outcomes for a few parameters [1].

Similarly, nowadays use of Internet of Things (IoT) based sensors to perform these tests is evidently increasing. However, there are several levels of security problems in IoT-based agricultural systems. Due to low security, the user may lose field parameter's data. IoT devices are at risk due to physical intrusion such as attacks by animals and predators [2, 3]. Many cost-related issues occur such as setup and running costs during Installation of IoT devices [4]. The lack of knowledge about technology becomes the main barrier in the IoT device implementation in farms [5, 6]. Before installing IoT infrastructure, farmers must also invest heavily in training. The IoT devices/Sensors are installed in an open environment so the harsh environmental conditions may cause communication failure hence to ensure the physical safety of deployed IoT devices/sensors to protect them from severe climate conditions is of prime importance [7]. The devices must provide functionality and support to the rest of the world without installing additional devices with overhead configuration [8]. Furthermore, in order to provide unhindered communication and information exchange between devices, it is imperative to choose the optimal deployment location. Integration challenges include connectivity, low-power communication devices etc. Furthermore, as IoT communication devices experience connectivity and latency concerns, monitoring and regulating processes have made IoT device data exchange increasingly complex. [9, 10].

The use of hyperspectral satellite imagery for soil nutrient measurement has also been the subject of several studies [11]. These multispectral photos are used to calculate the land's reflectance, which is a crucial characteristic for comprehending soil nutrients. A multi-spectral image's bands each have their own unique properties. [12, 13]. But the price is one of the biggest disadvantages. Because satellite photography is so costly, some researchers may find it prohibitive to acquire and process the data. Furthermore, satellite photography frequently has a low resolution, which makes it potentially incapable of identifying minute environmental changes [14, 15].

Recent advancements in the field of agriculture have also resulted in the use of technology based smart solutions to automate many aspects of farming. Major use of Artificial Intelligence (AI), Machine Learning (ML), Deep Learning (DL) and Internet of Things (IoT) has increased over the past few years [16]. Apart from performing laboratory tests, sensors data can be shared to these automated solutions using IoT and similarly satellite imaging data can also be utilized by these automated solutions. Also, we would like to highlight that in rare scenarios these values may get miscalculated or interpreted by the system. And use of assessment methodology like laboratory test, sensor, or image information; might not be possible due to various constraints as discussed above. Therefore, this wrong understanding about the soil composition may lead to applying inadequate quantity of fertilizers or wrong crop selection affecting the farms' yield. This directly impacts the sustainable growth of the plant, farm, and crop yield management.

Thus, through this research we aim to investigate the application supervised ML methods for imputing these missing values. We propose an automated solution focused on 12 nutrients and ML algorithm. Additionally, we compare and conclude the best possible solution for this problem. This can be instrumental in understanding correlation of 12 nutrients, as specific to the state of Maharashtra, India.

2 Literature Review

Throughout the literature we found that most of the studies have considered macronutrients only. Madhumathi, R., et al. have applied multiple linear regression to predict the values of soil's NPK. They used a statistical technique which can be used to explain the relationships between independent and dependent variables. Their proposed solution achieved an accuracy of approximately 80%. However, they mentioned that micronutrient analysis is the future scope for research in this domain [17]. Ahmed, Usman, et al. have used an improved genetic algorithm, which is an evolutionary computation for proposing a nutrient recommendation system for soil fertilization [18]. Dipto, Shakib Mahmud, et al. proposed a suitable crop recommendation system based on N, P and K values using various ML algorithms with accuracy as mentioned; Support Vector Machine (SVM) 98%, Random Forest (RF) 95%, AdaBoost 90% and logistic regression 88% [19]. Similarly, another study done by Gosai, Dhruvi, et al. have also used ML algorithms for the crop recommendation system with accuracy ranging between 10% by SVM and 99.31% by XGBoost [20]. Ratnaparkhi, Sanika, et al. have proposed a smart agriculture aspect by utilizing sensor information for automation [21].

Kaur et al., studied the impact of N over carbon (C), sulfur (S), lime, moisture, calcium (Ca), magnesium (mg), P, and K. They have considered total nine parameters for their prediction study [22]. Dash et al., research focused on establishing and interrelating the micronutrients and the weather parameters and classifying the type of crop based on the micronutrients using a support vector machine and decision tree algorithm. Based on the soil's features and input parameters, the developed model forecasts a suitable crop with a 92% accuracy level (Fitness Score). SVM with Linear Kernel shows better performance compared to other algorithms such as linear SVM and decision tree [23].

Sr No	Paper title	Year	Author details	Algorithm/Model used	Result/Accuracy in %
1	Soil NPK prediction using multiple linear regression. In 2022 8th international Conference on Advanced Computing and Communication Systems (ICACCS) (Vol. 1, pp. 542–546). IEEE	2022	Madhumathi, R., Arumugnathan, T., Shruthi, R., & Raghavendar, S	Multiple regression (MLR) technique	Approximately 80%
2	A nutrient recommendation system for soil fertilization based on evolutionary computation. Computers and Electronics in Agriculture, 189, 106,407	2021	Ahmed, U., Lin, C. W., Srivastava, G., & Djenouri, Y.	Improved genetic algorithm	NA
3	Suitable crop suggesting based on npk using machine learning models IEEE Asia–Pacific Conference Computer Science Data Engineering (CSDE) (pp. 1–6). IEEE	2021	Dipto, S. M., Iftekher, A., Ghosh, T., Reza, M. T., & Alam, M.A. (2021, December)	SVM, adboost, forest logistic regression	Maximum of 98% accuracy using SVM
4	Crop recommendation system using machine learning. International Journal of Scientific Research in Computer Science, Engineering and Information Technology, 7(3), 558–569	2021	Gosai, D., Raval, C., Nayak, R., Jayswal, H., & Patel, A.	Decision tree, Naïve Bayes, Support vector machine, logistic regression, Random Forest and XGBoost	99.31%

(continued)

(continued)

Sr No	Paper title	Year	Author details	Algorithm/Model used	Result/Accuracy in %
5	Kulkarni, N. H., Srinivasan, G. N., Sagar, B. M., & Cauvery, N. K.(2018, December). Improving crop productivity through a crop recommendation system using ensembling technique. In 2018 3rd International Conference on Computational Systems and Information Technology for Sustainable Solutions (CSITSS) (pp. 114–119). IEEE	2018	Kulkarni, N. H., Srinivasan, G. N., Sagar, B. M., & Cauvery, N. K.	The independent base learners used in the ensemble model are Random Forest, Naïve Bayes, and Linear SVM	99.91%
6	Pande, Chaitanya B., et al. "Prediction of soil chemical properties using multispectral satellite images wavelettransforms methods." Journal of the Saudi Society of Agricultural Sciences 21.1 (2022): 21–28	2022	Pande, C. B., Kadam, S. A., Jayaraman, R., Gorantiwar, S., & Shinde, M. (2022)	Four wavelets models like Daubechies, Symlet, Biorthogonal and Coiflet	NA
7	Dhiman, Gaurav, Jhilik Bhattacharya, and Sangita "Soil textures nutrients estimation using remote sensing data in north india-Punjab region." Procedia computer science 218 (2023):2041–2048	2023	Dhiman, Gaurav, Jhilik Bhattacharya, and Sangita Roy	Gradient Boosting (GB), Extreme Gradient Boosting (XGB), Random Forest Regressing (RFR), and Adaptive Boosting (ADA)	NA

(continued)

(continued)

Sr No	Paper title	Year	Author details	Algorithm/Model used	Result/Accuracy in %
8	Riad, Sumsuttibriz, et al. "Prediction of soil nutrients using hyperspectral satellite imaging." Proceedings of International Conference on Fourth Industrial Revolution and Beyond 2021. Singapore: Springer Nature Singapore, 2022	2022	Riad, S., Ahmed, M. S., Himel, M. H., Ahmed, M. R., hasan, M. M., Mim, A. H., & Mukta, M. S. H.	Machine learning based classification model	NA
9	Ahmed, Usman, et al. "Nutrient recommendation system for soil fertilization based on evolutionary computation." Computers and Electronics in Agriculture 189(2021):106,407	2021	Ahmed, U., J. C. W., Srivastava, G., & Djenouri, Y.	Improved genetic algorithm (IGA)	More efficient
10	Kaur, Supreet, and Kamal Malik. "Predicting and estimating the major nutrients of soil using machine learning techniques." Soft Computing for Intelligent Systems: Proceedings of ICSCIS 2020. Springer Singapore, 2021	2021	Kaur, Supreet, and Kamal Malik	Simple linear regression and multiple linear regression-based models	The results obtained by these three models justified interdependency of vital nutrients over other

(continued)

(continued)

Sr No	Paper title	Year	Author details	Algorithm/Model used	Result/Accuracy in %
11	Dash, R., Dash, D. K., & Biswal, G. C. (2021). Classification of crop based on macronutrients and weather data using machine learning techniques. Results in Engineering, 9, 100, 203	2021	Dash, R., Dash, D. K., & Biswal, G. C.	Support vector machine learning (SVM), SVM with Kernel and Decision tree	92% Accuracy with linear Kernel

3 Materials and Methods

In the following section, we have discussed the complete ML pipeline. Exploratory data analysis is the first step, followed by data preparation methods to enhance data quality for improved prediction outcomes, model training, and model testing. Since we have used multiple ML algorithms to impute these macro and micronutrient values. We have added a discussion on these techniques as well.

3.1 Data Set

We have used the dataset of Maharashtra, India. It consists of values of soil nutrients like N, P, K, S, Fe, Zn, Cu, Mn, B, pH, OC, and EC for 25 different villages. The second data is for crops and specific values of N, P, K required for these crops. Oil samples were collected from 25 separate villages in Ambegaon Taluka, located in the Pune District of Maharashtra. The samples were taken at a depth of 0–30 cm and each village had its own unique chemical and textural characteristics. The purpose of collecting these samples was for laboratory examination.

The process flow diagram below depicts the solution design, and the following sections elaborate them in detail (Fig. 1).

3.2 Exploratory Data Analysis (EDA)

EDA is a very important aspect of any data analysis research. This process is performed to investigate and understand the data well. It helps in summarizing the

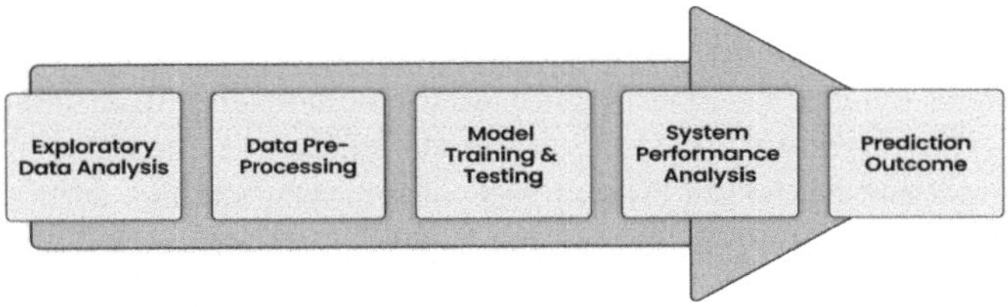

Fig. 1 Process flow diagram of the system

main characteristics from statistical perspective by using some visualization techniques. It is used to understand the further data preprocessing which can be applied on the data. We have performed the following EDA on our dataset.

3.3 Data Preprocessing

This process is another crucial aspect, where to prepare the data to make it ready for any prediction system. Accurately anticipating the primary aspect of interest requires preprocessing. While maintaining an increase in the quality of the data, several methods for improving the data can assist in eliminating particular kinds of mistakes. It is recommended to use various complementing preprocessing approaches in order to remove artifacts that would be left behind by employing a single technique.

We implemented the following technique for improvising our data—feature selection, handling missing values and null values, outlier detection and removal, removing extra characters for data using python libraries and packages.

1. Feature Selection

The goal of the feature selection procedure is to find the most instructive and pertinent characteristics for a given analysis or model. By evaluating the impact of individual features on the desired outcome, we can determine which ones contribute the most to achieving our objectives.

Initially, the dataset had a shape of (12,398, 14), indicating 12 soil features (PH, EC, OC, nitrogen, phosphorus, potassium, boron, Sulphur, zinc, iron, copper, and magnesium) along with two additional columns for longitude and latitude. However, for the subsequent analysis, the longitude and latitude columns were dropped, resulting in a revised shape of (12,361, 12). This step was likely taken because the focus shifted to analyzing the soil features rather than the geographical coordinates.

2. Removal of null values

After dropping the null values, the dataset was further refined, resulting in a shape of (12,361, 12). This step ensures that the analysis is performed on complete and valid data points.

3. Outlier detection and removal

To recognize and comprehend data points that substantially differ from the anticipated patterns or behavior within a dataset, outlier identification is an essential first step in data analysis and modelling. The process of removing outliers ensures that the subsequent analysis and modelling are not skewed or influenced by extreme data points, leading to more accurate and reliable insights and predictions.

Box plots and scatter plots were made to obtain information about the distribution and connections between the soil properties. Box plots illustrate the median, quartiles, and any possible outliers in a graphic depiction of the distribution of each attribute. Conversely, scatter plots make it possible to analyze the connections between feature pairs. A z-score test was conducted to detect outliers in the dataset. The z-score determines how much a data point deviates from the mean in standard deviations. Data points that surpass a threshold (often established at a z-score of 3) might be categorized as outliers.

It was discovered that there were outliers in the provided dataset after doing the outlier identification analysis. Data points that drastically depart from typical patterns or norms are known as outliers.

To address these outliers, a decision was made to remove them using the z-score test. By removing the outliers, the dataset was further refined, resulting in a new shape of (10,653, 12).

4. Data cleaning

Overall, the dataset underwent several data cleaning steps, including dropping the longitude and latitude columns, handling null values, identifying outliers through box plots and scatter plots, and finally removing outliers using the z-score test. These steps help prepare the data for further analysis and modeling, ensuring that the remaining data points accurately represent the soil features of interest.

5. Data Visualization

Heatmap: A heatmap is a visualization tool that represents data using colors on a two-dimensional grid. In the context of agriculture and soil analysis, a heatmap can be used to visualize the relationships and patterns among different features or variables. By plotting a heatmap, we can gain insights into the correlations between the soil features mentioned earlier, such as pH, EC, OC, nitrogen, phosphorus, potassium, boron, sulfur, zinc, iron, copper, and magnesium. The heatmap allows us to observe how these variables interact with each other and identify any significant associations. Analyzing the heatmap can help identify relationships between soil features that may influence plant growth, nutrient availability, or other important factors for the recommendation of the crops for that particular soil. Other plots are scatter plot and box plot for numerical values.

A two-dimensional data representation with different values shown in different colors is called a heatmap. A straightforward heat map lets users rapidly determine which data points are most crucial or pertinent by providing a visual summary of the information on two axes (Fig. 2).

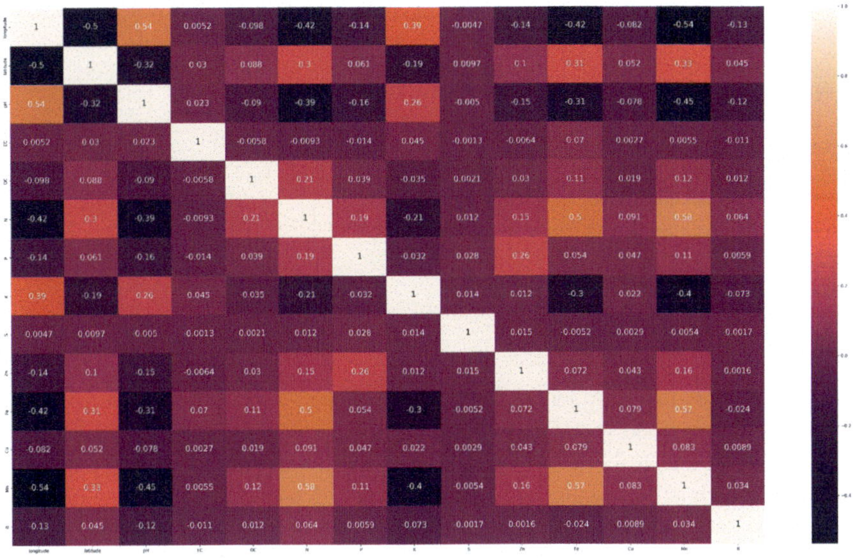

Fig. 2 Heat map for data visualization

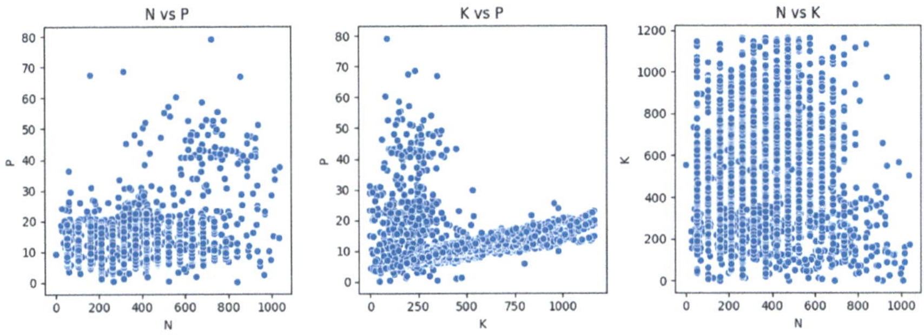

Fig. 3 **a** Scatter plot for N and P **b** Scatter plot for K **c** Scatter plot for N and K

- Scatter plots:

The values of two numerical variables are represented by points in a scatter plot, sometimes referred to as a scatterplot or scatter plot. A single data point's value is represented by its position on the horizontal and vertical axes. Examining correlations between variables, such as positive or negative, strong or weak, linear or nonlinear, is done via scatter plots (Fig. 3).

- Box plots:

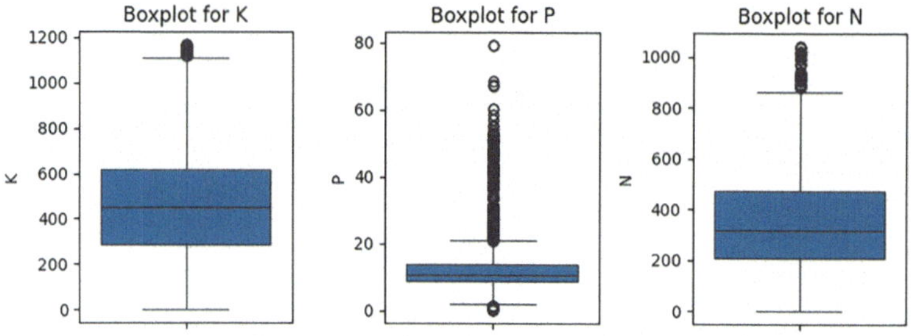

Fig. 4 Box plots

This kind of chart displays a set of numerical data along with their respective quartiles. Here's a quick method to see how our data is shaped. This facilitates cross-class comparison of data attributes with great ease. A box gives the data set's five-number summary, which is: Minimum: This is the data set's lowest value, omitting outliers. 25% of the data are below the first (lower) quartile, or the first quartile (Q1). The midpoint of the data set is designated as the median (Q2). It is exceeded by half of the values and is below the other half. 75% of the data are below the third (upper) quartile, which corresponds to the third quartile. Maximum: The highest value in the dataset without any outliers is this (Fig. 4).

3.4 Methodology Supervised ML Models

There are several machine learning algorithms that can be used to impute missing values in a dataset. Some of the commonly used algorithms which we investigated in our research study are mentioned below:

- Linear Regression:

As mentioned earlier, linear regression can be used to predict the missing values in a dependent variable based on the values of other independent variables. A multiple linear regression model is a linear equation that has the general form:

$$y = b1x1 + b2x2 + \ldots + c \tag{1}$$

where y is the dependent variable, x1, x2... are the independent variable, and c is the (estimated) intercept.

- Decision Trees:

Regression and classification issues can both be solved with decision trees. Based on the values of other independent variables, they can also be used to forecast the missing values in a dependent variable. This method operates on the disjunctive normal form, commonly referred to as Sum of Product. The two most widely used methods are the

Gini Index and Information Gain. To determine the root node and other branches, they are employed in attribute selection. We took information acquisition into account in our analysis. The basic formula for the same is:

$$\text{Gain} = \text{Entropy (Parent)} - \text{Entropy (Child)} \quad (2)$$

Where Entropy characterizes the impurity of any given collection of instances, and is a measure of uncertainty for a random variable. There is more information content when the entropy is higher. Assume that D is a dataset, A is an attribute, Values (A) is the set of all possible values for A, and Dv is the subset of D where A = v, then

$$\text{Gain (D, A)} = \text{Entropy (D)} + \Sigma v\varepsilon \text{ values } (A)|Dv| / |D| \cdot \text{Entropy (Dv)} \quad (3)$$

The ensemble ML approach combines the predictions of these distinct models, known as base learners, to arrive at a conclusion. The goal of ensemble methods is to increase prediction accuracy and generalization performance by utilizing the diversity and capabilities of the base learners. Among the several kinds of ensemble approaches are stacking, boosting, and bagging. We employed Random Forests (RF), ADABoost, and XGBoost. RF combines multiple decision trees. We investigated their use to impute missing values in our study.

It is important to note here, that the choice of the algorithm for imputing missing values will depend on the characteristics of the dataset and the nature of the missing values. For this research, multiple algorithms were implemented, and comparison of their performance was done to determine the best approach for imputing missing values for this domain and problem.

4 Result and Discussion

Model	Results	Pros	Cons
Linear regression	Moderately effective in predicting soil nutrient levels	Simplicity: Easy to interpret and implement Provides insights into linear relationships between features	Limited complexity: Struggles with capturing non-linear patterns in nutrient data Sensitivity: Vulnerable to outliers
Decision tree	Demonstrated strong performance, capturing non-linear relationships	Non-linearity: Effectively models complex interactions in nutrient levels Interpretability: Provides a clear decision-making structure	Overfitting: Prone to overfitting, particularly with deep trees Instability: Sensitive to small variations in the dataset

(continued)

(continued)

Model	Results	Pros	Cons
Random forest	High accuracy and robustness, especially in handling diverse nutrient variables	Ensemble learning: Reduces overfitting and enhances generalization Robust: Handles a large number of input variables well	Complexity: Ensemble of trees may complicate interpretation Computational Overhead: Resource-intensive for large datasets
XGBoost	Competitive accuracy and robustness to outliers	Boosting algorithm: High predictive accuracy with careful hyperparameter tuning Robustness: Tolerant to outliers and noise in the data	Hyperparameter tuning: Requires meticulous tuning for optimal performance Computational Cost: Can be resource-intensive, particularly during training
ADABoost	Effective in improving weak learners, showcasing versatility	Performance improvement: Enhances model performance through boosting Reduced Overfitting: Focuses on misclassified instances, reducing overfitting	Sensitivity to outliers: May be influenced by outliers in the dataset Complexity: The combination of weak learners may introduce complexity

To sum up, every machine learning model has its own set of benefits and limitations.

The objectives of the soil health analysis should direct which one to use, taking into account aspects like dataset type, interpretability, and computational capacity. Our work provides the groundwork for making well-informed choices when choosing the best model to automate nutrient estimates and enhance soil health evaluations.

Any machine learning methodology evaluates the system's performance by looking at its mistakes and accuracy levels. As a result, we have deliberately chosen to analyze the performance of our systems using the "R-Squared" number. R2 or r2 is the statistical coefficient of determination. In a regression model, it shows the percentage of a dependent variable's fluctuation that can be attributed to an independent variable. Likewise, the remaining parameters consist of Root Mean Square Error (RMSE) and Mean Absolute Error (MAE) (Table 1).

The average absolute difference between the actual and expected values is measured by the MAE. It is computed by averaging the absolute deviations between the values that were anticipated and those that were observed. The mean absolute error (MAE) provides a sense of the average magnitude of prediction errors and is given in the same unit as the target variable.

While the RMSE and MAE are comparable, the RMSE gives greater errors more weight. The square root of the average of the squared discrepancies between the

Table 1 R2 scores for combined village data

Model name	pH	EC	OC	N	P	K	S	Zn	Fe	Cu	Mn	B
Linear Regression	−1.136	0.02	0.114	−0.03	0.19	−1.06	0.0004	−0.06	−2.56	−0.06	0.45	−0.45
DecisionTree	0.7879	0.9486	0.9988	0.9915	0.3887	0.9651	0.9142	0.8357	0.9259	0.9668	0.9772	0.9999
Random Forest	0.9221	0.8979	0.9992	0.99936	0.50291	0.9561	0.95576	0.9694	0.9307	0.86303	0.9728	0.9979
XGBoost	0.8326	0.8527	0.9983	0.9993	0.3331	0.8918	0.995	0.8012	0.9662	0.9988	0.9907	0.9984
ADA-Boost	0.4763	0.2014	0.7874	0.9762	0.4451	0.6918	−0.9043	0.2958	0.6863	0.4506	0.8018	0.9969

expected and actual values is used to compute it. RMSE is represented in the same unit as the target variable, just like MAE. Greater mistakes are punished more severely than little ones. An expansion of R-squared that accounts for the number of predictors is called adjusted R-squared (Table 2).

5 Conclusion

In this research proposal, we have discussed the application of supervised machine learning methods to impute missing values for 12 soil nutrients in Maharashtra, India. An automated solution focused on understanding the correlation between these nutrients is proposed. We have used a dataset of soil nutrient values from 25 villages in Maharashtra, India.

The crops dataset with their specific N, P, and K values is also used. Our ML pipeline will consist of: Data preprocessing techniques for preparing data to enhance its quality and prepare it for modeling. Multiple ML algorithms to impute the missing soil nutrient values. We have tested the performance of the trained models on a held-out test set. By inputting missing soil nutrient values, our solution can help farmers to make better decisions about crop selection and fertilizer application. This can lead to increased crop yields and sustainable agriculture practices. Our solution useful in enhancing their comprehension of the connections between crop growth and soil nutrients, in addition to the advantages it offers farmers. This information could be used to develop new crop varieties and fertilization strategies.

In conclusion, our research focuses on leveraging machine learning to predict missing values for 12 essential soil nutrients in Maharashtra, India. Utilizing a dataset from 25 villages and crop data, our robust pipeline incorporates data preprocessing and five ML algorithms. Among these, the random forest algorithm outperforms others, offering an efficient solution. Our method helps farmers choose crops and apply fertilizer wisely, resulting in higher yields and more sustainable agriculture. Moreover, it presents valuable insights for agricultural researchers to refine fertilization strategies and develop new crop varieties, bridging the gap between data science and agricultural sustainability.

In the pursuit of advancing soil health analysis and automating the prediction of macro and micronutrients, our research employed several machine learning models, each contributing unique insights and trade-offs. The following is a discussion of the results obtained from these models, along with their respective pros and cons.

6 Future Work

Algorithms for machine learning (ML) have the power to completely change how we assess and forecast soil health. However, there is still much research to be done in this area. Here are some possible directions for future work:

Table 2 RMSE and MAE scores for combined village data

Model	Metric	pH	EC	OC	N	P	K	S	Zn	Fe	Cu	Mn	B
Random forest	RMSE	0.123	7.5462	0.0287	0.1762	10.1179	0.7379	0.1762	0.2964	4.7311	0.7379	4.3554	1.0939
	MAE	0.0419	0.0291	0.0028	0.0291	1.242	0.0207	0.0291	0.0665	1.0082	6.711	0.9111	0.0737
	Adjusted R2	0.877	0.7123	0.73	0.9272	0.9183	0.9988	0.8159	0.9073	0.972	–1.236	0.959	0.5854
XGBoost	RMSE	0.1104	0.2751	0.146	1.447	1.2371	4.7527	3.3232	0.3143	5.8364	0.5656	4.2615	2.5283
	MAE	0.0255	0.0203	0.0098	1.8187	0.9556	2.9687	0.6796	0.0379	0.8466	0.0713	1.1097	0.0795
	Adjusted R2	0.9093	0.8547	0.9936	0.9987	0.4876	0.926	0.4369	0.8958	0.9575	0.7052	0.9402	–1.214
ADABoost	RMSE	0.2835	0.9576	1.4985	5.8736	18.0156	3.0412	29.4137	0.9307	1.4964	0.7853	14.2284	0.8501
	MAE	0.1936	0.4403	0.383	2.5994	4.9078	20.647	5.5155	0.4971	1.1489	0.529	5.338	0.3813
	Adjusted R2	0.3725	–2.5117	0.1982	0.7418	0.8466	0.2333	0.8826	0.2331	0.5138	0.9769	0.6384	–23.7546
Decision tree	RMSE	0.1486	0.28105	0.1274	104.8201	24.5414	96.312	161.2106	0.1629	10.5015	0.2921	10.684	0.107
	MAE	0.028	0.0182	0.0033	4.0161	2.1756	0.8895	3.8138	0.03001	0.9196	0.0568	1.0148	0.01471
	Adjusted R2	0.8444	0.848	0.9951	0.8357	0.7328	0.8825	0.6397	0.9781	0.8593	0.9967	0.7947	0.9914
Linear regression	RMSE	0.3218	0.72108	1.7989	207.4729	46.9301	268.3815	268.7153	1.0025	23.5211	5.1055	18.9007	1.1939
	MAE	0.213	0.1536	0.1108	150.3	8.1222	205.1474	10.8451	0.4056	7.8202	0.6762	9.431	0.488
	Adjusted R2	0.2304	0.0022	0.0393	0.3563	0.0232	0.1156	–0.001	0.0816	0.3062	0.0048	0.3577	–0.0631

More accurate and robust ML models can be developed by using larger and more diverse datasets, as well as by developing new ML algorithms that are specifically designed for soil health analysis and prediction. Models can be developed to predict the spatial and temporal distribution of macro and micronutrients in soil. This information could be used to create precision agriculture maps, which would allow farmers to apply fertilizers and other inputs more precisely and efficiently. It could also be used to predict how soil nutrient levels will change over time, which could help farmers develop more sustainable management practices. ML models can be developed that can be used to assess the impact of different agricultural practices on soil health. This information could be used to develop more sustainable and environmentally friendly agricultural practices.

Other ML algorithms like singular value decomposition which is a matrix factorization technique and K-Nearest Neighbors a non-parametric technique may be explored as a future direction.

Along with these other time series techniques like, Autoregressive integrated moving averages (ARIMA) utilizing other environmental aspects and crop cultivation information generated through IoT devices, can also be added to make these predictions more useful for industrial support [23–25].

Acknowledgements We would like to thank Mr Rawindra Sudhakar Wadkar Soil Testing Officer Pune Department of Agriculture Government of Maharashtra and Ministry of Agriculture Government of India for providing data sets consisting of soil chemical analysis details, for our research work.

References

1. Shukre VA, Patil SS (2020) Comparative study of different methodologies used for measuring soil parameters: A review. In: International Journal Of Engineering Research & Technology (IJERT) ICSITS—2020 (Vol 8—Issue 05)
2. Varga P, Plosz S, Soos G, Hegedus C (2017) Security threats and issues in automation IoT. In: Proceedings of the 2017 IEEE 13th International Workshop on Factory Communication Systems (WFCS), Trondheim, Norway, 31 May–2 June 2017, pp 1–6
3. Duan J, Gao D, Yang D, Foh CH, Chen HH (2014) An energy-aware trust derivation scheme with game theoretic approach in wireless sensor networks for IoT applications. IEEE Internet Things J 1:58–69
4. Elijah O, Rahman TA, Orikumhi I, Leow CY, Hindia MN (2018) An overview of Internet of Things (IoT) and data analytics in agriculture: Benefits and challenges. IEEE Internet Things J 5:3758–3773
5. Elijah O, Orikumhi I, Rahman TA, Babale SA, Orakwue SI (2017) Enabling smart agriculture in Nigeria: Application of IoT and data analytics. In: Proceedings of the 2017 IEEE 3rd International Conference on Electro-Technology for National Development (NIGERCON), Owerri, Nigeria, 7–10 November 2017, pp 762–766
6. Asikainen M, Haataja K, Toivanen P (2013) Wireless indoor tracking of livestock for behavioral analysis. In: Proceedings of the 2013 9th International Wireless Communications and Mobile Computing Conference (IWCMC), Sardinia, Italy, 1–5 July 2013, pp 1833–1838

7. Al-Fuqaha A, Guizani M, Mohammadi M, Aledhari M, Ayyash M (2015) Internet of Things: A survey on enabling technologies, protocols, and applications. IEEE Commun Surv Tutor 17:2347–2376
8. Biral A, Centenaro M, Zanella A, Vangelista L, Zorzi M (2015) The challenges of M2M massive access I wireless cellular networks. Digit Commun Netw 1:1–19
9. Perera C, Liu CH, Jayawardena S, Chen M (2014) A survey on internet of things from industrial market perspective. IEEE Access 2:1660–1679
10. Xiaojing Z, Yuanguai L (2012) Zigbee implementation in intelligent agriculture based on internet of things. In: Proceedings of the 2nd International Conference on Electronic Mechanical Engineering and Information Technology, Shenyang, China, 7 September 2012
11. Pande, Chaitanya B, et al (2022) Prediction of soil chemical properties using multispectral satellite images and wavelet transforms methods. J Saudi Soc Agric Sci 21(1):21–28
12. Dhiman G, Bhattacharya J, Roy S (2023) Soil textures and nutrients estimation using remote sensing data in north india-Punjab region. Procedia Comput Sci 218:2041–2048
13. Sumsuttibriz R, et al (2022) Prediction of soil nutrients using hyperspectral satellite imaging. In: Proceedings of International Conference on Fourth Industrial Revolution and Beyond 2021. Singapore: Springer Nature Singapore
14. Mochamad Firman G, et al (2020) Generating soil salinity, soil moisture, soil pH from satellite imagery and its analysis. Inf Process Agric 7(2):294–306
15. Solmaz F, et al (2021) Effect of multi-temporal satellite images on soil moisture prediction using a digital soil mapping approach. Geoderma 385:114901
16. Bhoite S, Patil CH, Thatte S, Magar VJ, Nikam P (2023) A data-driven probabilistic machine learning study for placement prediction. In: 2023 International Conference on Intelligent Data Communication Technologies and Internet of Things (IDCIoT), Bengaluru, India, pp 402–408. https://doi.org/10.1109/IDCIoT56793.2023.10053523
17. Madhumathi R, et al (2022) Soil NPK prediction using multiple linear regression. In: 2022 8th International Conference on Advanced Computing and Communication Systems (ICACCS), vol 1, IEEE
18. Usman A, et al (2021) A nutrient recommendation system for soil fertilization based on evolutionary computation. Comput Electron Agric 189:106407
19. Shakib Mahmud D, et al (2021) Suitable crop suggesting system based on NPK values using machine learning models. In: 2021 IEEE Asia-Pacific Conference on Computer Science and Data Engineering (CSDE). IEEE
20. Dhruvi G, et al (2021) Crop recommendation system using machine learning. Int J Sci Res Comput Sci, Eng Inf Technol 558–569
21. Sanika R, et al (2020) Smart agriculture sensors in IOT: A review. In: Materials Today: Proceedings
22. Ritesh D, Ku Dash D, Biswal GC (2021) Classification of crop based on macronutrients and weather data using machine learning techniques. Results Eng 9:100203
23. Tikiwala V, Khule S, Nadgauda C, Thatte S (2023) Automatic prediction of egg production in poultry farm system. In: S Kumar, S Hiranwal, S Purohit, M Prasad (eds) Proceedings of International Conference on Communication and Computational Technologies. ICCCT 2023. Algorithms for Intelligent Systems. Springer, Singapore. https://doi.org/10.1007/978-981-99-3485-0_12

24. Muthumanickam D, et al (2022) Smart farming: Internet of Things (IoT)-based sustainable agriculture. Agriculture 12(10):1745
25. Supreet K, Malik K (2021) Predicting and estimating the major nutrients of soil using machine learning techniques. In: Soft Computing for Intelligent Systems: Proceedings of ICSCIS 2020. Springer Singapore

Intelligent Systems for Disease Prediction and Assessment-II

A Deep CNN Model for Analytical Prediction of Tomato Diseases Using Multiple Disease Dataset

Akash S. Kumar, Aman Sahil, Aniket Prasad, Aditya Singh, and Sumit Kumar

Abstract Tomatos are one of the most important crops in global agriculture, responsible for many of the world's vegetable production.Tomatoes are one of the most important crops in global agriculture, responsible for many of the world's vegetable production. However, various diseases and pests often affect tomato plant health and yield. Early detection of these problems is essential to avoid crop loss and to optimize agricultural practices. It helps farmers take action on time for healthier tomato production. This paper uses a deep learning approach to predict leaf diseases by building a deep Convolutional Neural Network (CNN) model. Many different data sets are used to train the proposed model, including healthy tomato leaves and ones affected by diseases. The dataset that is used contains 1 class of healthy images consisting of 154 images and 10 classes of plant disease images consisting of 2869 images for the detection of diseases such as Bacterial spot, early blight, late blight, leaf-mold, powdery mildew, septoria leaf-spot, spidermitted two-spotted, target spot, Mosaic virus, yellow leaf-curl virus. The data is collected from the plant village. Multiple layers of convolution and Max pooling are used to build the model. The model accuracy is found to be 96.25%.

1 Introduction

In an agricultural country like India, farming and agriculture are practiced in a vast area, and most of the population depends on agriculture [1]. Many problems farmers face today include low productivity, increased food demand, irrigation, climate change, pest and rodent attacks, plant diseases, rising prices, and soil erosion. It is very important to discuss the problems and challenges that farmers face. One of the major problems that the farmers face is that the cultivated crop has been infected or affected by diseases that lead to their damage, and most of the time, these diseases are not detected at an early stage. So, solving this issue and detecting the diseases as early

A. S. Kumar · A. Sahil · A. Prasad · A. Singh · S. Kumar (✉)
Symbiosis Institute of Technology, Pune Campus, Symbiosis International (Deemed University), Pune 412115, Maharashtra, India
e-mail: er.sumitkumar21@gmail.com

A. J. Kulkarni and M. Hajiaghaei-Keshteli (eds.), *Advances in Intelligent Systems for Sustainable Agriculture*, Lecture Notes in Networks and Systems 1201,
https://doi.org/10.1007/978-981-97-9839-1_9

as possible is very important. Tomato is one of the major crops grown worldwide and essential for meeting people's nutritional needs [2]. Different types of fungi, bacteria, and viruses can cause discolored leaves, low crop yields, and bad crop quality. To get rid of these problems, it is very important to detect them early and accurately predict the disease. Usually, you need to hire a human expert to do a visual inspection, but that is time-consuming and prone to mistakes [3]. That is why modern technology and approaches are useful for detecting and diagnosing plant diseases. It can help reduce the need for pesticides by allowing targeted treatments [4]. This could be a huge help for farmers and researchers alike, as it can help keep an eye on tomato crop health and give them more control over disease management so they can get better crop quality and yield [5]. These diseases and problems faced by the farmers tend to give rise to more problems like less crop yield, rise in the price of the crop, and increase in food demand, which in turn gives rise to even further problems and the chain continues [6]. So, this work aims to solve the problem by proposing a system that can predict tomato leaf-related diseases using leaf images. This approach addresses the farmers' difficulties without relying on a third party [7]. Doing so will allow the farmers to potentially address the plant disease and treat it, leading to an increase in the quality and quantity of the crop and hence earning the farmers profits [8]. A Deep CNN model is introduced, which can detect diseases by comparing them with an available dataset of images. The model's performance was assessed using different parameters, including training accuracy, validation accuracy, testing accuracy, and the number of trainable parameters compared to a pre-trained model [9]. The literature survey is discussed in Sect. 2. The details of the Methodology for this model's work are mentioned in Sect. 3. The results observed are mentioned in Sect. 4, followed by the conclusion in Sect. 5.

2 Literature Survey

Deep learning has advanced a lot in this rapidly moving technological world, providing ease of work and development in many sectors. Agriculture is one of them [10]. It is crucial to recognize the research that has already been done to advance in the right way, and plant disease has always been a major area for research where deep learning is used widely for its accurate classification [11]. Different types of CNN have been used to detect different types of tomato leaf diseases. N. Panuganti et al. [12] used some of the most popular CNN to meet the specific needs of disease classification. It includes VGG (Veolia Green Gamma), ResNet (Inception), and MobileNet. Research has shown that deeper and more complicated networks are better at picking up fine-grain disease patterns on tomato leaves. P. Pradhan et al. [13] showed Transfer learning is a new way to detect tomato leaf disease. It works by using pre-trained CNN on bigger data sets like ImageNet and then fine-tuning them on smaller, tomato-specific data sets. It is a great way to get high accuracy with small data sets, which is especially useful in agriculture, where data can be limited [14]. Understanding how CNN models make predictions is important for building trust and confidence in agriculture [15]. Karen Simonyan et al. [16] have examined ways to explain CNN decisions, like creating heatmaps to show disease-caused areas in images. XAI methods are getting more and more important for understanding model behavior. While much of the research is focused on model design and testing, the practical implementation of CNN-based detection systems in agriculture is still a work in progress [17]. Researchers and practitioners are still figuring out how to set up the model, connect it to farming equipment, and make it easy for farmers to use [18]. Future research on how CNNs can detect tomato leaf disease should focus on making sure they are scalable, practical, and easy to use. This means creating easy-to-use applications, making them easier to understand, and ensuring they are integrated into precision agriculture processes [19].

3 Methodology

3.1 Dataset

The Data set is taken from Plantvillage [20] with a total of 3009 images of tomato leaves in which 11 classes contain 2855 images of unhealthy leaves and 1 class of 154 images of healthy leaves. Figure 1 shows the samples of healthy and infected leaves taken from each class. Early Blight (Alternaria solani), this fungal disease affects tomato leaves, causing dark brown to black lesions with concentric rings. It often starts at the bottom of the plant and can spread to the fruit.Tomato Yellow Leaf Curl Virus (TYLCV) is a viral disease transmitted by whiteflies. It causes curling, yellowing, and stunting of tomato leaves. Infected plants often have reduced fruit production [21]. Septoria Leaf Spot (Septoria lycopersici) is a fungal disease characterized by small, circular lesions with a grayish center and dark borders on tomato leaves. It can lead to the premature defoliation of plants. Powdery Mildew (Oidium spp) is a fungal disease that appears as a white, powdery substance on tomato leaves [22]. It can reduce photosynthesis and weaken the plant. Cucumber Mosaic Virus (CMV)—a viral disease that can affect tomato plants, causing mottled or mosaic-like patterns on leaves and stunted growth [23]. Figure 2 is the data visualization of the data set for better representation.

3.2 Data Preprocessing

A diverse and representative dataset of tomato leaf images from Plant Village containing healthy and diseased leaves is gathered. The data set is split into training, validation, and testing sets (typically 70–80% for training, 10–15 for validation, and 10–15 for testing) [24]. After Resizing all images to a consistent resolution, i.e., 256×256, to ensure uniformity, the training data is augmented with techniques

Fig. 1 Classwise example of dataset

like rotation, flipping, and zooming to increase data set diversity and improve model generalization [25].

3.3 *Convolutional Neural Network*

The CNN deep learning architecture stands for Convolutional Neural Network. Architecture can predict or classify something, classifying the images concerning their matching characteristics. With the help of activation operations/functions, CNN works out potential maps and does not require supervision to monitor the working. It is very efficient in image recognition and classification. It is very helpful in handling large-scale datasets. Figure 3 explains the basic structure of CNN using parameters such as lining, surface depth, height, and thickness. The model is classified using a set of layers through which the images are passed. As the model is trained, it learns the features of the images, which are used to detect the disease [26].

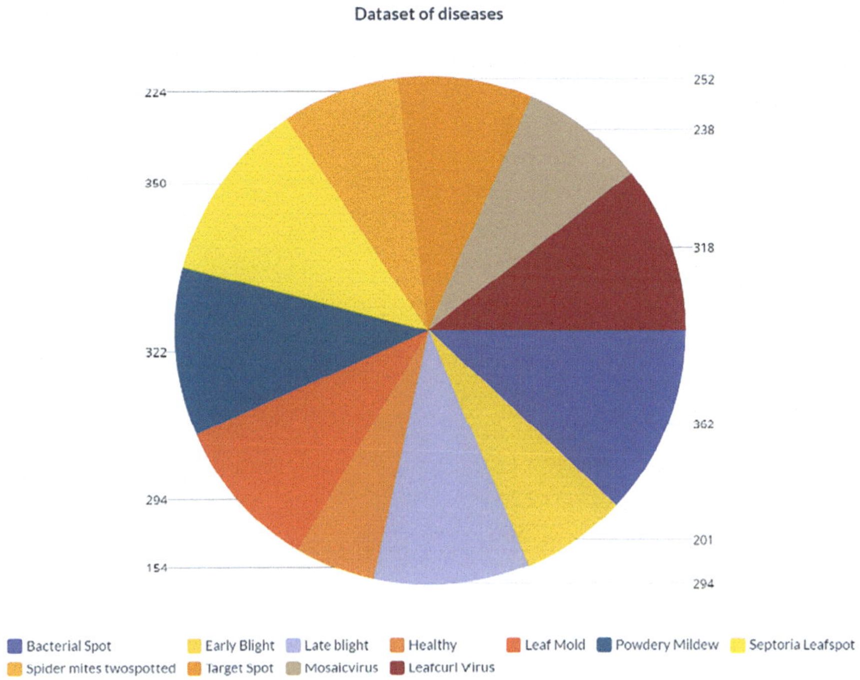

Fig. 2 Classwise number of dataset

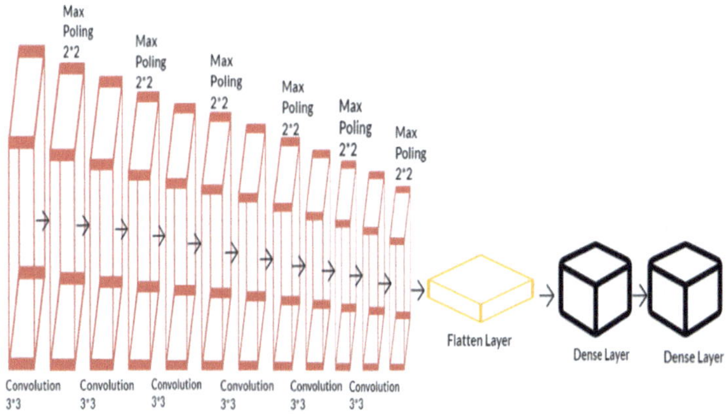

Fig. 3 Pictorial representation of CNN

3.4 Model Training

It begins by initializing a Convolution Neural Network (CNN) model with random weights [27]. The CNN is the "brain" that will learn to recognize diseases in tomato

Table 1 Parameters of CNN

Layer (type)	Output shape	Param
Sequential	(32,256,256,3)	0
Conv2D	(32,254,254,32)	896
Max Pooling 2D	(32,127,127,32)	0
Conv2D 1	(32,125,125,64)	18496
Max Pooling 2D 2	(32,62,62,64)	0
Conv2D 2	(32,60,60,64)	36928
Max pooling 2D 3	(32,30,30,64)	0
Conv2D 3	(32,28,28,64)	36298
Max pooling 2D 4	(32,14,14,64)	0
Conv2D 4	(32,12,12,64)	36928
Max pooling 2D 4	(32,6,6,64)	0
Conv2D 5	(32,4,4,64)	36928
Max pooling 2D 5	(32,2,2,64)	0
Flatten	(32,256)	0
Dense	(32,64)	16448
Dense1	(32,11)	725

leaf images. The data set was acquired and divided into training, validation, and test data. This way, the model could learn from one part of the data set, fine-tune its learning on the other, and then test on the third, invisible part [28]. Providing the model with the training data, it could make predictions on the tomato leaf images, matching those predictions to the real labels (healthy or disease) [29]. It kept tweaking its parameters to get better and better at it. Let us say you are teaching your friend a new board game. As they get better at the game, you want them to be able to play well, even on a board they have never seen before. Just like the model (computer) will be able to learn the rules of the game, the model will be able to spot diseases not just from the training pictures it has seen but from new unseen pictures, too. To do this, a set of pictures is called for the validation of data set [13]. Table 1 shows the layers and parameters used in the model training. Every time the model shows a new set of pictures, it asks to look at them and explain what it sees. It repeats the training and validation steps for a set number of training cycles or until the model converges, meaning it reaches its best possible performance. In this model, training was set to be 0.8, which shows that 80% of the dataset was used to train the model and, similarly, 0.1 (10%) for both validation and testing the model. The training, validation, and testing were done by shuffling the dataset after each EPOCH [30] (Figs. 4 and 5).

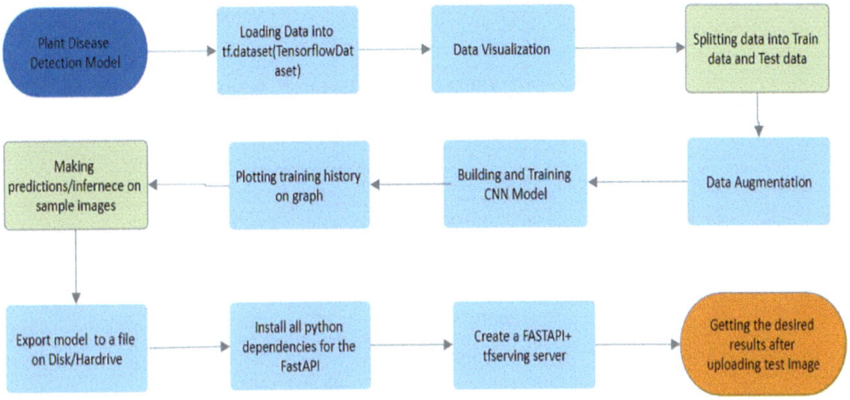

Fig. 4 Flow of proposed model

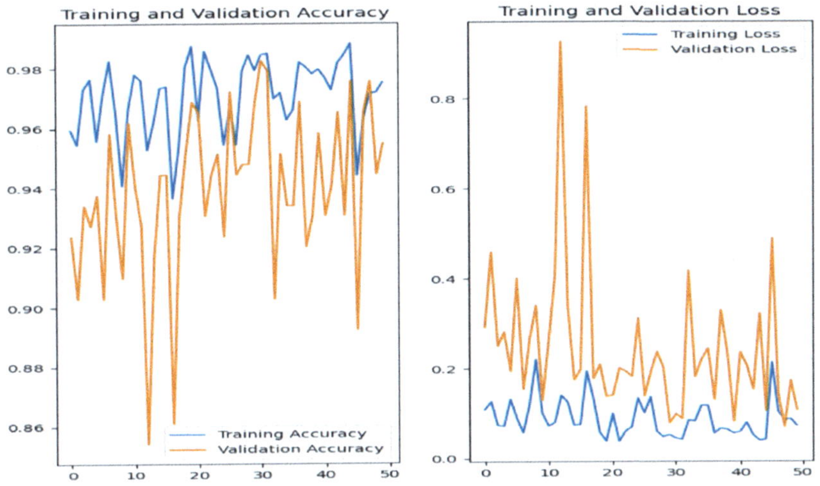

Fig. 5 Training and validation graph

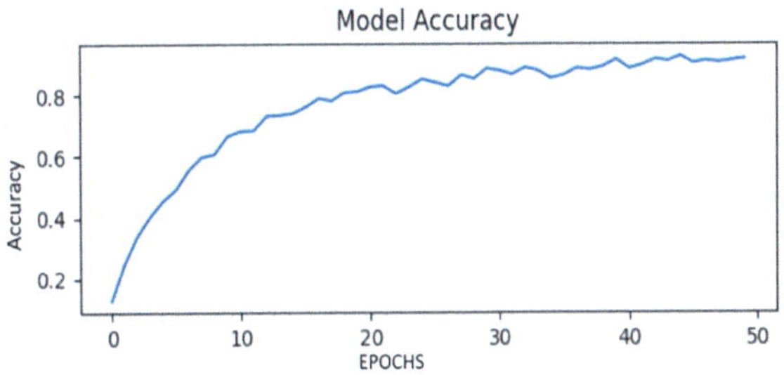

Fig. 6 Proposed model accuracy

Table 2 Hyperparameters of the proposed model

Type	Value
No of convolution layer	6
No of max pooling layer	6
Flatten layer	1
Dense layer	2
Batch size	32
Image size	256
Channels	3
EPOCHS	50
Training size	0.8
Validation on size	0.1
Testing size	0.1

4 Results

The results in Fig. 6 demonstrate the performance of the trained Deep-learning model for detecting tomato leaf disease. A total of 5 models were trained by trying out different hyper-parameters and shuffling the images, and based on the accuracy of each trained model, the best accuracy model was chosen as the proposed model. Table 2 shows the Hyper-parameters and their values in the proposed model. The performance of the proposed model was evaluated based on two metrics: Loss and Accuracy. Loss measures how well the model predictions match the actual data. A lower loss value indicates that the model is producing more accurate predictions. Accuracy measures the percentage of correct predictions that the model makes. Higher accuracy values indicate how well the model can distinguish between healthy tomato leaves and diseased ones. The loss value of the Proposed Model is 0.1992, and the accuracy is 0.9625. While comparing the proposed model with pre-trained models (Mobilenet [31], VGG 16 [32], Inception V3 [33]), shown in Table 3. The proposed model achieved better results than the pre-trained models (Fig. 7).

Table 3 Comparison of time and space complexity with pre-trained models

Sr.No.	Model	Accuracy (%)	Storage	Trainable parameters	Non-trainable parameters
1	Mobilenet	63.75	82,498 kb	20,507,146	559,808
2	VGG 16	77.2	94,452 kb	9,449,482	14,714,668
3	Inception V3	63.4	1,63,734 kb	41,247,146	490,880
4	Proposed model	96.25	2533,112 kb	1,84,267	0

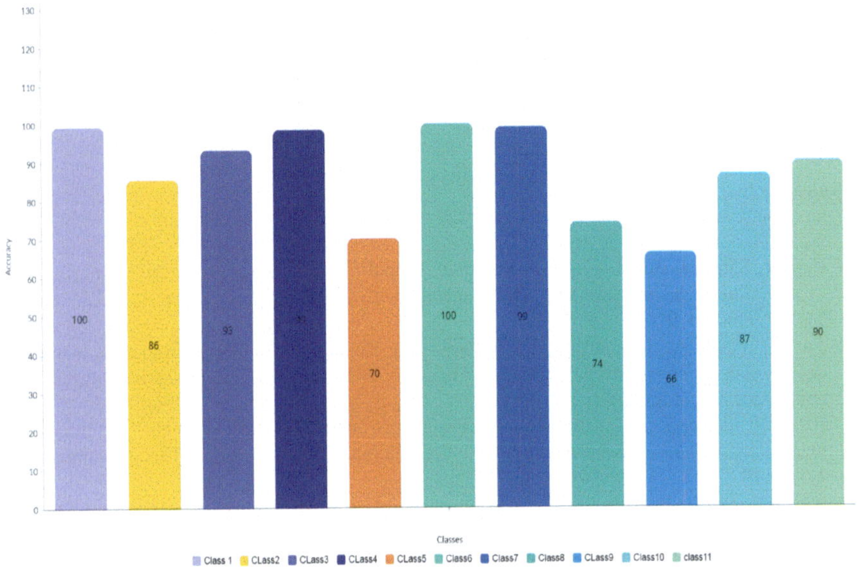

Fig. 7 Per-class accuracy

5 Conclusion

In summary, implementing the Convolutional Neural Network (CNN) to detect tomato leaf disease could revolutionize modern agriculture. This technology can provide farmers with a powerful, AI-driven tool to identify when their plants need assistance. CNNs have demonstrated remarkable proficiency in detecting various tomato leaf diseases, including Early Blight, Late Blight, and Bacterial Spot. When these diseases are identified promptly, farmers can take appropriate measures to prevent their spread, resulting in increased yields and reduced costs. Furthermore, CNNs are compatible with various devices, including smartphones and drones, allowing farmers to use this technology regardless of location. However, it is important to remember that this technology can only do so much based on the data they have been trained on. Scientists, farmers, and tech experts must work together to improve these systems. In the future, CNNs will be a key part of precision agriculture, helping farmers make smarter decisions about how to care for their crops. That means using resources more wisely, reducing the use of chemicals, and farming in an environmentally friendly way. To put it another way, using such techniques to detect tomato leaf diseases is revolutionary. It is like having a tech-savvy assistant for farmers who can help them grow more food efficiently and sustainably. As it continues to develop this technology, the model is closer to making smart farming the norm and providing better food for everyone. Experimentally, the Accuracy achieved by the model is 96.25%, and the storage size is 2.41 MB, which is much less than the pre-trained models available, which makes it better to implement and gives more accurate results.

In the future, the model can be loaded with more data sets from various other crops to detect diseases of more crops. Further, this model can be implemented in an App to click images and directly detect diseases from the app.

References

1. Rani R, Sahoo J, Bellamkonda S, Kumar S, Pippal SK (2023) Role of artificial intelligence in agriculture: an analysis and advancements with focus on plant diseases. IEEE Access 11:137 999–138 019
2. Tm P, Pranathi A, SaiAshritha K, Chittaragi NB, Koolagudi SG (2018) Tomato leaf disease detection using convolutional neural networks. In: 2018 Eleventh international conference on contemporary computing (IC3), pp 1–5
3. Patil RR, Kumar S (2022) Priority selection of agro-meteorological parameters for integrated plant diseases management through analytical hierarchy process. Inte J Electr Comput Eng 12(1):649
4. Patil RR, Kumar S, Rani R (2022) Comparison of artificial intelligence algorithms in plant disease prediction. Revue d'Intelligence Artificielle 36(2)
5. Agarwal M, Singh A, Arjaria S, Sinha A, Gupta S (2020) Toled: tomato leaf disease detection using convolution neural network. Procedia Comput Sci 167:293–301, 2020, international Conference on Computational Intelligence and Data Science. [Online]. Available: https://www.sciencedirect.com/science/article/pii/S1877050920306906
6. Patil RR, Kumar S (2021) Predicting rice diseases across diverse agro-meteorological conditions using an artificial intelligence approach. Peer J Comput Sci 7:e687
7. Kumar S, Patil RR, Rani R (2023) Smart iot-based pesticides recommendation system for rice diseases. In: Kulkarni AJ, Mirjalili S, Udgata SK (eds) Intelligent systems and applications. Springer Nature Singapore, Singapore, pp 17–25
8. Patil RR, Kumar S (2020) A bibliometric survey on the diagnosis of plant leaf diseases using artifificial intelligence. Library Philos Practice, 1–26
9. Kumar S, Patil RR, Kumawat V, Rai Y, Krishnan N, Singh S (2021) A bibliometric analysis of plant disease classification with artificial intelligence using convolutional neural network. Libr Philos Pract 5777:1–14
10. Agnihotri S, Gupta J, Garg N, Khatri P (2023) Comparative analysis of tomato leaf disease detection using machine learning. In: 2023 6th International conference on information systems and computer networks (ISCON), pp 1–5
11. Nivethitha T, Vijayalakshmi P, Jaya J, Shriram S (2022) A review on coconut tree and plant disease detection using various deep learning and convolutional neural network models. In: 2022 International conference on smart and sustainable technologies in energy and power sectors (SSTEPS), pp 130–135
12. Panuganti N, Ranjan P, Batra KS, Rai JK (2022) Automation in agriculture and smart farming techniques using deep learning. In: IEEE Conference on interdisciplinary approaches in technology and management for social innovation (IATMSI), pp 1–5
13. Pradhan P, Kumar B (2022) Automatic detection of tomato diseases using fine-tuned pre-trained deep learning models. In: 2022 3rd International conference for emerging technology (INCET), pp 1–5
14. de Luna RG, Dadios EP, Bandala AA (2018) Automated image capturing system for deep learning-based tomato plant leaf disease detection and recognition. TENCON 2018–2018 IEEE region 10 conference, pp 1414–1419
15. Kim T, Kim H, Baik K, Choi Y (2022) Instance-aware plant disease detection by utilizing saliency map and self-supervised pre-training. Agriculture 12(8):1084
16. Simonyan K, Vedaldi A, Zisserman A (2014) Deep inside convolutional networks: Visualising image classification models and saliency maps

17. Elhassouny A, Smarandache F (2019) Smart mobile application to recognize tomato leaf diseases using convolutional neural networks. In: International conference of computer science and renewable energies (ICCSRE), pp 1–4
18. David H, Ramalakshmi K, Gunasekaran H, Ramachandran V (2021) Literature review of disease detection in tomato leaf using deep learning techniques, pp 274–278
19. Tripodi P, Nicastro N, Pane C, Cammarano D (2020) Artificial intelligence based efficient phenotyping for agronomics. In: 2020 3rd International conference on intelligent sustainable systems (ICISS), pp 484–488
20. Priyadharshini G, Judie Dolly DR (2023) Comparative investigations on tomato leaf disease detection and classification using cnn, r-cnn, fast r-cnn and faster r-cnn. In: 2023 9th International conference on advanced computing and communication systems (ICACCS), vol 1, pp 1540–1545
21. Moriones E, Navas-Castillo J (2000) Tomato yellow leaf curl virus, an emerging virus complex causing epidemics worldwide. Virus Res 71(1):123–134, [Online]. Available: https://www.sciencedirect.com/science/article/pii/S0168170200001933
22. Patil RR, Kumar S, Chiwhane S, Rani R, Pippal SK (2023) An artificial-intelligence-based novel rice grade model for severity estimation of rice diseases. Agriculture 13(1), [Online]. Available: https://www.mdpi.com/2077-0472/13/1/47
23. Kodali RK, Gudala P (2021) Tomato plant leaf disease detection using cnn. In: 2021 IEEE 9th Region 10 humanitarian technology conference (R10-HTC), pp 1–5
24. Gunarathna M, Rathnayaka R (2020) Experimental determination of cnn hyper-parameters for tomato disease detection using leaf images. In: 2020 2nd international conference on advancements in computing (ICAC), vol 1, pp 464–469
25. Mezenner A, Nemmour H, Chibani Y, Hafiane A (2022) Tomato plant leaf disease classification based on cnn features and support vector machines. In: 2022 2nd International conference on advanced electrical engineering (ICAEE), pp 1–5
26. Patil RR, Kumar S (2022) Rice-fusion: a multimodality data fusion framework for rice disease diagnosis. IEEE Access 10:5207–5222
27. Ranjana V, Mundada MR, Sawkar SP, Bellad RS, Keerthi PS (2020) Design and development of efficient techniques for leaf disease detection using deep convolutional neural networks. In: 2020 IEEE International conference on distributed computing, VLSI, electrical circuits and robotics (DISCOVER), pp 153–158
28. Kibriya H, Rafique R, Ahmad W, Adnan S (2021) Tomato leaf disease detection using convolution neural network. In: International Bhurban conference on applied sciences and technologies (IBCAST), pp 346–351
29. Alkaff AK, Prasetiyo B (2022) Hyperparameter optimization on cnn using hyperband on tomato leaf disease classification. In: IEEE International conference on cybernetics and computational intelligence (CyberneticsCom), pp 479–483
30. Mahilraj J, Sivaram P, Sharma B, Lokesh N, Bobinath B, Moriwal R (2023) Detection of tomato leaf diseases using attention embedded hyper-parameter learning optimization in cnn. In: 2023 6th International conference on information systems and computer networks (ISCON), pp 1–6
31. Howard AG, Zhu M, Chen B, Kalenichenko D, Wang W, Weyand T, Andreetto M, Adam H (2017) Mobilenets: efficient convolutional neural networks for mobile vision applications
32. Simonyan K, Zisserman A (2014) Very deep convolutional networks for large-scale image recognition. arXiv preprint arXiv:1409.1556
33. Szegedy C, Vanhoucke V, Ioffe S, Shlens J, Wojna Z (2016) Rethinking the inception architecture for computer vision. In: IEEE Conference on computer vision and pattern recognition (CVPR), pp 2818–2826

Synergistic Solutions: Imagen-Based Dataset Augmentation and Yolo-V8-Driven Disease Detection for Enhanced Cherry Tree Health

Sagar Dhanraj Pande

Abstract Agriculture is one of the most functional ancient cultivation forms that is now updating itself with many modern techniques. Its contribution to the economy might be a bit low in some countries but its impact on society always remains at the top, it is impossible to imagine a day on worldwide scale with no food or agricultural products. By saying the importance of agriculture and knowing that plants are the atomic units in agriculture, the discussed importance is reflected same as the importance of detection of disease in plants, with the help of state of arts technology like deep learning. Thus, this paper proposed a multi-staged deep learning approach for detection of disease in plants with the help of methods like diffusion model like Imagen and detection model like You Only Look Once (YOLO-V8). The proposed model utilizes the self-collected images of different plant parts which are further augmented and increased with the help of imagen and then that dataset is further processed and YOLO-V8 was assigned responsibility to detect the diseased parts in image. The proposed has achieved training accuracy of 96.8% and testing accuracy of 95.9%.

Keywords Deep learning · Elastic Clout Compute 2 · Diffusion models · Imagen · Detection · Agriculture

1 Introduction

Within the dynamic field of agriculture, a long-standing custom is embracing modernity by integrating cutting-edge methods and tools to improve productivity and sustainability. Even if it isn't always the main force behind a country's economic growth, agriculture has a lasting impact on society. We cannot live without it, and the global community is largely dependent on the steady supply of food and agricultural

S. D. Pande (✉)
School of Engineering and Technology, Pimpri Chinchwad University (PCU), Pune, Maharashtra, India
e-mail: sagar.pande@pcu.edu.in

A. J. Kulkarni and M. Hajiaghaei-Keshteli (eds.), *Advances in Intelligent Systems for Sustainable Agriculture*, Lecture Notes in Networks and Systems 1201,
https://doi.org/10.1007/978-981-97-9839-1_10

goods. This paper suggests that plants, the basic components of agriculture, have a crucial function to play in this setting. The need for early plant disease identification looms even larger than the relevance of agriculture [1]. The research presents a multi-staged deep learning technique that utilizes state-of-the-art technologies, including the detection model YOLO-V8 and the diffusion model Imagen, to handle this crucial element.

The methodology commences with the collection of self-captured images of various plant components. These images undergo augmentation and enrichment using Imagen, ultimately resulting in a comprehensive and diversified dataset. Subsequently, this dataset undergoes further refinement, with the pivotal responsibility of disease detection being entrusted to YOLO-V8 [2]. This paper signifies the remarkable potential of deep learning in reshaping the landscape of agriculture. It presents a method that not only harnesses state-of-the-art technology but also underscores the vital importance of plant health in safeguarding our agricultural future. Join us in the exploration of the intricate details of this multi-staged approach for plant leaf disease detection and the promise it holds for enhancing agricultural sustainability on a global scale.

Agriculture, deeply rooted in our history, has witnessed a gradual transformation over the years. The paper delves into the dynamics of this transformation, which has been accelerated by the integration of modern techniques and technology. While the economic contribution of agriculture may vary among nations, its societal impact remains universally significant [3]. This paper underscores the essential role played by agriculture in sustaining our daily lives, highlighting the critical nature of uninterrupted access to agricultural products. Moreover, it emphasizes the intrinsic connection between agriculture and plant health, positioning the detection of plant diseases as a pivotal aspect to ensure a thriving agricultural ecosystem.

The paper introduces a comprehensive multi-staged approach of deep learning tailored to the task of plant disease detection. The methodology begins with the collection of self-acquired images, showcasing various plant parts. These images are subjected to a sophisticated augmentation process facilitated by Imagen, expanding the dataset's scope and diversity. It is during this stage that the dataset is fortified, laying the groundwork for subsequent analysis [4]. YOLO-V8, a robust detection model, assumes the role of identifying diseased portions within the images. This multifaceted approach holds promise for improving the efficiency of and accuracy plant disease diagnosis.

The contributions of the paper are as follows:

- An extensive dataset of different plant parts is collated for the purpose of implementing the proposed methodology.
- A regressive survey on the previous works is conducted for gaining knowledge on the topics like gaps, model's problem.
- A novel two staged deep learning-based methodology is proposed for the detection of the diseases in plant parts image.

Section 2, which gives a summary of previous research, and Sect. 3, which contains details on the dataset and recommended methodology, form the basis for the rest of the study. Section 4 presents the results of the recommended techniques. The paper's conclusion and future scope are described in Sect. 5.

2 Literature Survey

Prior to putting the suggested multi-staged deep learning methodology for plant disease detection into practice, it is critical to recognize how important it is to have a thorough awareness of the agricultural landscape. The background knowledge and perspectives offered by this fundamental comprehension enable the identification of the most important obstacles and needs in the field of plant disease detection. The selection and modification of deep learning algorithms can be guided by knowledge of the plant species, illnesses, and environmental circumstances. This ensures that the methodology chosen is in line with the practical requirements of agriculture. Table 1 represents the previous work done on detection topic.

Plant disease detection techniques now in use frequently have drawbacks, including low accuracy, poor generalization to new datasets, and susceptibility to picture noise. This work suggests a novel two- stage method that combines the advantages of conventional image processing approaches and deep learning to overcome these restrictions [5]. While the second step uses a conventional image processing technique to partition the leaf area and identify unhealthy patches, the first stage uses a deep learning model to extract high-level properties from plant photos. Compared to earlier techniques, this two-stage method has various advantages, such as more generalization, decreased sensitivity to image noise, and increased accuracy.

3 Methodology

3.1 Dataset Description

2460 photos of both healthy and sick plant leaves made up the dataset at first. Using the diffusion model Imagen, data augmentation techniques were used to increase the dataset's size and diversity. As a result, an enhanced dataset of 4100 photos was produced, offering a more thorough depiction of plant leaf variations and disease symptoms. Every image in the collection was assigned a label, with 0 denoting healthy leaves and 1 denoting diseased leaf. The dataset was divided into training and testing sets in an 8:2 ratio, with 80% of the photos utilized for training and 20% for testing, to enable efficient model training and evaluation. Figure 1 shows the distribution of the dataset. Figure 1 shows the information provided in the previous paragraph.

Table 1 Represents the literature review on the previous works done on the topic

References	Year	Dataset	Methodology	Advantage	Gap of the paper
[1]	2023	Pre-existing leaves dataset	CNN	Feature extraction using classic and customized CNN is very powerful and easy	The paper has not compared the results with other methods, and it did not use the new methodologies
[2]	2023	Pre-existing plant village dataset	Alexnet and mobilenet	Bothe Alexnet and Mobilenet are complex convolutional networks that can facilitate the transferring of weights in training process	The issue of image-level noise, which might provide a serious challenge in the identification of plant diseases, is not addressed in the paper. An image of a damaged plant leaf, for instance, might have been captured in dim light or with other items partially blocking it
[3]	2023	Histogram of plant leaves	Random forest	RF models can handle large datasets with many features without overfitting. This is because the algorithm averages the predictions of multiple decision trees, which helps to reduce the impact of individual trees that overfit to the training data	The issue of image-level noise, which might provide a serious challenge in the identification of plant diseases, is not addressed in the paper an image of a damaged plant leaf, for instance, might have been captured in dim light or with other items partially blocking it
[4]	2023	Pre- existing datasetof different plants	GoogleNet, Alexnet	The capacity of a model to function successfully on data that it has not been trained on is known as generalization. Enhancing the model's generalization might be achievable bymerging two models with dissimilar architectures	The topic of intra-class variation, which can also provide a serious challenge to the diagnosis of plant diseases, is not covered in the paper. For instance, the severity of the illness or the type of plant may have a significant impact on how a diseased leaf looks

(continued)

Table 1 (continued)

References	Year	Dataset	Methodology	Advantage	Gap of the paper
[5]	2023	Pre-existing plantvillage dataset	EfficientNet-B7, modified u- net	EfficientNet-B7 is a scalable model, meaning that it can be easily scaled up or down in size to fit the specific needs of a particular application. For example, a smaller version of EfficientNet-B7 can be used for applications that require low latency, while a larger version can be used for applications that require high accuracy	The paper does not address the vulnerability of the models to adversarial attacks. Adversarial attacks are a type of attack that can be used to fool a machine learning model into making a wrong prediction. In the context of plant disease detection, an adversarial attack could be used to create an image of a healthy tomato plant that the model would mistake for a diseased plant
[6]	2023	Pre-existing plantvilla dataset	Imagenet	Several effective deep learning models, including VGG16, ResNet50, and InceptionV3, have been trained on ImageNet and have achieved state-of-the-art performance on a variety of image classification tasks. The amount of time and effort needed for training can be greatly decreased by fine-tuning these pre-trained models for plant disease detection	Crop disease detection systems must measure the degree of uncertainty in addition to offering precise forecasts. We can find situations where the model is less certain by using uncertainty estimate approaches, which opens the door to more investigation or data collecting. Furthermore, it is imperative that deep learning models exhibit resilience against fluctuations in image quality, lighting, and other environmental factors that may impact the visibility of crop diseases

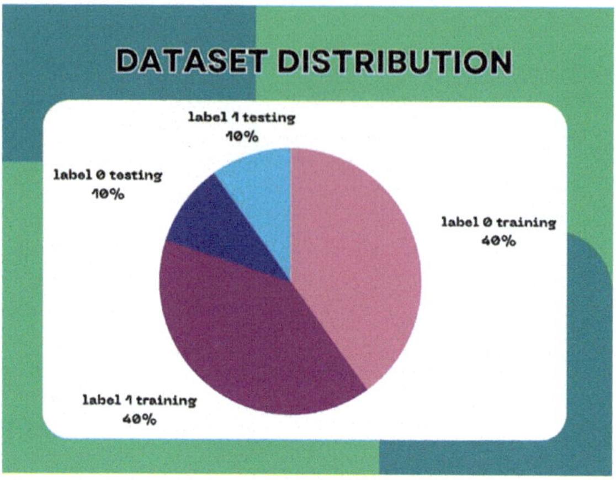

Fig. 1 The dataset distribution description

3.2 Imagen

In the intricate process of creating images from text, the T5-XXL encoder serves as the language maestro. The model takes the complexity of text and transforms it into a unique code, capturing the intricate details and nuances embedded in the words. From this linguistic foundation, the visual realm unfolds with an initial image of pure chaos—a canvas of random noise akin to the fuzz on a malfunctioning TV screen, devoid of recognizable patterns [7].

The magic unfolds as the model embarks on a step-by-step journey to clean up this visual commotion. Guided by the wisdom encoded from the text, the model predicts and refines colors and shapes, gradually unveiling a coherent image that aligns with the underlying textual description. The secret lies in the diffusion time horizon—the number of cleanup steps [8]. More steps mean a more polished and detailed final image, but the trade-off is an increased demand for computational muscle.

As the computational symphony plays out, the output image emerges. What once was a chaotic blend of noise transforms into a tangible, realistic picture, a visual manifestation of the encoded textual prompt. For those craving even greater visual fidelity, the journey continues with text-conditional super-resolution diffusion models. These sophisticated models take the output image and elevate it to higher resolutions, all while maintaining fidelity to the original text [9]. It's an extra layer of refinement, resulting in high-resolution images that not only meet the demands of the task but also faithfully embody the essence of the initial textual description.

3.3 *YOLO-V8*

YOLO v8 represents a leap forward in early disease detection, potentially saving crops and livelihoods. As AI and sensor technologies evolve, even more sophisticated models can emerge, offering real-time monitoring, disease prediction [10], and personalized treatment recommendations. Imagine farmers equipped with AI-powered tools, making informed decisions, and safeguarding their harvests with greater efficiency and sustainability [11]. The future of plant health looks bright, with YOLO v8 paving the way for a new era of intelligent crop protection.

YOLO v8 starts by learning. A massive dataset of healthy and diseased leaves feeds the algorithm, teaching it subtle visual cues of each condition. This training phase fine-tunes its ability to recognize patterns and differentiate between healthy and sick tissues. Presented with a new leaf image, YOLO v8 swings into action. It scans the entire image at once, not pixel by pixel, making it super-fast. Its trained "eyes" identify potential disease regions, marking them with bounding boxes. Beyond just detection, YOLO v8 can even classify the disease type. This requires an additional learning stage with labeled examples of different diseases. With this extra knowledge, it can pinpoint not just "diseased," but also "rust," "blight," or "mildew," empowering targeted treatment.

Beyond simply identifying diseased leaves, YOLO v8 can unearth valuable insights hidden within images. By analyzing the size, shape, and color of diseased regions, it can estimate the severity of the infection, guiding farmers towards the most effective interventions. Additionally, the model can track disease progression over time, helping predict outbreaks and optimize treatment strategies. This data-driven approach empowers farmers to move beyond reactionary measures and adopt preventative tactics, fostering healthier crops and maximizing yields. Imagine the possibilities: targeted application of fungicides only where needed, reduced reliance on broad-spectrum chemicals, and ultimately, a more sustainable and productive agricultural landscape (Fig. 2).

4 Results and Analysis

Google Colab was used in the construction of the models. The outcomes were assessed following the addition of the dataset to the model. The Google Colab laptop, namely the professional variant, which has an Intel i10 7th generation 8-core CPU clocking at 5.30 GHz and 16 GB of RAM, was used for the tests.

The results obtained from Imagen's image generation process are in, unveiling a remarkable transformation from initial noise to detailed, realistic images. The model, guided by a large T5-XXL encoder, effectively cleanses the visual chaos through progressive denoising steps. The output images reflect a faithful interpretation of the input textual prompts, demonstrating the power of this sophisticated approach to generate compelling visual content. Shifting gears to the realm of disease detection,

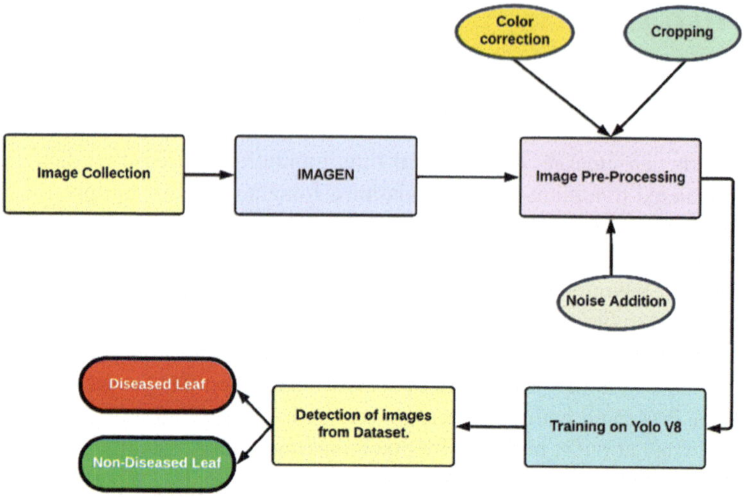

Fig. 2 The working diagram of the proposed architecture

Yolo-v8 emerges as a pivotal player in the analysis of cherry images and the result of it are depicted in Figs. 4 and 5. Leveraging YOLO-V8 model's efficiently processes a dataset of cherry images, employing object detection algorithms to identify and localize signs of disease. The scalability and computational prowess of YOLO-V8 enable seamless adaptation to varying demands, making it an ideal platform for the precise and efficient detection of diseases affecting cherry trees. The results obtained from the YOLO-V8 are compared with other methodologies on the same dataset, the results of the experiment were represented in Table 2.

Algorithm with that different deep learning architectures are used for the generation of dataset to compare the capability of the model the results are represented as graph, in that each model is tested to its capability by making it generate various types and kinds of leaver with variety of shapes and colors, the results obtained are depicted in Fig. 3. From that we can see how well the Imagen outperforms all other models.

The achievement of the Yolo V8 results, that took over the other trending models, are of two factors the primary reason was that it is the latest version in the Yolo family and the neck head architecture with revised bottle neck connection has been affecting the rate of detection capability. The second reason is the corelative effect of

Table 2 The result comparison of Detection

Model	mAP (%)	F1score (%)	Recall (%)
YOLO V8	**95.78**	**89.85**	**90.95**
RCNN	70.56	73.83	70.52
YOLO V6	82.68	89.62	73.46
YOLO V7	91.01	79.32	95.13

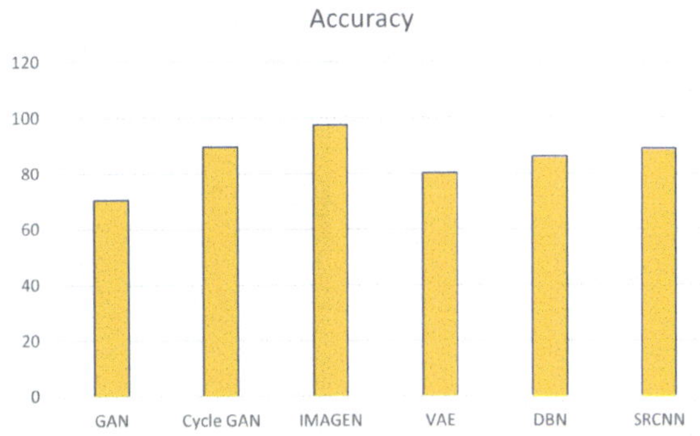

Fig. 3 Accuracy comparison of the generative models

Fig. 4 Detection of the Diseased Leaf

hybrid model the hybrid nature has covered the weakness of each other to enhance the rate of achieving accurate results. This also serves as a reason Imagine achieving good results.

5 Conclusion and Future Scope

In conclusion, the integration of Imagen for dataset augmentation and the utilization of the YOLO-V8 for image detection in the context of cherry trees present a powerful synergy. Imagen's capacity to generate diverse and realistic images contributes significantly to expanding the dataset, facilitating a robust training ground for detection

Fig. 5 Detection No disease Leaf

models. The incorporation of two classes, one representing diseased cherries and the other non-diseased, enhances the model's ability to discern between healthy and compromised specimens. The efficiency and scalability of the YOLO-V8 prove invaluable in processing this augmented dataset, ensuring swift and accurate identification of disease markers. Looking forward, the project holds promising prospects, with potential advancements in leveraging cutting-edge image generation models like Imagen and cloud computing technologies such as YOLO-V8 for enhanced precision in agricultural disease detection, ultimately contributing to the improvement of crop health and yields.

References

1. Panchal AV, Patel SC, Bagyalakshmi K, Kumar P, Khan IR, Soni M (2021) Image-based plant diseases detection using deep learning. Mater Today Proc,https://doi.org/10.1016/j.matpr.2021.07.281
2. Zhang P, Liu Q, Chen Y, Sun Y (2016) Using deep learning for image-based plant disease detection. Frontiers Plant Sci 7(1114), https://doi.org/10.3389/fpls.2016.01114
3. Ramesh S et al. (2018) Plant disease detection using machine learning. 2018 International conference on design innovations for 3cs compute communicate control (ICDI3C), Bangalore, India, pp 41–45, https://doi.org/10.1109/ICDI3C.2018.00017
4. Ferentinos KP (2018) Deep learning models for plant disease detection and diagnosis. Comput Electron Agric 145:311–318. https://doi.org/10.1016/j.compag.2018.01.009
5. Chowdhury MEH et al (2021) Automatic and reliable leaf disease detection using deep learning techniques. AgriEngineering 3(2):294–312. https://doi.org/10.3390/agriengineering3020020

6. Kulkarni O (2018) Crop disease detection using deep learning. 2018 Fourth international conference on computing communication control and automation (ICCUBEA), Pune, India, 2018, pp 1–4, https://doi.org/10.1109/ICCUBEA.2018.8697390
7. Singh D, Jain N, Jain P, Kayal P, Kumawat S, Batra N (2020) PlantDoc: a dataset for visual plant disease detection. Proceedings 7th ACM IKDD CoDS 25th COMAD, pp 249–253
8. Chohan M, Khan A, Chohan R, Katpar SH, Mahar MS (2020) Plant disease detection using deep learning. Int J Recent Technol Eng 9(1):909–914
9. Sahu SK, Pandey M, Geete K (2021) Classification of soybean leaf disease from environment effect using fine tuning transfer learning. Ann Romanian Soc Cell Biol 25:2188–2201
10. Bhagwat R, Dandawate Y (2021) A framework for crop disease detection using feature fusion method. Int J Eng Technol Innov 11(3)
11. Lee SH, Goeau H, Bonnet P, Joly A (2021) Conditional multi-task learning for plant disease identification. Proceedings 25th international conference pattern recognition (ICPR), pp 3320–3327.

Advanced Detection of Diseases in Tomato Plant Leaf Using Deep Learning

G. S. Madhan Kumar, S. N. Chandrakanth, H. D. Akash Patil, M. R. Sanjay Kumar, and K. Vamshi Ram

Abstract India, the world's second-largest tomato producer, plays an essential role in feeding millions and providing substantial income for farmers. However, the backbone of Indian agriculture, the tomato, faces a significant challenge in early disease detection. Traditional disease detection methods, relying on expertise, prove time-consuming and prone to human error. The presence of undetected diseases poses a continual threat to the health and productivity of tomato plants, resulting in reduced yields and profits. To surpass these constraints and expedite the detection process, this paper employs a deep learning model built using a Convolutional Neural Network (CNN). The model follows a workflow involving image acquisition, pre-processing, extraction of features, and CNN training to identify diseases in tomato leaf plants. The proposed model exhibits an average score of detecting all tomato diseases categories with 98.27% of accuracy, 98.27% of precision, 97.81% of recall, and 98.06% of F1 score surpassing traditional approaches like SVM, KNN, and backpropagation.

Keywords Convolutional Neural Network (CNN) · Tomato leaf plant · Early blight · Bacterial spot · Tomato mosaic virus

1 Introduction

Agriculture is a commercial sector that involves cultivating crops for food, fiber, and other products. It plays an essential role in supporting global economies and ensuring food security. India ranked second in the cultivation of tomato globally. Tomato cultivation plays a crucial role in global agriculture, serving as a main ingredient and contributing significantly to the world's food production. However, the tomato crop faces persistent threats from various diseases like early blight, bacterial spot, late blight, tomato mosaic virus, leaf mold, and yellow leaf curl virus. These diseases not only degrade the health of the tomato plants but also pose a substantial risk to overall

G. S. Madhan Kumar (✉) · S. N. Chandrakanth · H. D. Akash Patil · M. R. Sanjay Kumar · K. Vamshi Ram
Maharaja Institute of Technology Thandavapura, NH 766, Nanjangud Taluk, Mysuru 571302, India
e-mail: gsmadhanmitt@gmail.com

A. J. Kulkarni and M. Hajiaghaei-Keshteli (eds.), *Advances in Intelligent Systems for Sustainable Agriculture*, Lecture Notes in Networks and Systems 1201,
https://doi.org/10.1007/978-981-97-9839-1_11

productivity. Symptoms may include minor changes in the color, shape, or function of the plant as it reacts to the presence of the pathogen. In this context, by addressing these disease symptoms that should be noted if there is a perceived decline in plant growth. The precise classification and identification of leaf diseases play a pivotal role in preventing agricultural losses. Various plant leaves exhibit distinct diseases, each characterized by different symptoms as shown in Fig. 1.

- **Bacterial spot**: Bacterial spot, triggered by the bacterium Xanthomonas campestris pv. vesicatoria, is a prevalent ailment impacting tomato plants. It predominantly affects the leaves and fruits of the plant, exhibiting symptoms characterized by small, dark lesions with a water-soaked appearance on the leaves. With lesion growth, they typically acquire a raised, scabby texture and may merge, resulting in notable harm. The disease can rapidly escalate in warm and humid environments.
- **Early Blight (Alternaria)**: Another disease affecting tomato plants is caused by the Alternaria fungus, which triggers leaf spot or early blight. On lower leaves, noticeable brown or black spots develop, characterized by dark edges resembling a target pattern. In certain instances, the stem ends of fruits are also affected, exhibiting significant sunken black areas adorned with concentric rings. Typically, this fungus affects tomato plants after fruit setting has occurred.
- **Late blight**: A tomato plant disease caused by the Phytophthora infestans fungus emerges during cool, rainy periods, typically toward the end of the growing season. Its appearance mirrors frost damage on leaves, forming irregular green-

Fig. 1 Various symptoms and diseases

black patches. Fruits are also affected, displaying large, irregular-shaped brown patches that quickly deteriorate. Importantly, this fungal infection has the potential to spread to potatoes and be transmitted from them. It is advised to employ similar controls used for managing septoria leaf spot to effectively handle this tomato plant disease.

- **Mosaic virus**: Mosaic virus is widespread among different plants, including tomatoes. While it doesn't cause the plant's death, it notably diminishes both the quantity and quality of fruits. The virus gets its name from the characteristic markings resembling a mosaic pattern of light green and yellow on the leaves, accompanied by mottling on the fruits of infected plants. Furthermore, the leaves may display abnormal growth, resembling misshapen ferns.
- **Yellow leaf curl virus**: Whiteflies transmit the tomato yellow leaf curl virus, being minuscule insects that extract plant sap. Infected tomato plants display several discernible symptoms. Initially, the leaves exhibit an upward curl, appearing twisted and unusual. Moreover, the leaves adopt a yellow hue, particularly on the upper surface, originating from the veins and extending throughout the entire leaf. This yellowing imparts an unhealthy appearance to the plant. Furthermore, the virus can inhibit plant growth, resulting in overall smaller plants.
- **Leaf mold**: The pathogen Pseudocercospora fuligena induces a prevalent fungal disease recognized as leaf mold in tomato plants. It is distinguished by the emergence of unique lesions on the leaves. Symptoms usually comprise yellow or pale spots on the upper surface of the leaf, along with a fuzzy, mold-like growth on the underside of the leaves. As the disease advances, the affected regions may transition to brown, rendering the leaves a papery texture. Unlike certain other tomato ailments, leaf mold flourishes in environments of elevated humidity and moderate temperatures.

In the early days, the traditional methods of disease detection primarily involve manual inspection by agricultural experts, a process that is both time-consuming and susceptible to human error. The need for a more efficient, accurate, and timely approach to disease detection in tomato plants has motivated the exploration of technological solutions. In automated machine learning, Support Vector Machine (SVM) [1] autonomously separates classes, boasting an accuracy of 87.6%. Backpropagation [2], a neural network training technique, achieves a remarkable 98.3% accuracy and K-Nearest Neighbors (KNN) [3] classifies data without human intervention, achieving 78.76% accuracy in automated predictions. In contrast, the approach endeavors CNNs, a deep learning algorithm, to analyze and interpret different patterns present in tomato plant leaf images. The model is trained on diverse instances of each disease by utilizing an extensive dataset, making it capable of understanding small differences and telling apart different problems. One of the distinguishing factors of methodology is a notable improvement in accuracy in comparison to prior results models. The objective of this work is to set a new standard in the field of agricultural disease by addressing existing detection model limitations and elevating the overall productivity of tomato crops. This paper marks a significant changes toward a more resistent and productive tomato farming industry, emphasizing the crucial role of cutting-edge technology in securing global food sustainability. In light of these challenges, this research aims to revolutionize the early detection of diseases in tomato plants, providing a significant solution to find a way to reduce impact of infections. The primary goal is to enhance the productivity of tomato crops by using sophisticated deep learning methods, particularly Convolutional Neural Networks (CNN), to achieve precise and timely disease identification.

1.1 Novelty and Contribution

This paper mainly aims to develop a CNN model for detecting diseases in tomato plants, addressing challenges such as variability in symptoms, where diseases manifest diverse signs. Environmental factors impact the expression of diseases, and maintaining high image quality is crucial for accurate analysis. Real-time detection poses a significant hurdle, requiring efficient algorithms to process data promptly.

1.1.1 Contribution

The contributions are as follows:

- To collect a diverse and representative dataset of tomato leaf images, covering various diseases and healthy images, to enable robust training and validation of the automated detection system.
- To create an automated system for detecting early signs of plant diseases in tomato plants, The goal is to identify at the initial stages, offering farmers a comprehensive diagnostic solution that can handle the complexity of multiple diseases affecting a single tomato plant simultaneously.

2 Related Work

Ashok et al. [4] explore the advanced techniques introducing a robust method for early Tomato Leaf Plant disease detection, employing processing of images and CNNs. With a remarkable 98.12% accuracy, the approach outperforms alternatives, providing a valuable tool for preemptive agricultural protection and economic preservation in India. Kibriya et al. [5] proposed a robust system, which explores early identifying diseases in tomato leaves using GoogLeNet and VGG16 CNN models, achieving remarkable accuracies of 99.23 and 98%. Trained on the Plant Village dataset, the system provides an effective solution for minimizing crop loss in tomato fields. Sherly Puspha Annabel and Muthulakshmi [6] explore an AI-powered system that presents efficient tomato leaf disease detection. It integrates RGB to grayscale conversion, thresholding, GLCM, and a random forest classifier, achieving a notable 94.1% accuracy. Outperforming SVM and MDC, it provides automated early disease identification, empowering farmers and enhancing crop yield while reducing losses. Adhikari et al. [7] exploring automated classification and detection, mainly focuses on tomato plant diseases. The system employs a CNN-based model with Raspberry Pi as the main computing unit. Trained on a dataset of 520 images, it achieves an 89% accuracy, successfully identifying diseases like Late Blight, Gray Spot, and Bacterial Canker. Abbas et al. [8] evaluate the conducting novel avenues; this method presents the utilization of a Conditional Generative Adversarial Network (C-GAN) for tomato plant disease detection. By conducting training on a combination of real and computer-generated pictures, the DenseNet121 model achieves impressive accuracy rates, surpassing existing methodologies with 99.51, 98.65, and 97.11% for 5, 7, and 10-class classifications, respectively.

de Luna et al. [9] proposed an investigating innovative method; the study introduces an Automatic Picture-Taking System for Tomato Plant Disease Detection. Utilizing a motor-controlled box and Convolutional Neural Network, it achieves 95.75% accuracy in disease recognition. A Faster R-CNN, with an 80 confidence score, detects anomalies, resulting in a practical testing accuracy of 91.67%. Das et al. [1] explore automated detection methods, the study employs SVM, Random Forest, and Logistic Regression for identifying tomato leaf diseases. Through processing of image and feature extraction, SVM demonstrates superior results. The model facilitates early disease identification, potentially benefiting crop production and farmers' well-being in India. Nagamani and Sarojadevi [2] explore advanced techniques in tomato leaf disease detection. The study utilizes Fuzzy SVM, CNN, and Region-based Convolutional Neural Network. Leveraging image processing methods like color thresholding, flood filling, and deep learning, the study achieves an impressive accuracy of 96.73%, enabling effective early diagnosis and classification of plant diseases. Mokhtar et al. [10] propose a robust approach for tomato leaf disease detection. It employs Gabor wavelet transform and Support Vector Machines (SVMs) with various kernel functions. Leveraging a dataset of 200 real tomato plant leaf images, the SVM-based method achieves high accuracy, particularly with Cauchy and Laplacian kernels (100 and 98, respectively). Jiang et al. [11] explore innovative techniques, the study introduces a deep learning-based method for tomato leaf disease classification. Utilizing an enhanced Resnet-50 model with Leaky-ReLU activation and an 11×11 convolution kernel, the approach achieves notable Precision, showing its capability for practical crop disease detection applications.

Prajwala et al. [12] proposed a work that introduces a simplified Convolutional Neural Network, a variation of LeNet, for tomato leaf plant disease detection. By utilizing minimal computing resources, the model succeeds a 94–95% accuracy. The methodology involves thorough investigation into data acquisition, pre-processing, and classification. Results indicate the model's effectiveness for accurate disease identification in agriculture. Vetal and Khule [13] explore processing of image techniques. The paper presents a Tomato Plant Leaf Disease Detection system with segmentation and Multi-class SVM. Implemented in MATLAB, the method achieves a 93.75% classification accuracy for four diseases, surpassing other techniques. This exploration highlights its effectiveness in early disease detection. Balakrishna and Rao [3] proposed a two-stage system to tell which tomato leaves are good and which ones are not. In the first stage, KNN classifies leaves, and in the second stage, PNN and KNN categorize unhealthy leaves based on GLCM, Gabor, and color features. The focus on a fusion approach with PNN proves superior, outperforming other methods in disease detection. Agarwal et al. [14] present "ToLeD," a Tomato Leaf Plant Disease Detection model utilizing Convolutional Neural Networks. Having a typical precision of 91.2%, it outperforms VGG16, InceptionV3, and MobileNet. The study conducts an in-depth exploration, addressing the critical issue of tomato crop diseases and emphasizing the model's potential for timely and accurate disease identification. Harakannanavar et al. [15] explore diverse techniques. The paper proposes a leaf disease detection model using Visual computing and automated learning algorithms. Employing methods such as K-means clustering, Discrete Wavelet Transform, and Convolutional Neural Network, the model achieves high accuracies of 88, 97, and 99.6% for SVM, KNN, and CNN, respectively.

Paymode et al. [16] developed the AgroDeep mobile app to collect real leaf images for disease detection. The proposed advanced learning methods, specifically Convolutional Neural Network (CNN), achieved an unprecedented 97% accuracy in classifying diseases on the leaves of tomatoes. This research significantly contributes to increasing food production and motivating farmers for sustainable agriculture. Gadade et al. [17] explore the machine learning-based method introduced in this work, the goal is to identify and classify tomato leaf diseases. The study employs SVM, KNN, Naive Bayes, Decision Trees, and LDA, achieving robust disease severity grading. The exploration delves into feature extraction, incorporating shape, color, and texture features. Notably, SVM classification demonstrates enhanced accuracy in disease identification. Mohanty et al. [18] developed a comprehensive system for the timely identification of tomato diseases using machine learning algorithms, such as logistic regression and random forest, and support vector machine. Feature extraction via Histogram of Oriented Gradients and assessment through F1 score, precision, and recall reveal the superior performance of the support vector machine. The study culminates in deploying image classification via the Streamlight application. Langar et al. [19] explore the innovative approach outlined in this work, employing Convolutional Neural Networks for tomato leaf disease detection. Emphasizing agriculture's economic impact, the study covers image processing, disease symptoms, and achieves a notable Mean Average Precision of 0.76. The authors envision future applications, proposing mobile apps to assist farmers in disease identification. Chandak et al. [20] explore the economic impact on India's agriculture, this delves into the automated detection of diseases in early stages within tomato crops. Utilizing machine learning and examining leaf attributes, the study explores the identification of diseases transmitted through air or soil. The research contributes to preserving agricultural productivity and bolstering economic stability.

The state of art is showcased in Table 1. From the table, it is clear that most of the research has been conducted in detecting single tomato plant leaf disease. This paper contributes in detecting multiple diseases and thus enhances the method of detecting tomato leaf diseases using the CNN model.

Table 1 State of the art

SI. No.	Authors	Algorithm	Proposed methodology	Results
1	Das et al. [1]	SVM	Image acquisition Image pre-processing Masking Feature extraction Classification and result Output	Identifies the diseased or affected area on the tomato leaf
2	Usama mokhtar et al. [10]	SVM	Image acquisition Pre-processing Image segmentation Feature extraction Classification	It identifies only 2 tomato leaf viruses with the highest classification accuracy of 91.5%
3	Prajwal et al. [12]	CNN	Data acquisition Classification	Accuracy of disease detection of proposed methodology is 94–95% Identifies the diseased or affected
4	Langar et al. [19]	CNN	Image acquisition Image pre-processing Image segmentation Classification and detection of disease	It recognizes whether crop is infected or not and which type of disease

3 Problem Statement

The inefficiency and reliance on manual methods in order to detect the tomato plant leaf diseases pose a significant impediment to timely intervention and the prevention of widespread infections. Addressing this challenge is essential for enhancing the overall health and productivity of tomato crops, ensuring sustainable agricultural practices and nutrition security.

4 System Model

Step 1: Image Acquisition

$$\text{Dataset} = \{(X_1, y_1), (X_2, y_2), \ldots, (X_N, y_N)\} \tag{1}$$

where X_i is the i-th image and y_i is the corresponding label indicating if the leaf is healthy or diseased.
By collecting a dataset (Dataset) of tomato leaf images, each labeled with whether leaf is healthy or diseased. The dataset plays a crucial role in training and evaluating the CNN model.

Step 2: Pre-processing

$$X_{\text{normalized}} = \frac{X - \mu}{\sigma} \tag{2}$$

where X is the input image, μ is the mean of pixel values, and σ is the standard deviation.
The collected images (X) undergo pre-processing, where it calculates the mean (μ) and standard deviation (σ) of the pixel values across entire dataset. Then, normalize each image ($X_{\text{normalized}}$) by eliminating the mean and dividing the standard deviation. This ensures consistency and better convergence during training.

Step 3: Feature Extraction

$$\text{Texture\_features} = \text{Apply\_Texture\_Analysis}(X) \tag{3}$$

$$\text{Color\_histogram} = \text{Compute\_Color\_Histogram}(X) \tag{4}$$

$$\text{Edges} = \text{Apply\_Edge\_Detection}(X) \tag{5}$$

where Apply_Texture_Analysis represents a texture analysis algorithm.
In this step, the system model extracts features from the preprocessed images:
- Texture features using a dedicated texture analysis algorithm (Apply_Texture _Analysis(X)), – Color histogram computed with (Compute_Color_Histogram(X)), – Edges detected through an edge detection algorithm (Apply_Edge_Detection(X)).

Step 4: Deep Neural Network Architecture

$$Y_i = \sigma\left(\sum_{j=1}^{n} W_{ij} * X_j + b_i\right) \tag{6}$$

where Y_i is output of the i-th feature map, W_{ij} is the filter weight, X_j is input image or feature map, b_i is the bias term, and σ is the activation function (e.g., ReLU).

$$\text{Pooled}ij = \text{MaxPooling}(Xij) \tag{7}$$

where MaxPooling extracts the maximum value from a set of values in a specified region.

$$Y_k = \sigma\left(\sum_{j=1}^{m} W_{kj} \cdot X_j + b_k\right) \tag{8}$$

where Y_k is the output neuron, W_{kj} is the weight, X_j is the input from the previous layer, b_k is the bias term, and σ is the activation function.
System model constructs a CNN to learn hierarchical features extracted from the images of tomato leaves. The input is the preprocessed image, and the structure involves convolutional layers (Y_i), activation functions (σ), and max pooling (MaxPooling) to down-sample spatial dimensions.

Step 5: Training

$$\text{Loss} = -\frac{1}{N}\sum_{i=1}^{N}\left[y_i \log(\hat{y}_i) + (1 - y_i)\log(1 - \hat{y}_i)\right] \tag{9}$$

where N is the number of samples, y_i is the true label, and $\hat{y}_i$ is the predicted probability.
For training the CNN, we need to provide the preprocessed and feature-extracted images along with corresponding labels from the dataset. The CNN learns to classify healthy and diseased tomato leaves by adjusting its internal parameters based on a cross-entropy loss function (Loss).

Step 6: Evaluation

$$\text{Accuracy} = \frac{\text{Number of Correct Predictions}}{\text{Total number of predictions}} \tag{10}$$

After training, evaluate the performance of approach using a separate dataset that it has not seen before. The accuracy is calculated by comparing the model's predictions to the true labels, providing insights into its effectiveness in disease classification.

Step 7: Diesese Detection
Upon successful training and evaluation, the trained CNN model is prepared for real-time disease detection in tomato plant leaves. The algorithm provides a prediction indicating the identified disease, which may include bacterial spot, early blight, late blight, tomato mosaic virus, septoria leaf spot, leaf mold, or yellow leaf curl virus, along with a confidence score, providing the disease label for effective and targeted intervention in tomato plant leaves.

5 Proposed Methodology

Figure 2 shows the Proposed methodology for Detection of Diseases in Tomato Plant Leaf.

- **Image Acquasition**: The initial phase in the proposed work involves acquiring images, a critical step in any image classification system. Here, the focus lies on obtaining high-quality images of tomato plant leaves, laying the groundwork for subsequent disease detection analysis. The dataset utilized in this method comprises a total of 7000 samples, meticulously organized into eight separate folders. Among these folders, seven exclusively contain images depicting various diseased tomato leaves, offering a diverse array of examples representing different leaf diseases. The remaining folder contains images of healthy tomato leaves, serving as comparison reference for analysis. The structured dataset guarantees that the model encounters an equitable representation of both healthy and diseased instances, fos-

Fig. 2 Proposed methodology

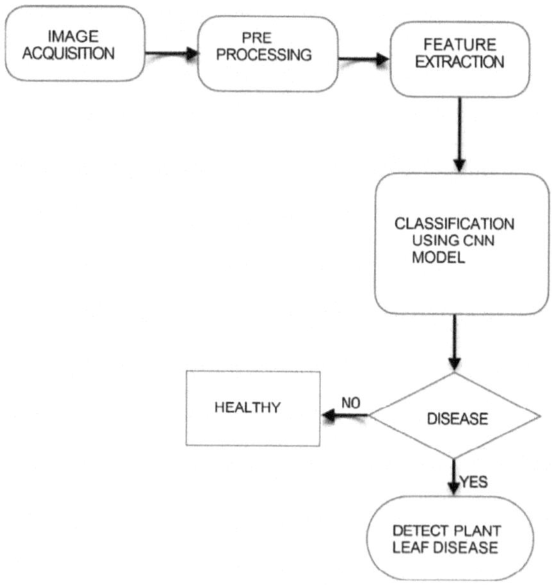

tering balanced learning, thereby improving its capacity to accurately classify and discern various conditions during subsequent analysis and classification stages.

- **Pre processing**: It increases the efficiency of the classification, a pre-processing phase should be considered to enhance the quality of the input tomato leaves images and to make the feature extraction phase more reliable. Pre processing can be done in two steps:

  Extracting image of tomato leaves from the acquired images:

  As previously mentioned, many acquired images contain multiple leaf images and other elements. Given the research's focus on identifying the types of viruses affecting tomato leaves, it is necessary to process single tomato leaves individually. To accomplish this, the images were manually cropped to extract single leaves, which were then saved in PNG format.

  Image Resizing:

  First, the images are resized in order to match the size of the input layer of CNN. All images must be with the same size and equal dimensions.

- **Feature Extraction**: Regarding feature extraction for the detection of tomato plant leaf diseases, three key techniques are applied: Texture Analysis, Color Histogram computation, and Edge Detection.

  Texture Analysis: This entails utilizing a texture analysis algorithm on the input image. Texture analysis is centered on examining the spatial configuration of pixels within an image, detecting patterns and fluctuations that contribute to its visual texture. Within the realm of tomato plant leaves, this method can capture nuances pertaining to surface attributes like smoothness, roughness, or the presence of patterns indicative of disease-related irregularities.

  Color Histogram: A color histogram serves as a depiction of the color distribution within an image. In the context of detecting diseases in tomato plant leaves, this method aids in capturing details about the prevalent colors found in the leaf.

  Edge Detection: The last technique involves Edge Detection, which accentuates the edges of structures within the image. Within the context of tomato leaves, these edges may signify crucial features such as veins or irregularities present on the leaf surface.

- **CNN Model**: Figure 3 shows a Convolutional Neural Network (CNN) model suitable for developing a computational approach that works on the unstructured image inputs and converts them to corresponding classification output labels. They belong to the category of multi-layer neural networks which can be trained to learn the required features for classification purposes.

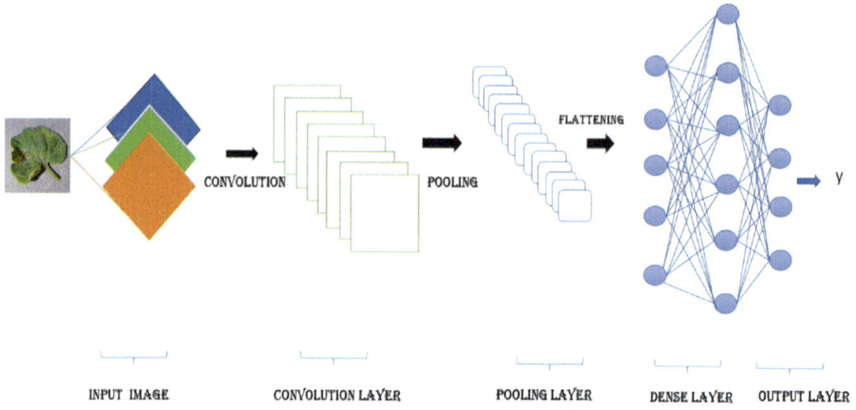

Fig. 3 CNN model

Convolutional Layer: Function: The convolutional layer serves as the fundamental component of a CNN, conducting convolution operations on the input image by utilizing filters (also referred to as kernels) to extract local features and identify patterns.

Purpose: Convolutional layers are effective in capturing features like edges, textures, and more complex patterns.

Pooling Layer: The pooling layer follows the convolutional layer and reduces the spatial dimensions of the available volume, effectively down sampling the information. Typical pooling operations consist of max pooling and average pooling.

Dense Layer (Fully Connected Layer): Dense layers establish connections between each neuron in one layer and every neuron in the subsequent layer. These layers often follow the convolutional and pooling layers in a CNN architecture.

Output Layer: The output layer produces the final prediction or classification.

6 Results and Discussion

6.1 Dataset

Figure 4 shows common diseases affecting tomato leaves. Creating a high-quality dataset is a critical aspect of developing and evaluating a system for automated detection of tomato leaf diseases using image processing and deep neural networks.

Fig. 4 Common diseases affecting tomato leaves

Early Blight, Late Blight, Bacterial Spot, etc.: include images representing a range of prevalent diseases that impact tomato plants. This ensures that the approach is trained to recognize different manifestations, enhancing its ability to generalize.
Healthy Tomato Leaves: Including pictures of normal tomato leaves so the model can compare and understand the difference between healthy and diseased ones. And Fig. 5 shows healthy tomato leaves.

6.2 *Experimental Setup*

The experimental setup employs an 11th Gen Intel processor CPU running at 2.40 GHz, accompanied by 8 GB of RAM, and operates on the Windows OS. Python version 3.7 serves as the primary programming language, supported by essential libraries including SciPy 1.6.3, pandas 1.2.3, NumPy 1.20.1, and Matplotlib 3.4.1. Additionally, a Deep Learning Framework such as TensorFlow or PyTorch is utilized. The scikit-learn library, version 0.24.2, is integrated for machine learning and data analysis tasks. This comprehensive configuration forms the foundation for conducting the detection of tomato leaf disease experiment, ensuring a robust and well-equipped framework for analysis and results interpretation in the conference paper.

Fig. 5 Healthy tomato leaves

6.3 Model Evaluation

In evaluating the efficiency of the proposed model in detecting Tomato Leaf Disease, which utilized Convolutional Neural Networks (CNN), the model was implemented in Python and subjected to a rigorous training regimen spanning 100 epochs. The evaluation of the model's performance was carried out using key metrics such as accuracy, recall, and precision, and F1 score, providing a comprehensive evaluation of its capability to discern and classify various diseases affecting tomato leaves.

6.4 Result

The effectiveness of the CNN model proposed is as shown in Table 2. The Approach achieved an 98.27% of accuracy, 98.27% of precision, 97.81% of recall, and 98.06% of F1 score for both disease and healthy plants.

Comparative Analysis: The table below offers a thorough overview of the performance of the proposed Tomato Leaf Disease Detection Model through comparison in contrast to three alternative models: Support Vector Machines (SVM), Backpropagation, and k-Nearest Neighbors (KNN). The evaluation metrics used include accuracy and precision.

The comparative examination encompasses diverse disease categories, such as Early Blight and Late Blight, Tomato mosaic virus, Bacterial spot, Septorial leaf spot, Target spot, Gray mold, and Healthy leaves, reveals insightful findings across the proposed system, Support Vector Machines (SVM), Backpropagation, and k-Nearest Neighbors (KNN) (Table 3).

The results showcase the remarkable performance of proposed model, achieving an accuracy of 0.92 and a precision of 0.91. This showcases its exceptional capabil-

Table 2 Proposed CNN performance metric

Disease category (%)	Accuracy (%)	Precision (%)	Recall (%)	F1 score (%)
Early blight	97.8	97.8	98.2	98.0
Late blight	98.5	98.5	97.6	98.0
Tomato mosaic virus	99.1	99.1	98.7	98.9
Bacterial spot	97.3	97.3	96.9	97.1
Yellow leaf curl virus	98.7	98.7	98.3	98.5
Leaf mold	97.1	97.1	96.2	96.6
Healthy leaves	99.4	99.4	99.2	99.3

Table 3 Comparison table

Disease	Algorithm (%)	Accuracy (%)	Precision (%)	Recall (%)	F1 score (%)
Early blight	Proposed system	97.8	97.8	98.2	98.0
	SVM	94.5	94.2	95.0	94.6
	Backpropagation	96.0	95.8	96.5	96.1
	KNN	92.3	91.5	93.2	92.7
Late blight	Proposed system	98.5	98.5	97.6	98.0
	SVM	95.7	95.3	94.8	95.0
	Backpropagation	97.2	96.8	97.0	96.9
	KNN	93.8	93.2	93.5	93.3
Tomato mosaic virus	Proposed system	99.1	99.1	98.7	98.9
	SVM	96.0	96.2	95.3	96.5
	Backpropagation	98.2	97.0	97.3	97.2
	KNN	93.0	93.2	93.6	93.3
Bacterial spot	Proposed System	97.3	97.3	96.9	97.1
	SVM	93.8	93.5	94.2	94.0
	Backpropagation	96.0	95.7	95.5	95.8
	KNN	91.5	91.0	91.8	91.3
Yellow leaf curl virus	Proposed system	98.7	98.7	98.3	98.5
	SVM	96.8	95.5	95.7	96.6
	Backpropagation	97.8	97.5	97.7	97.6
	KNN	92.5	93.8	92.6	92.8
Leaf mold	Proposed System	97.1	97.1	96.2	96.6
	SVM	96.5	96.2	95.8	95.9
	Backpropagation	96.2	95.8	95.5	95.7
	KNN	95.8	95.5	95.0	95.3
Healthy leaves	Proposed system	99.4	99.4	99.2	99.3
	SVM	98.0	97.8	98.0	97.9
	Backpropagation	98.2	98.0	98.5	98.1
	KNN	97.8	97.5	97.0	97.3

ity to precisely detect and categorize tomato leaf diseases. While Backpropagation, SVM, and Traditional AI also exhibit commendable performances with accuracies ranging from 0.78 to 0.88. The proposed model distinguishes itself with a notable advantage, emphasizing its potential as an advanced tool for precision agriculture. The visual representation of these results in the diagram below further underscores the comparative superiority of proposed Tomato Leaf Disease Detection Model. Figure 6 shows the average performance metric of detecting all tomato disease categories.

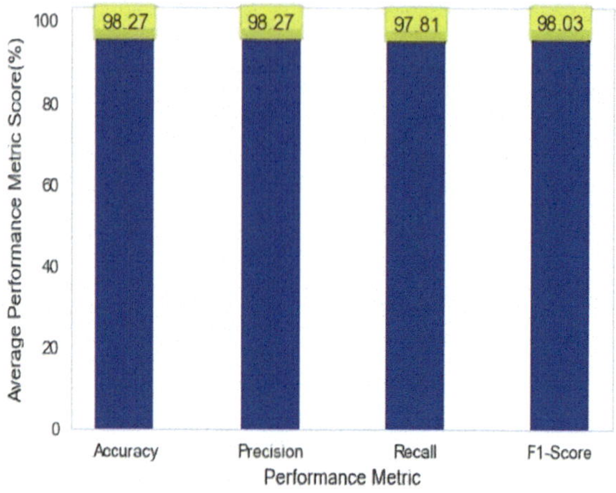

Fig. 6 Average performance metric of all tomato disease categories

7 Conclusion and Future Work

In this paper, a Convolutional Neural Network (CNN) is used to identify tomato leaf diseases in India, the world's second-largest tomato producer. The proposed model follows a streamlined workflow, including image acquisition, pre-processing, feature extraction, and CNN training, achieving an impressive average score of 98% accuracy, precision, recall, F1 score. This surpasses the effectiveness of traditional methods, including SVM, KNN, and backpropagation. The proposed approach shows potential to transform the management of tomato leaf diseases in Indian agriculture., addressing the critical need for timely intervention and ultimately safeguarding crop productivity. In the future, the dataset could be expanded to encompass a broader range of diseases and environmental conditions. This diversification aims not only to enhance the model's accuracy in recognizing various threats but also to ensure its efficacy under varying growth circumstances. Furthermore, there is potential to develop an easy-to-use, mobile application-based interface, holding substantial promise in making this technology accessible to farmers. This user-friendly interface is instrumental for disease detection and management, empowering farmers by seamlessly integrating it into their daily agricultural routines, regardless of their technological proficiency. Sophisticated algorithms like Recurrent Neural Networks (RNN) and Long-Short Term Memory (LSTM) can be utilized to enhance the overall effectiveness.

References

1. Das D, Singh M, Mohanty SS, Chakravarty S (2020) Leaf disease detection using support vector machine. In: 2020 international conference on communication and signal processing (ICCSP), Chennai, India, pp 1036–1040. https://doi.org/10.1109/ICCSP48568.2020.9182128
2. Nagamani HS, Sarojadevi H (2022) Tomato leaf disease detection using deep learning techniques. Int J Adv Comput Sci Appl 305–311. https://doi.org/10.14569/IJACSA.2022.0130138
3. Balakrishna K, Rao M (2019) Tomato plant leaves disease classification using KNN and PNN. Int J Comput Vis Image Process 51–63. https://doi.org/10.4018/IJCVIP.2019010104
4. Ashok S, Kishore G, Rajesh V, Suchitra S, Sophia SGG, Pavithra B (2020) Tomato leaf disease detection using deep learning techniques. In: 2020 5th international conference on communication and electronics systems (ICCES), Coimbatore, India, pp 979–983. https://doi.org/10.1109/ICCES48766.2020.9137986
5. Kibriya H, Rafique R, Ahmad W, Adnan SM (2021) Tomato leaf disease detection using convolution neural network. In: Proceedings of 2021 18th international Bhurban conference on applied sciences & technology (IBCAST), pp 346–351. https://doi.org/10.1109/IBCAST51254.2021.9393311
6. Sherly Puspha Annabel L, Muthulakshmi V (2019) AI-powered image-based tomato leaf disease detection. In: Proceedings of the third international conference on I-SMAC (IoT in social, mobile, analytics and cloud) (I-SMAC 2019), pp 506–511. https://doi.org/10.1109/I-SMAC47947.2019.9032621
7. Adhikari S, Shrestha B, Baiju B, Saban Kumar KC (2018) Tomato plant diseases detection system using image processing. In: Proceedings 1st KEC conference, pp 81–86
8. Abbas A, Jain S, Gour M, Vankudothu S (2021) Tomato plant disease detection using transfer learning with C-GAN synthetic images. J Comput Electron Agric 0168–1699. https://doi.org/10.1016/j.compag.2021.106279
9. de Luna RG, Dadios EP, Bandala AA (2018) Automated image capturing system for deep learning-based tomato plant leaf disease detection and recognition. In: TENCON 2018–2018 IEEE region 10 conference, Jeju, Korea (South), pp 1414–1419. https://doi.org/10.1109/TENCON.2018.8650088
10. Mokhtar U, Alit MAS, Hassenian AE, Hefny H (2015) Tomato leaves diseases detection approach based on support vector machines. In: Proceeding of 978-1-5090-0275-7/15/31.00 2015 IEEE, pp 246–250. https://doi.org/10.1109/ICENCO.2015.7416356
11. Jiang D, Li F, Yang Y, Yu S (2020) A tomato leaf diseases classification method based on deep learning. In: Proceeding of 2020 Chinese control and decision conference (CCDC 2020), pp 1446–1450. https://doi.org/10.1109/CCDC49329.2020.9164457
12. Vetal S, Khule RS (2017) Tomato plant disease detection using image processing. J Int J Adv Res Comput Commun Eng 293–297. https://doi.org/10.17148/IJARCCE.2017.6651
13. Vetal S, Khule RS (2017) Tomato plant disease detection using image processing. J Int J Adv Res Comput Commun Eng 293–297. https://doi.org/10.17148/IJARCCE.2017.6651
14. Agarwala M, Singhb A, Arjariac S, Sinhad A, Gupta S (2020) Tomato leaf disease detection using convolution neural network. In: Proceeding of international conference on computational intelligence and data science (ICCIDS-2019), pp 293–301. https://doi.org/10.1016/j.procs.2020.03.225
15. Harakannanavar SS, Rudagi JM, Puranikmathb VI, Siddiquaa A, Pramodhini R (2022) Plant leaf disease detection using computer vision and machine learning algorithms. In: Proceeding publishing services by Elsevier B.V. on behalf of KeAi Communications Co. Ltd., pp 306–310. https://doi.org/10.1016/j.gltp.2022.03.016
16. Paymode AS, Magar SP, Malode VB (2021) Tomato leaf disease detection and classification using convolution neural network. In: 2021 international conference on emerging smart computing and informatics (ESCI), Pune, India, pp 564–570. https://doi.org/10.1109/ESCI50559.2021.9397001

17. Gadade HD, Kirange DK (2021) Machine learning based identification of tomato leaf diseases at various stages of development. In: Proceeding of the fifth international conference on computing methodologies and communication (ICCMC 2021), pp 814–819. https://doi.org/10.1109/ICCMC51019.2021.9418263
18. Mohanty R, Wankhede P, Singh D, Vakhare P (2022) Tomato plant leaves disease detection using machine learning. In: 2022 international conference on applied artificial intelligence and computing (ICAAIC), Salem, India, pp 544–549. https://doi.org/10.1109/ICAAIC53929.2022.9793302
19. Langar G, Jain P, Panchal N (2020) Tomato leaf disease detection using artificial intelligence and machine learning. J Int J Adv Sci Res Eng Trends 1–5
20. Chandak A, Sharma A, Khandelwal A (2022) Detection of diseases in tomato plant using machine learning. Int J Next Gener 943–952. https://doi.org/10.47164/ijngc.v13i5.941

Intelligent Systems for Climate Prediction

Application of CNN for Separation Between No Rain and Rain from Remote Sensing Data

Mourad Lazri, Youcef Attaf, Karim Labadi, Noureddine Bensafi, Slimane Hameg, and Soltane Ameur

Abstract We present in this chapter a strategy for classifying meteorological images using the convolution neural network (CNN). The CNN is applied to images from the Meteosat Second Generation (MSG) meteorological satellites. The study region is divided into several zones considered as considered small image. All zones are assigned to a class, either no rain class or rain class. The processing is carried out on all the images collected during the rainy period from October 2009 to March 2010. The processing consists of introducing these zones extracted from MSG images of different sizes to the CNN and obtaining as output the rain class or the no rain class. The classification results are compared to the results obtained by a standard artificial neural network (ANN). To show the contribution of the CNN model, a comparison of the results with the standard ANN model is performed. The results show better performance in the case of CNN. Indeed, for the POD, for the CNN indicates 95% against 91% for the MLP. The POFD and FAR show 06 and 20% for CNN compared to 09 and 27% for MLP, respectively. The same trends are observed in the case of CSI and PC. 88 and 96% for CNN versus 82 and 92% for MLP, respectively. As for Bias, CNN shows a very slight overestimation which is better than MLP.

Keywords Rainfall · CNN · Meteorological satellite · Image

1 Introduction

Water plays a very important role in the agricultural sector. Land irrigation systems take over when precipitation rates are insufficient. In this context, knowledge of precipitation rates in time and space helps farmers plan irrigation periods and determine, based on these rates, the sowing period to ensure a better harvest. Precipitation

M. Lazri (✉) · Y. Attaf · N. Bensafi · S. Hameg · S. Ameur
Laboratory LAMPA, University Mouloud MAMMERI, Tizi Ouzou, Algeria
e-mail: m_lazri@yahoo.fr

K. Labadi
Laboratory LR2E, ECAM-EPMI/Quartz-Lab, Cergy Pontoise, France

A. J. Kulkarni and M. Hajiaghaei-Keshteli (eds.), *Advances in Intelligent Systems for Sustainable Agriculture*, Lecture Notes in Networks and Systems 1201,
https://doi.org/10.1007/978-981-97-9839-1_12

is measured using rain gauges or weather radars. However, these measures are one-off and cannot cover large areas. For this reason, data from meteorological satellites are used to estimate precipitation. This type of data makes it possible to cover a wider field. Meteorological satellites are therefore becoming an essential means for observing the atmosphere and monitoring clouds. Precipitation estimates can be made from satellite observation series. The use of multispectral information from the same satellite or from several satellites allows a better estimate. This involves finding a relationship between satellite observations and precipitation. In this sense, the diversity of existing satellite precipitation estimation techniques depends, on the one hand, on the multispectral combination and the characteristics of the sensors used. On the other hand, it depends on the differentiation in satellite orbits, low orbit terrestrial, such as National Oceanic and Atmospheric Administration–Advanced Very High Resolution Radiometer (NOAA-AVHRR) and geostationary orbits, such as Geostationary Operational Environmental Satellite—Visible and Infrared Spin Scan Radiometer (GOES-VISSR), Meteosat Second Generation (MSG). Indeed, the satellites of type GEO have optical sensors and passive microwave sensors are placed on satellites of type LEO. From data issued satellite, precipitation contained in clouds is estimated using mathematical models, in particular models based on machine learning. A number of works have been published using neural networks [1, 2, 3], SVMs ([4], [5]), random forest [6], or the combination of these models [7, 8, 9].

Despite the level of precision in precipitation estimates provided by machine learning, improvements can be made. Ensemble methods are also implemented to optimize mathematical models [4]. Recently, deep learning such as CNN (Convolutional Neural Networks) and DNN (Deep Neural Networks) are increasingly used. These models have revolutionized object recognition in image classification [10].

In precipitation estimates, classification is an essential step. Indeed, the estimation results strongly depend on the quality of classification. With this in mind, in this chapter, we will apply convolutional neural networks for image classification. Areas of pixels are fed into the CNN model for classification. Two classes are considered, namely the rain class and the no rain class.

The chapter is structured as follows. In Sect. 2, Study region and Data is described. Section 3 contains the methodology in which the CNN model is introduced. The classification results and their interpretation are presented in Sect. 4. The conclusion is given in Sect. 5.

2 Study Region and Data

The CNN model is used for image classification into two classes, rain class and no rain class. The input data are MSG images and for training and validation, the output data are radar images. In the following sections, we present the database used which is composed of MSG images and Radar images as well as the study region.

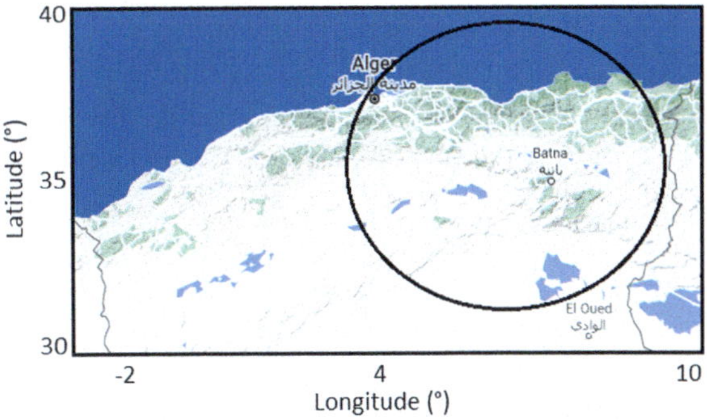

Fig. 1 Study area indicated by the Radar coverage of Sétif (North-East of Algeria)

2.1 Study Region

The study region is the northern region of Algeria (Fig. 1). Precipitation in this region is a mixture of mid-latitude and subtropical precipitation.

Coincidences pairs of images, namely, MSG image and Radar image used initially for the learning of CNN. These images are collected during the period from October 2009 to April 2010. Period from October 2010 to April 2011 is used for the evaluation of the classification by CNN.

2.2 MSG Image

Multispectral images in 12 channels at spatial resolutions of 3×3 km^2 (VSH at 1×1 km^2) are collected by MSG satellite every 15 min. A pixel can take values between 0 and 1023. These digital counts can be converted into radiance and then into brightness temperature in the case of the infrared channels and into reflectance in the case of the visible channels [11]. Infrared, such as IR10.8, IR8.7, IR3.9, IR12.1, WV6.2, WV7.3, and visible channels, such as VIS0.6, VIS1.6 are used. The spatial resolution in the study region reaches 4×5 km^2.

Radar images used here are collected by the Sétif radar every 15 min. The spatial resolution is 1×1 km^2. A pixel is coded on four bits giving 16 classes (dBZ). Levels below 12 dBz are considered no rain, while levels above are considered rain.

In order to have the best match between the MSG image and the Radar image, the radar images have been reduced to get the same spatial resolution of the images from MSG. The averages of 4×5 pixels in the radar image coincide with one pixel of the MSG image. In the case of temporal resolution, the radar and satellite do not require processing because the time difference is very weak.

2.3 Selection of Input Attributes

A set of channels in the infrared and in the visible are used. They allow to provide information on the temperature of the cloud tops (CTT), the cloud vertical extension (CTH), and the cloud water path (CWP) [12, 1, 13] (Table 1).

- The TIR10.8 parameter is used to extract information about the CTT.
- The ΔTIR10.8-IR12.1 parameter provides information on the cloud phase (CP). The two channels show opposite similarities in the case of ice particles.
- The parameter ΔTIR8.7-IR10.8 like ΔTIR10.8-IR12.1 used to separate "ice" and "water", which therefore provides information on the CP.
- The parameter ΔTIR3.9-WV7.3 and ΔTIR3.9-IR10.8 provides information on precipitation droplet radii and cloud optical thickness (CWP) which are decisive in precipitation characterization.
- The ΔTWV7.3-IR12.1 parameter provides information on CTT and CTH.
- The parameter ΔTWV6.2-IR10.8 like ΔTWV7.3-IR12.1 is used to eliminate non-precipitating cirrus clouds and indicates information on the presence of convective systems.
- The parameter VIS0.6 and NIR1.6 available only during the day are used to have information on the CWP.

3 Methodology

The model applied in this paper is the CNN. The study region is divided into several zones in the form of n × n pixel windows on the MSG images. Each zone is injected at the input of CNN. After training, the CNN model generates an output response

Table 1 Input attributes and the corresponding information

Input parameters	Possible values		Information on
	Daytime	Nighttime	
$T_{10.8}$ (Kelvin)	207.2 to 283.9 k	205.3 to 282.4 k	CTT and CTH
$\Delta T_{10.8-12.1}$ (Kelvin)	−0.3 to 7.4 k	−0.3 to 7.1 k	CP
$\Delta T_{8.7-10.8}$ (Kelvin)	−4.6 to 1.3 k	−4.8 to 1.7 k	CP
$\Delta T_{6.2-10.8}$ (Kelvin)	50.1 to 6.4 k	−51.8 to 5.1 k	CTT and CTH
$\Delta T_{7.3-12.1}$	−50.3 to 6.6 k	−52.0 to 5.7 k	CTT and CTH
$R_{0.6}$ (μm)	0.02 to 1 μm	No used	CWP
$R_{1.6}$ (μm)	0.03 to 1 μm	No used	CWP
$\Delta T_{3.9-10.8}$ (Kelvin)	No used	−10.3 to 15.1 k	CWP
$\Delta T_{3.9-7.3}$ (Kelvin)	No used	−4.9 to 25 k	CWP

to indicate whether the area is precipitating or non-precipitating. The mathematical concept of CNN is presented in the next section.

3.1 Convolution Neural Network

We present one of the most efficient Deep Learning algorithms, namely, CNN. These are powerful programming models used especially in image recognition. In this study, CNN is applied to an image area, the area is assigned to the rain or no rain class.

CNN are neural networks that use convolution to extract features. One of the particularities of CNNs is that they process images as input. Their architecture is made up of two blocks.

- The first works as a feature extractor. By applying convolution operations with several convolution kernels, the image is filtered and returns "feature maps", that are normalized using an activation function. The operation is repeated several times. Indeed, the feature maps are filtered again and again with other kernels that gives new feature maps to resize and normalize. The last feature maps are concatenated in order to get a vector which will be the input of the second block.
- The second block receives the values of the vector from the first block which transforms into output using linear combinations and activation functions permitting to return a new vector as output. The vector obtained at the last stage is a set of probabilities and each probability corresponds to a class.

The layer parameters are determined by backpropagation of the gradient: during the learning phase the cross entropy is minimized. These parameters designate in particular the features of the images. The CNN has four layers: the convolution, the pooling, the ReLU correction, and the fully connected.

- Convolution layer

This layer that can be seen as hidden layer is the act of applying a mathematical filter to an image. The operation permits to find areas of the image, which could be interesting. A neuron is only connected to a neuron that can be found in the next layer.

- ReLu layer

After convolution, the ReLu function should be applied to each pixel. Each negative value is replaced with a 0. ReLu is extensively used in CNN, it corresponds to max(0, y).

- Pooling

Pooling is used in convolution layers: it gets several feature maps as input, and applies the pooling to each of them. It reduces the size of images, preserving their important characteristics. The image is divided into regular cells. The maximum value is retained within each cell.

The 2 × 2 pixels slides over the image, as for convolution with a step of 2. For each step, we take the maximum. This is called Max Pooling. The pooling reduces the number of parameters what leads to simply the calculations in the network. This operation thus allows the efficacy of the network while avoiding overlearning.

- Couche fully connected

The last step corresponds to the fully connected layer of the CNN. All the features obtained from the previous layers are put into a vector which is injected at the input of the last layer. It produces an output results vector to indicate the final class. It applies an activation function and a linear combination to the values taken as input. The last fully connected layer therefore classifies the image input to the network thanks to the results which is a vector of size N, where N represents the number of classes in the image classification problem. It is a probability set where the greatest probability corresponds to the assignment class.

3.2 *Learning and Application*

The CNN is first trained before being applied to MSG images. The reference data for learning is weather radar data. Two classes were considered in this study, the no rain class and the rain class. The objective of this work is to classify non-precipitating areas and precipitating areas on MSG images. In order to implement our model, images are subdivided into several areas. Areas are inputs to CNN and rain and no rain classes are outputs. The CNN first allows characteristics to be extracted from the area of size n × n pixels, then these characteristics are used as input to a fully connected neural network (see Fig. 2).

The inputs are multispectral images. We therefore combined in the feature extraction part the images T10.8 (Kelvin), ΔT10.8-12.1 (Kelvin), ΔT8.7-10.8 (Kelvin), ΔT6.2-10.8 (Kelvin), ΔT7.3-12.1 (Kelvin), R0.6 (μm), R1.6 (μm), ΔT3.9-10.8 (Kelvin), and ΔT3.9-7.3 (Kelvin). As noted above, areas are classified.

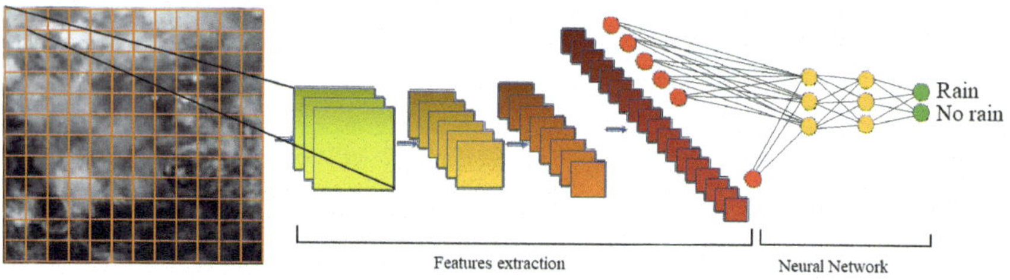

Fig. 2 Main architecture of CNN

4 Results and Interpretation

The CNN method was tested on MSG images collected during the period from October 2009 to March 2010. These images are therefore classified using the CNN. The choice of window size is determined experimentally. We varied this size and calculated a set of evaluation parameters. These parameters are the POD (Probability Of Detection) given by Eq. 1, the POFD (Probability Of False Detection) calculated by Eq. 2, the FAR (False Alarm Ratio) given by Eq. 3, the CSI (Critical Success Index) given by Eq. 4, the PC (Percentage of Correct) calculated by Eq. 5, and the Frequency BIAS index calculated by Eq. 6. These parameters are calculated using Table 2.

The POD shows the fraction of pixels detected by Radar that were correctly identified by the satellite using CNN:

$$\mathrm{POD} = \frac{\mathrm{a}}{\mathrm{a} + \mathrm{c}} \tag{1}$$

The POFD indicates the part of pixels which is incorrectly classified by the satellite using CNN.

$$\mathrm{POFD} = \frac{\mathrm{b}}{\mathrm{b} + \mathrm{d}} \tag{2}$$

The FAR corresponds to the part of pixels classified by satellite using CNN but does not correspond to the Radar data:

$$\mathrm{FAR} = \frac{\mathrm{b}}{\mathrm{a} + \mathrm{b}} \tag{3}$$

The BIAS shows the underestimation or the overestimation performed by CNN

$$\mathrm{Bias} = \frac{\mathrm{a} + \mathrm{b}}{\mathrm{a} + \mathrm{c}} \tag{4}$$

The CSI corresponds to the fraction of observed and/or identified pixels that were correctly retrieved.

Table 2 A contingency table

	Detected by radar			
Classified by model		Class (i)	No class (i)	Total
	Class (i)	a	b	a + b
	No class (i)	c	d	c + d
	Total	a + c	b + d	a + b + c + d = n

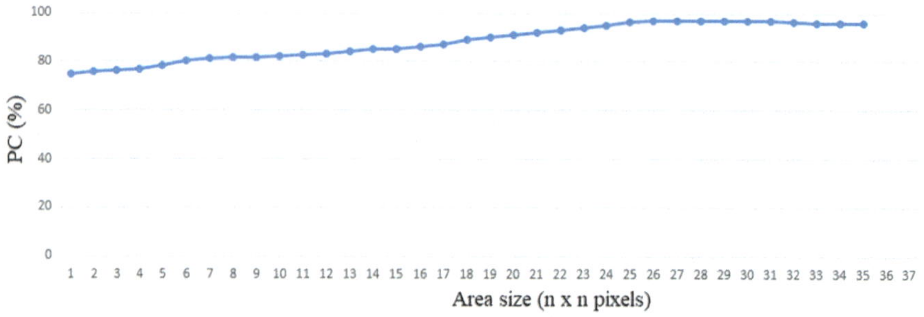

Fig. 3 Percent correct depending on the size of the zones

$$\mathrm{CSI} = \frac{\mathrm{a}}{\mathrm{a} + \mathrm{b} + \mathrm{c}} \tag{5}$$

The PC corresponds to the percentage of correct identification:

$$\mathrm{PC} = \frac{\mathrm{a} + \mathrm{d}}{\mathrm{n}} \tag{6}$$

The PC was used to show the best area to take in this study. The PC values as a function of the zone sizes are given in the following Fig. 3.

With large areas, we obtained very good classification results. However, the spatial resolution is too low. For example, a PC of 98% corresponds to areas of more than 10,000 km^2. The results remain in terms of unusable resolution. A compromise is found between the precision of the results and the spatial resolution. Thus, according to the comparisons, the input areas of size 16 × 16 pixels show a good compromise. Smaller areas do not have enough information to be used in a convolutional neural network. The results are not interesting. With larger areas, the spatial resolution decreases significantly. For an area of size 16 × 16, we made a comparison between the results obtained by applying the CNN and the results obtained by applying the ordinary neural network (multilayer perceptron). The individual pixels in the 16 × 16 pixel area are replaced by the average value. Table 3 gives the comparison results for the two models.

As these results show, the CNN gives better performance. Indeed, for the POD, for the CNN indicates 95 against 91% for the MLP. The POFD and FAR show 06 and 20% for CNN compared to 09 and 27% for MLP, respectively. The same trends are observed in the case of CSI and PC. 88 and 96% for CNN versus 82 and 92%

Table 3 Values obtained for the evaluation parameters for MLP and CNN

	POD (%)	POFD (%)	FAR (%)	Bias	CSI (%)	PC (%)
CNN	91	09	27	1.1	82	92
Optimal values	95	06	20	1.05	88	96
CNN	100	0	0	1	100	100

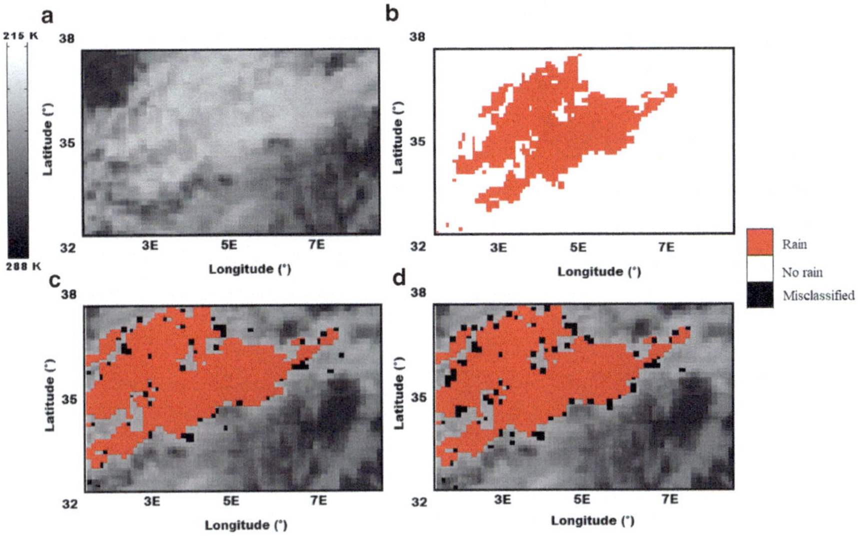

Fig. 4 Classification results, **a** IR10.8 image, **b** Radar image of reference, **c** result using CNN, **d** result using MLP

for MLP, respectively. As for Bias, CNN shows a very slight overestimation which is better than MLP.

The instantaneous scene taken by the MSG satellite in the IR10.8 infrared band is classified using the CNN and MLP models (see Fig. 4). The reference image from radar is also given.

As we can see, CNN model shows better performance compared to MLP. Misclassified pixels are more present in the case of the results obtained by the application of MLP. These pixels were reclassified correctly using the CNN. However, the application of CNN has a major drawback regarding spatial resolution. The results are better with low spatial resolutions. For this reason, a compromise is necessary between the quality of the results and the spatial resolution.

We also made precipitation estimates using CNN for the month of January of the rainy season. To compare, estimates using MLP were also performed (Fig. 5).

As these estimates show, the results of CNN are better than those operated by MLP. The estimates correlate well with radar measurements for CNN. The correlation coefficient of CNN is 0.95 compared to 0.84 for MLP.

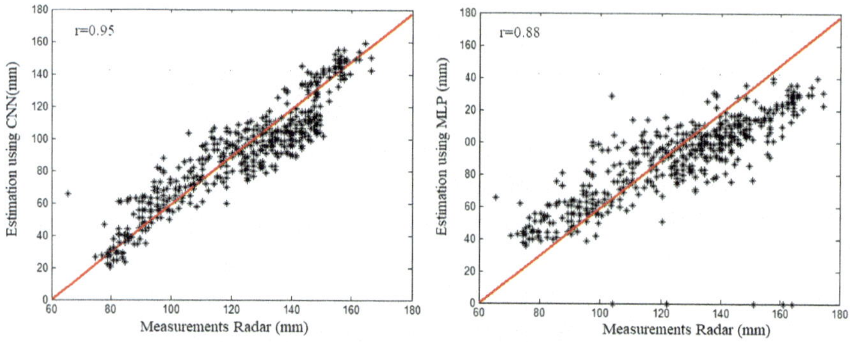

Fig. 5 Estimation using CNN and using MLP versus measurements radar respectively

5 Conclusion

The work presented in this chapter consisted of the classification of pixels from MSG images into classes, the no rain class and the rain class. Deep leaning such as CNN was applied. To do this, areas of different sizes on the images were tested. The results are highly dependent on the size of the area. The best results are obtained with larger zone sizes. For these areas, the spatial resolution is too low. A compromise is obtained for an area of size 16 × 16 pixels. The lower areas give better spatial resolution but the quality of the results is not good. For higher areas, the results are better but the resolution is very low.

To show the performance of the model used, a comparison was also carried out. Indeed, these results obtained by applying the CNN model are compared to the results obtained by applying the standard model of the MLP neural network. As we can see, CNN shows better performance compared to MLP. The classifications are better by applying the CNN. All calculated evaluation parameters indicated the superiority of CNN over MLP. Indeed, for the POD, for the CNN indicates 95% against 91% for the MLP. The POFD and FAR show 06 and 20% for CNN compared to 09 and 27% for MLP, respectively. The same trends are observed in the case of CSI and PC. 88 and 96% for CNN versus 82 and 92% for MLP, respectively. As for Bias, CNN shows a very slight overestimation which is better than MLP. This is justified by the fact that the CNN made it possible to extract considerable information but it shows a major drawback in terms of spatial resolution.

In this context and as a perspective of this work, it would be interesting to explore the deep neural network to extract information at the pixel level instead of taking areas of pixels. This will allow work with high spatial resolutions.

References

1. Lazri M, Ouallouche F, Ameur S, Brucker JM, Mohia Y (2012) Identifying convective and stratiform rain by confronting SEVERI sensor multispectral infrared to radar sensor data using neural network. Sens Transducers J 145(10):19–32
2. Lazri M, Labadi K, Ouallouche F, Ameur S (2023) Optimization of rainfall intensities classification based on artificial intelligence using recurrent neural network. In: Intelligent systems and applications: select proceedings of ICISA 2022. Springer Nature Singapore, Singapore, pp 39–48
3. Tebbi MA, Haddad B (2016) Artificial intelligence systems for rainy areas detection and convective cells' delineation for the south shore of Mediterranean Sea during day and nighttime using MSG satellite images. Atmos Res 178:380–392
4. Belghit A, Lazri M, Ouallouche F, Labadi K, Ameur S (2023) Optimization of One versus All-SVM using AdaBoost algorithm for rainfall classification and estimation from multispectral MSG data. Adv Space Res. https://doi.org/10.1016/j.asr.2022.08.075
5. Sehad M, Lazri M, Ameur S (2017) Novel SVM-based technique to improve rainfall estimation over the Mediterranean region (north of Algeria) using the multispectral MSG SEVIRI imagery. Adv Space Res 59(5):1381–1394
6. Mohia Y, Ouallouche F, Lazri M et al (2023) Classification of precipitation intensities from remote sensing data based on artificial intelligence using RF multi-learning. J Ind Soc Remote Sens. https://doi.org/10.1007/s12524-023-01665-5
7. Belmahdi F, Lazri M, Ouallouche F, Labadi K, Absi R, Ameur S (2023) Application of Dempster-Shafer theory for optimization of precipitation classification and estimation results from remote sensing data using machine learning. Remote Sens Appl Soc Environ 29:100906
8. Lazri M, Labadi K, Brucker JM, Ameur S (2020) Improving satellite rainfall estimation from MSG data in Northern Algeria by using a multi-classifier model based on machine learning. J Hydrol 584:124705
9. Lazri M, Ameur Z, Ameur S, Mohia Y, Brucker JM, Testud J (2013) Rainfall estimation over a Mediterranean region using a method based on various spectral parameters of SEVIRI-MSG. Adv Space Res 52:1450–1466
10. Zhang M, Liu L, Jin Y, Lei Z, Wang Z, Jiao L (2024) Tree-shaped multiobjective evolutionary CNN for hyperspectral image classification. Appl Soft Comput 152:111176
11. EUMETSAT (2004) Applications of Meteosat Second Generation—conversion from counts to radiances and from radiances to brightness temperatures and reflectance. http://oiswww.eumetsat.org/WEBOPS/msg_interpretation/index.html
12. Feidas H, Giannakos A (2011) Identifying precipitating clouds in Greece using multispectral infrared Meteosat Second Generation satellite data. Theor Appl Climatol 104:25–42
13. Thies B, Nauss T, Bendix J (2008) Precipitation process and rainfall intensity differentiation using Meteosat second generation spinning enhanced visible and infrared imager data. J Geophys Res 113
14. Hameg S, Lazri M, Ameur S (2016) Using naive Bayes classifer for classification of convective rainfall intensities based on spectral characteristics retrieved from SEVIRI. J Earth Syst Sci 125:945–955

Comparative Analysis of Ensemble-Based Nowcasting Models for Precipitation Prediction in the UAE

Achraf Tounsi and Safa Bhar Layeb

Abstract Accurate and timely nowcasting is vital in safeguarding public property and ensuring people's safety, particularly in regions with infrequent but potentially hazardous rainfall events like the United Arab Emirates (UAE). The UAE experiences rare rainfall occurrences that can cause flooding, making it crucial to address these weather challenges. In this study, we compared the predictability of two widely used ensemble-based nowcasting systems, STEPS and LINDA, to assess their performance during various precipitation events in the UAE. Our ROC analysis showed that both systems have a commendable detection rate of over 0.70, a low false alarm rate of less than 0.2, and a substantial area under the curve exceeding 0.75%. Additionally, both algorithms demonstrated their ability to produce reliable nowcasts for up to two hours, using a 5 mm/h threshold. These findings highlight the potential of these models in mitigating the impacts of rare rainfall events, such as flooding, and emphasize the importance of investing in advanced nowcasting technologies for improved weather prediction and public safety.

Keywords Nowcasting · Precipitation · Flooding · Public safety

A. Tounsi
Department of Civil, Environmental and Ocean Engineering, Stevens Institute of Technology, Hoboken, NJ, USA
e-mail: atounsi@stevens.edu

S. B. Layeb (✉)
LR-OASIS, National Engineering School of Tunis, University of Tunis El Manar Tunis, Tunis, Tunisia
e-mail: safa.layeb@enit.utm.tn

Centre Génie Industriel, Université Toulouse, IMT Mines Albi, Albi, France

A. J. Kulkarni and M. Hajiaghaei-Keshteli (eds.), *Advances in Intelligent Systems for Sustainable Agriculture*, Lecture Notes in Networks and Systems 1201,
https://doi.org/10.1007/978-981-97-9839-1_13

1 Introduction

Recently, rapid climate change has led to severe weather events around the world, causing unexpected precipitation and flash floods [1]. The importance of precise and dependable weather forecasting has become more evident with these extreme occurrences. Predicting rainfall in a specific area over a short period, usually up to six hours [2], is crucial for public safety from severe weather events like flash floods, tropical cyclones, thunderstorms, lightning, and strong winds that can impact large populations and economies [3]. In this context, the United Arab Emirates (UAE) is characterized by a very hard climate. Even with minimal rainfall all year round, flooding might still be experienced in places that are prone to urbanization [4, 5]. Due to this, most cities are now more prone to flooding than ever before; amplify runoff, increased peak discharges, raised flood depths and expanded flood-prone regions are common results in intense rainfall episodes [6, 7].

Predicting the weather in real-time (Nowcasting) is important for preventing flash floods, but it comes with its own set of difficulties because meteorological patterns can change unexpectedly. Using Numerical Weather Prediction (NWP) for short-term forecasts is not always reliable because it requires a lot of computing power, is easily influenced by inaccuracies, and is very sensitive to the starting conditions [8]. As a result, the current methods lead to delays in making short-term predictions because of the time it takes for data assimilation and simulation in weather prediction models. This results in forecasts that may already be outdated by the time they are produced. To overcome these challenges, nowcasting has typically relied on Radar Echo Extrapolation (REE). This technique involves using reflected electromagnetic waves from water particles in the atmosphere to provide precise details on storm characteristics such as intensity, speed, shape, and movement direction in real-time. The method of extrapolation has been shown to be more effective than NWP-based models [9] and is currently considered the best technique for predicting precipitation [10–13]. Computer vision techniques, like optical flow, are commonly used to extrapolate radar maps in the field of nowcasting [14, 15].

Satellite-based data has been utilized in various studies for precipitation nowcasting [16–18]. Data assimilation technique was employed by Otsuka et al. [18] for nowcasting from satellite images. Rivolta et al. [17] utilized a two-step approach to predict future infrared radiance data from satellites, followed by the use of an artificial neural network for further nowcasting. Liu et al. [19] used computer vision with optical flow to predict precipitation using satellite data. Son and Thong [16] developed a picture fuzzy clustering method to predict weather patterns from satellite images. While studies such as [16–18] have shown promise, they can be computationally intensive and involve multiple processing steps to generate nowcasting results. Moreover, the use of optical flow-based techniques in nowcasting has demonstrated only modest success, as they tend to treat the tracking of pixels and the extrapolation of values during the prediction phase as two distinct processes, with limited interconnection between the two [20, 21].

Within this study, the performance of the STEPS and LINDA ensemble-based nowcasting systems in the UAE is thoroughly evaluated using GPM (Global Precipitation Measurement) satellite precipitation data. GPM, as a state-of-the-art satellite-based measurement system, provides valuable and high-resolution information on precipitation patterns, to evaluate the nowcasting systems. This study investigates the unique weather conditions experienced in the UAE. Ultimately, this research endeavors to improve the precision and reliability of nowcasting in the region, reducing the potential impact of extreme rainfall events and facilitating better disaster management efforts.

2 Methodology

2.1 Study Area and Data

The UAE has national borders with Saudi Arabia to the southwest and Oman to the east and north. Its southeastern shore is next to the Arabian Gulf, as seen in Fig. 1, and a tiny portion of its eastern section has an eastward-facing coastline on the Gulf of Oman, an extension of the Arabian Sea in the North Indian Ocean.

The UAE's geography is dominated by desert plains, which represent nearly 95% of the country's total geographical area. The Hajar Mountains, on the other hand,

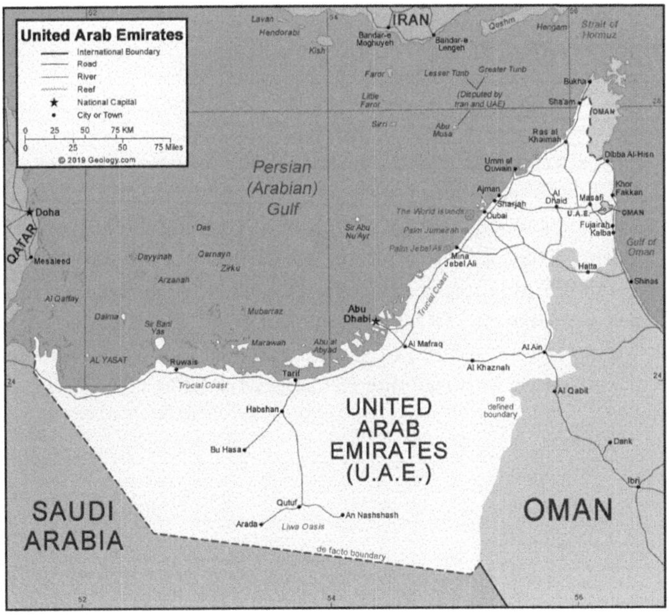

Fig. 1 Map of the United Arab Emirates

Table 1 Details of the precipitation events

Event	Date	Start time (UTC)	Duration (h)
1	01 January 2022	1430	9
2	26 May 2020	2300	11
3	09 January 2020	0230	13
4	28 October 2019	1330	5
5	16 December 2017	1530	4
6	25 March 2014	1900	7
7	15 March 2014	0130	20
8	21 November 2013	0030	5
9	17 November 2013	0900	7

rise in the country's northeastern section, reaching an altitude of approximately 2000 m. Since the mid-twentieth-century oil boom, the UAE has enjoyed tremendous economic expansion, particularly in the banking and tourist industries [4].

Rainfall occurrences are rare in the UAE's arid, desert-like environment, adding to the country's general dryness and lack of water. Despite limited water supplies and infrequent rainfall events, floods are uncommon due to limited water sources and infrequent rainfall occurrences. As a result, only 9 flood occurrences were researched in order to analyze and comprehend flooding dynamics in the region. The study used the rain events listed in Table 1 for the experiments.

2.2 *Enhanced Rainfall Estimation Through IMERG GPM Satellite Data*

The partnership between NASA and the Japan Aerospace and Exploration Agency (JAXA) on the Global Precipitation Measurement (GPM) mission, starting in March 2014, has greatly improved global precipitation forecasting. Following the Tropical Rainfall Measuring Mission (TRMM), the GPM mission focuses on maintaining satellite-based rainfall monitoring [22, 23]. This mission, succeeding the Tropical Rainfall Measuring Mission (TRMM), is dedicated to the continuity of satellite-based rainfall observations. Within the GPM framework, a plethora of products is available, with Integrated Multisatellite Retrievals for GPM (IMERG) standing out among them [22].

2.3 Ensemble-Based Nowcasting

It is well-known that ensemble-based nowcasting uses multiple weather models to predict weather conditions ahead of time, averaging or analyzing different scenarios to provide a more reliable forecast. In this work, two state-of-the-art models are investigated, namely, the LINDA model and the Short-Term Ensemble Prediction System (STEPS) model are investigated. The LINDA model was initially created to forecast heavy rainfall in specific areas, focusing on Autoregression [22, 23]. Besides, the Lagrangian extrapolation of radar image data by the STEPS model incorporates random changes to handle forecast uncertainty [22, 24].

2.4 Evaluation Metrics

Hamill [25] introduced the rank histogram as a tool for evaluating the alignment between ensemble spread and observed uncertainty. This histogram is constructed by sorting the predicted values from ensemble members (labeled as 1, 2, …, m) in ascending order for each grid cell. It comprises $m + 1$ bins, with each bin representing the range between two consecutive ensemble members where a verifying observation falls. The first and last bins correspond to observations that are lower or higher than all ensemble members, respectively. In a well-calibrated ensemble forecast, the rank histogram will display a flat distribution, indicating that no observations fall into the extreme bins.

In addition, Bröcker and Smith [26] were the first to create the reliability diagram, which is an important tool for assessing how well predicted probabilities match up with actual frequencies. The diagram should ideally show a perfect alignment along the diagonal, indicating perfect agreement. In addition to the diagram, there is a histogram showing the number of samples within different probability intervals, which is crucial for its accuracy. A reliable forecast is characterized by predicted probabilities that are mostly either very close to zero or very close to one. This indicates a strong tendency for clear-cut yes/no predictions, helping to reduce uncertainty.

In their study [27], Jolliffe and Stephenson explained the ROC curve, which is a way to show how well a forecast can predict whether it will rain enough to surpass certain rainfall thresholds. This curve shows the relationship between the probability of detecting rain (POD) and the likelihood of false alarms (POFD) at different probability levels. A perfect forecast would result in a curve that goes through the top-left corner of the graph, indicating perfect performance.

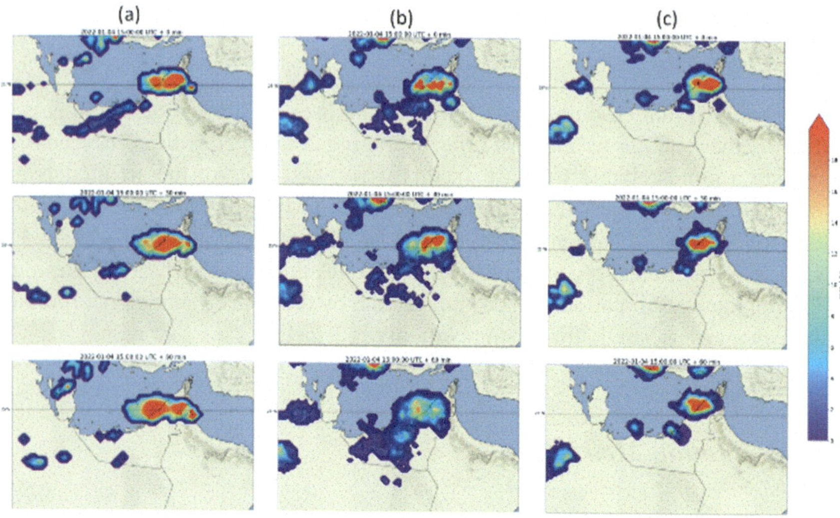

Fig. 2 Qualitative comparison of observations on January 4th 2022

3 Results and Discussions

3.1 Qualitative Analysis

Figure 2 presents a selection of members from both the 30-min LINDA and STEPS nowcast ensembles across the UAE. Upon meticulous examination of the vertical panels, it becomes evident that both approaches adeptly replicate the progression and dissipation dynamics of convective systems. Nonetheless, Fig. 2 displays discernible disparities that emerge between the forecasted outcomes and the authentic observations. The nowcasts tended to predict a decaying process, whereas the observed convective cloud system appeared to be still at the mature stage. Furthermore, the LINDA nowcast demonstrated a higher frequency of small rain cells, while the STEPS nowcast amplified the intensity of these cells. Moreover, the smoothed reflectivity fields generated by the STEPS ensemble exhibited deficiencies in accurately portraying the most extreme values within a quasi-linear convective system. On the other hand, the LINDA nowcasts exhibited more pronounced high rain rates, implying a potentially larger ensemble spread in their predictions. These observations align with the ensemble-based exceedance probabilities depicted in Fig. 3. Specifically, the exceedance probabilities generated by LINDA displayed sharper distributions, with a greater fraction of values being either very close to zero or one, indicating more confidence in those particular predictions.

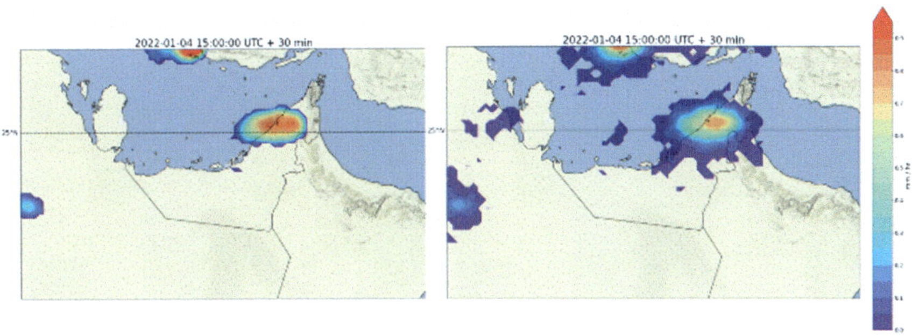

Fig. 3 Comparative analysis of 30 min exceedance probability nowcasts for 5 mm/h rain-rate threshold on 04 Jun 2020: LINDA (**a**) versus STEPS (**b**)

3.2 Quantitative Resuls

In the charts presented in Fig. 4, we can see how different nowcasting approaches compare in terms of reliability and accuracy. The horizontal dashed line on each chart represents a forecast that offers no improvement over a climatological forecast, showing the percentage of occurrences that surpass a certain threshold. The shaded area's lower boundary signifies the halfway point between a climatological forecast and a perfect forecast, indicating the range where forecast skill is useful. It is worth noting that the reliability diagrams in Fig. 4 show significant improvements in both reliability and accuracy for the LINDA approach compared to the STEPS approach, particularly for rainfall rates of 5–20 mm h^{-1}. With LINDA, there is a noticeable increase in reliability, as indicated by a higher concentration of values near one, pointing to enhanced accuracy. However, it should be noted that both approaches have their drawbacks. In the case of assessing 30 min nowcasts against a threshold of 20 mm h^{-1}, the reliability curves closely follow the line of no improvement, similar to a climatological forecast. This ultimately leads to a decrease in accuracy.

In Fig. 5, it can be seen that LINDA outperforms STEPS based on the ROC curves. These curves are created by dividing the probability range from 0 to 1 into 20 evenly spaced sections. The best threshold, which maximizes the Probability of Detection minus the Probability of False Detection (POD-POFD), is shown in black. The skill score is determined by calculating 2 times the area under the curve (AUC) minus 0.5 for evaluation purposes.

Both methods show a comparable probability of detection at the optimal threshold, but STEPS has a higher false alarm rate (0.26–0.32) compared to LINDA (0.2–0.22). Figures 4 and 5 suggest that probabilistic LINDA nowcasts, especially with the 20 mm h^{-1} threshold and optimal probability threshold for binary decisions, have potential to offer valuable skill for at least 30 min.

A rank histogram of STEPS and LINDA nowcasts for 30 min is shown in Fig. 6. Using the grid cells in the ensemble and the corresponding verification observations, the histograms were calculated without grid cells with rainfall rates below 5 mm h^{-1}.

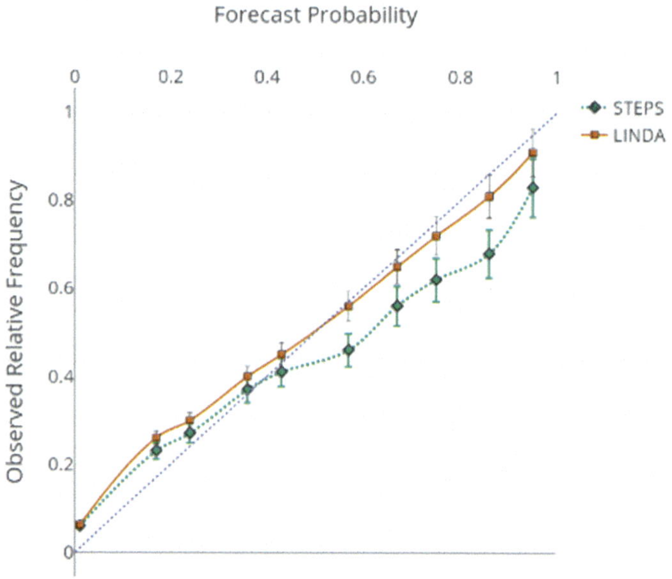

(a)

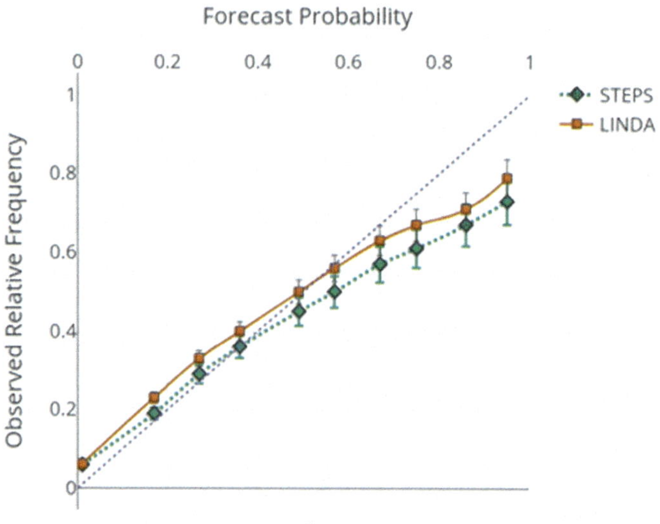

(b)

Fig. 4 Comparative analysis of reliability and sharpness in 30 min nowcasts by STEPS and LINDA for rain-rate thresholds of 5 mm h^{-1} (**a**), 10 20 mm h^{-1} (**b**), and **c** 20 mm h^{-1}

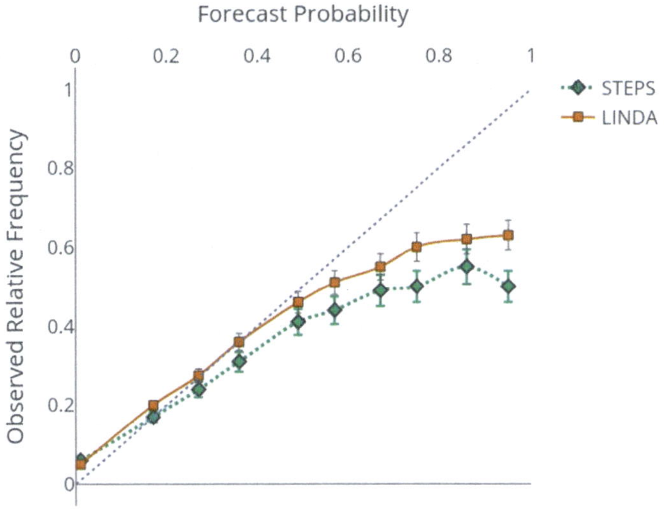

(c)

Fig. 4 (continued)

In spite of their somewhat conservative nature, both approaches are highly effective in capturing outliers. The number of observations exceeding the highest ensemble member remains below 4%. In contrast to STEPS, LINDA displays a marginally broader range than does STEPS, as depicted in Fig. 6. However, both ensembles show biases. Histograms of STEPS are right-skewed, indicating overestimation, while those of LINDA are left-skewed, indicating underestimation.

Furthermore, both techniques exhibit limitations in capturing the complete range of extreme values depicted on the right side of the histograms. In the STEPS method, the skewness is evident after the 110th member, while in the LINDA method it is discernible after the 90th member. These findings highlight the relative strengths and limitations of both nowcasting techniques in representing rainfall distribution, with a particular focus on the portrayal of extreme events.

4 Conclusion

Stochastic nowcasting was employed in the UAE region using satellite data (GPM) in combination with the LINDA and STEPS models. The GPM satellite data provided valuable information on precipitation, especially as the UAE is situated near the Arabian Sea and the Persian Gulf, where weather systems often approach from these bodies of water. The results of this satellite-based nowcasting approach demonstrated good accuracy for all study events. The ROC analysis indicated a high detection rate (more than 0.70), a low false alarm rate (less than 0.2), and a large area under the

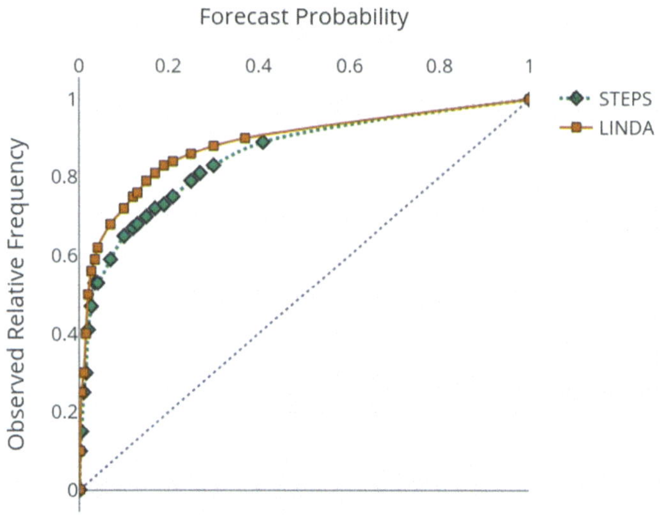

(a)

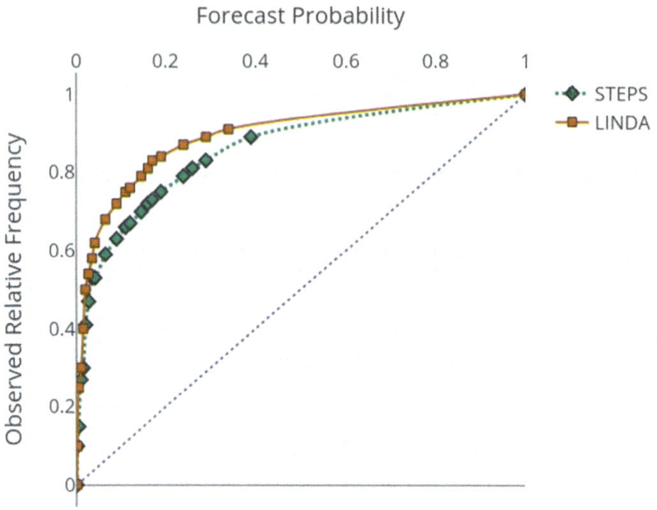

(b)

Fig. 5 STEPS and LINDA nowcasts for rain-rate thresholds: 5 mm h^{-1} (**a**), 10 20 mm h^{-1} (**b**), and **c** 20 mm h^{-1} (ROC Curves and Skill Scores)

curve (more than 0.75%). Both algorithms proved capable of generating reliable nowcasts for up to two hours with a 5 mm/h threshold.

Some avenues of research could be explored in future work. Indeed, the complexity of the algorithm schema requires attention to make it more user-friendly and accessible. Overall, this study lays the groundwork for future research and

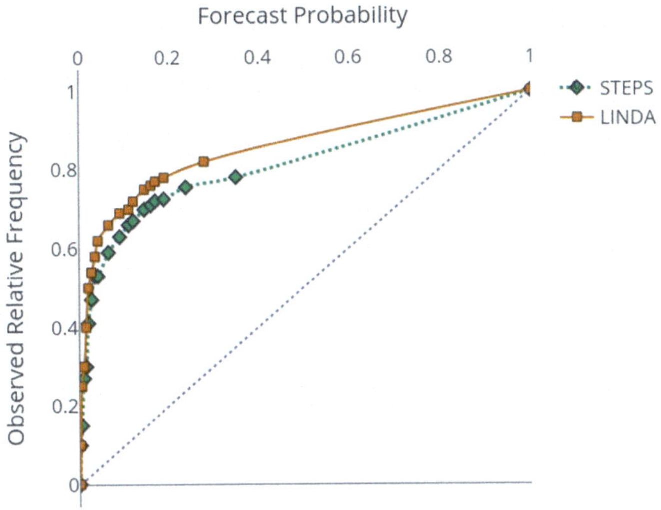

(c)

Fig. 5 (continued)

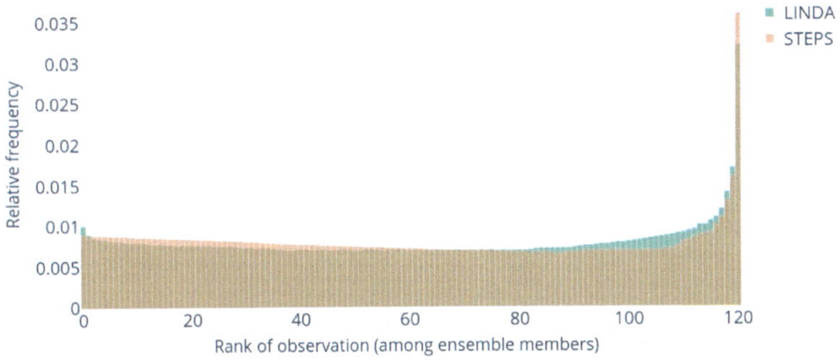

Fig. 6 Rank histograms for 30 min STEPS and LINDA-P nowcasts with a rain-rate threshold of 5 mm h^{-1}

improvements in satellite-based nowcasting in the UAE, with the potential to enhance extreme weather warning systems and provide valuable information for decision-makers and the public alike.

References

1. Douglas I, Alam K, Maghenda M, Mcdonnell Y, McLean L, Campbell J (2008) Unjust waters: climate change, flooding and the urban poor in Africa. Environ Urban 20(1):187–205

2. Smith K, Austin G (2000) Nowcasting precipitation—a proposal for a way forward. J Hydrol 239(1–4):34–45
3. Lebedev V et al (2019) Precipitation nowcasting with satellite imagery. In: Proceedings of the 25th ACM SIGKDD international conference on knowledge discovery & data mining, pp 2680–2688
4. Terry JP, Al Ruheili A, Almarzooqi MA, Almheiri RY, Alshehhi AK (2023) The rain deluge and flash floods of summer 2022 in the United Arab Emirates: causes, analysis and perspectives on flood-risk reduction. J Arid Environ 215:105013
5. Hussein K, Alkaabi K, Ghebreyesus D, Liaqat MU, Sharif HO (2020) Land use/land cover change along the Eastern Coast of the UAE and its impact on flooding risk. Geomat Nat Haz Risk 11(1):112–130
6. Booth DB, Bledsoe BP (2009) Streams and urbanization. The water environment of cities, pp 93–123
7. Douglas I, Kobold M, Lawson N, Pasche E, White I (2007) Characterisation of urban streams and urban flooding. Adv Urban Flood Manag. Taylor & Francis, New York, USA, pp 29–58. https://doi.org/10.1201/9780203945988
8. Heye A, Venkatesan K, Cain J (2017) Precipitation nowcasting: leveraging deep recurrent convolutional neural networks. In: Proceedings of the cray user group (CUG). Cray User Group Washington
9. Jensen DG, Petersen C, Rasmussen MR (2015) Assimilation of radar-based nowcast into a HIRLAM NWP model. Meteorol Appl 22(3):485–494
10. Germann U, Zawadzki I (2002) Scale-dependence of the predictability of precipitation from continental radar images. Part I: description of the methodology (in English). Mon Weather Rev 130(12):2859–2873. https://doi.org/10.1175/1520-0493(2002)130<2859:Sdotpo>2.0.Co;2
11. Germann U, Zawadzki I, Turner B (2006) Predictability of precipitation from continental radar images. Part IV: Limits to prediction. J Atmos Sci 63(8):2092–2108
12. Li L, Schmid W, Joss J (1995) Nowcasting of motion and growth of precipitation with radar over a complex orography. J Appl Meteorol Climatol 34(6):1286–1300
13. Mecklenburg S, Joss J, Schmid W (2000) Improving the nowcasting of precipitation in an Alpine region with an enhanced radar echo tracking algorithm. J Hydrol 239(1–4):46–68
14. Bowler NE, Pierce CE, Seed A (2004) Development of a precipitation nowcasting algorithm based upon optical flow techniques. J Hydrol 288(1–2):74–91
15. Cheung P, Yeung H (2012) Application of optical-flow technique to significant convection nowcast for terminal areas in Hong Kong. In: The 3rd WMO international symposium on nowcasting and very short-range forecasting (WSN12), pp 6–10
16. Son LH, Thong PH (2017) Some novel hybrid forecast methods based on picture fuzzy clustering for weather nowcasting from satellite image sequences. Appl Intell 46:1–15
17. Rivolta G, Marzano FS, Coppola E, Verdecchia M (2006) Artificial neural-network technique for precipitation nowcasting from satellite imagery. Adv Geosci 7:97–103
18. Otsuka S, Kotsuki S, Miyoshi T (2016) Nowcasting with data assimilation: a case of global satellite mapping of precipitation. Weather Forecast 31(5):1409–1416
19. Liu Y, Xi DG, Li ZL, Hong Y (2015) A new methodology for pixel-quantitative precipitation nowcasting using a pyramid Lucas Kanade optical flow approach. J Hydrol 529:354–364. https://doi.org/10.1016/j.jhydrol.2015.07.042
20. Ayzel G, Heistermann M, Winterrath T (2019) Optical flow models as an open benchmark for radar-based precipitation nowcasting (rainymotion v0.1). Geosci Model Dev 12(4):1387–1402. https://doi.org/10.5194/gmd-12-1387-2019
21. Shi X, Chen Z, Wang H, Yeung DY, Wong WK, Woo WC (2015) Convolutional LSTM network: a machine learning approach for precipitation nowcasting. Adv Neural Inf Process Syst 28
22. Tounsi A, Temimi M, Abdelkader M, Gourley JJ (2023) Assessment of deterministic and probabilistic precipitation nowcasting techniques over New York metropolitan area. Environ Model Softw 168:105803
23. Pulkkinen S, Chandrasekar V, Niemi T (2021) Lagrangian integro-difference equation model for precipitation nowcasting (in English). J Atmos Ocean Technol 38(12):2125–2145. https://doi.org/10.1175/jtech-d-21-0013.1

24. Bowler NE, Pierce CE, Seed AW (2006) STEPS: a probabilistic precipitation forecasting scheme which merges an extrapolation nowcast with downscaled NWP. Q J R Meteorol Soc 132(620):2127–2155. https://doi.org/10.1256/qj.04.100
25. Hamill TM (2001) Interpretation of rank histograms for verifying ensemble forecasts (in English). Mon Weather Rev 129(3):550–560. https://doi.org/10.1175/1520-0493(2001)129<0550:Iorhfv>2.0.Co;2
26. Bröcker J, Smith LA (2007) Increasing the reliability of reliability diagrams (in English). Weather Forecast 22:651–661. https://doi.org/10.1175/WAF993.1
27. Jolliffe IT, Stephenson DB (2003) Forecast verification: a practitioner's guide in atmospheric science. John Wiley and Sons, p 296

Intelligent Analytical and Recommendations Systems

AI-Driven Approach for Optimal Soil-Based Crop Recommendations

Suvarna Patil, Deepali Hajare, Shivganga Gavhane, Nayan Panchal, Sahil Shelke, and Aditya Nikam

Abstract As agriculture remains a primary source of income in India, the rising instances of farmer distress, often associated with crop failure and low yields, emphasize the immediate need to promote agricultural sustainability. This challenge extends its implications to both the nation's economy and the global agricultural sector, becoming a significant concern for countries worldwide. The key issue revolves around the precise prediction of crop yields under diverse climatic conditions in India, considering various attributes. The overarching objective is to equip farmers with essential tools for informed decision-making, enabling them to choose the most suitable crops and optimize fertilizer usage. In the proposed system, Machine Learning (ML) model is used to address a variety of agricultural challenges, including soil classification and crop recommendation. Addressing these challenges, a proposed model aims to categorize and forecast suitable crops based on soil nutritional levels, employing diverse ML models termed as K-Nearest Neighbors (KNN), Logistic Regression (LR), Support Vector Machine (SVM), Stochastic Gradient Boosting (SG), and XG-Boost (XGB). The research dataset encompasses seven distinct soil parameters characterized by seven nutrient attributes. Training and testing of the models are conducted using 80 and 20% of the dataset, respectively. Notably, the outcomes reveal that XG-Boost surpasses other models, achieving an accuracy of 98.64%, precision of 98.49%, recall of 98.76%, and an F-score value of 98.59%.

Keywords Soil parameters · Crop recommendation · Soil classification · Agriculture · Sustainable agricultural practices · Machine learning

S. Patil (✉) · D. Hajare · S. Gavhane · N. Panchal · S. Shelke · A. Nikam
D Y Patil Institute of Engineering, Management and Research, Akurdi, Pune, India
e-mail: suvarna.patil@dypiemr.ac.in

D. Hajare
e-mail: deepali.hajare@dypiemr.ac.in

S. Gavhane
e-mail: shivganga.gavhane@dypiemr.ac.in

S. Patil · D. Hajare · S. Gavhane
D Y Patil International University, Akurdi, Pune, India

A. J. Kulkarni and M. Hajiaghaei-Keshteli (eds.), *Advances in Intelligent Systems for Sustainable Agriculture*, Lecture Notes in Networks and Systems 1201,
https://doi.org/10.1007/978-981-97-9839-1_14

1 Introduction

A substantial segment of the Indian populace (approximately two-thirds), depends directly on agriculture for their sustenance. Consequently, agriculture has historically served as a fundamental pillar of the Indian economy, constituting 20% of the country's Gross Domestic Product (GDP). In India, farmers often referred to as the nation's "Anna data" (Food Provider), hold a central position in the agricultural sector, currently facing several challenges:

1. Farmers face challenges in choosing the most appropriate and economically viable crop for their soil, specific circumstances, and geographical location, primarily due to the diverse soil types prevalent across the country. This variability frequently results in financial losses for farmers.
2. Farmers struggle to predict their harvest and potential profit during a specific planting season due to unpredictable weather patterns.
3. The "farm to market" system, involving numerous intermediaries in the transportation and sale of crops, results in farmers receiving minimal returns for their labor.

The success of agriculture relies significantly on soil, as good soil structure supports the health of both the soil and crops. A soil structure facilitates the free circulation of water and air throughout its profile, which is vital for the growth of crops in agriculture. Moreover, soil serves as a reservoir for water crucial for plant development and facilitates the movement of machinery and animals in farming operations. The various components necessary for plant growth are predominantly acquired through root uptake from the soil [1].

Agriculture has been a fundamental cultural practice since ancient times, and the significance of soil in this context cannot be overstated. Therefore, understanding the specific soil type suitable for agricultural cultivation is vital for achieving optimal crop yields. Various methods, including traditional approaches, expert knowledge, and advanced technology, can be employed to identify soil types [2]. Understanding soil classification is valuable for predicting soil behavior, which, in turn, helps forecast how well a particular soil can support the growth of crops. Soil classification, as a method, serves to determine the specific type of soil [3].

Modern agriculture heavily relies on AI and ML, offering solutions such as crop recommendation systems, precision farming methods, and fertilizer prediction techniques. Implementing AI systems has the potential to provide a much-needed boost to the struggling agriculture industry. Machine learning, as an emerging agricultural technology, can elevate crop quality and yield by recognizing patterns in agricultural data and prioritizing crucial information. This entails the collection, processing, and training of data before conducting tests on samples.

To forecast soil and crop results using historical patterns and soil types, algorithms like Support Vector Machines (SVM) can be utilized. Essential datasets for these projects include a soil dataset with various chemical attributes and a crop dataset with geographical information. These datasets serve as foundational elements for implementing machine learning approaches in agriculture. In the proposed system, five algorithms were selected for experimentation: KNN, SVM, LR, SG, and XGB, to identify the most effective models.

1.1 Contribution of the Proposed System

The primary contributions of the proposed system are outlined as follows:

- The system proposes an innovative method for soil classification by leveraging machine learning algorithms, addressing the constraints associated with conventional soil classification approaches.
- Evaluate the efficacy of machine learning algorithms, including KNN, SVM, LR, Stochastic Gradient, and XG-boost, in the context of soil classification. It offers insights into the effectiveness of these algorithms in recommending appropriate crops for distinct soil types.
- The system outlines a methodological approach for gathering, analyzing, and applying soil data to train and assess machine learning models. This contributes to the advancement of data-driven methodologies in the field of agriculture.
- The research underscores the significance of soil classification in agricultural decision-making and illustrates how machine learning can aid farmers in making informed decisions regarding crop selection and overall farm management.

1.2 Organization of the Paper

The structure of the remaining sections is illustrated in this section. Section 2 conducts a thorough exploration of relevant literature on crop prediction, offering essential research context and background. The article outlines the framework of the proposed system, presenting the methodology and design of our research in Sect. 3. Section 4 provides insights into the collected data for the experiments, the experimental setup, and the outcomes, essential for validating our findings and conclusions in Sect. 5.

2 Related Work

In their paper, Raj et al. [4] underscore the need for a standardized language among scientists engaged in diverse land areas, especially when categorizing various soil types. The authors employ machine learning and established algorithms in their research to foster effective communication within the scientific community. Given the diverse soils worldwide, each with distinct characteristics, the suitability for growing different crops varies. To classify soils, the authors utilize Neural Networks, Decision Trees, Naïve Bayes, and SVM approaches for comparative analysis.

In a related study by Mohammad Diqi et al. [5], the focus centers on classifying ebony and red soil types. Here, the authors employ Convolutional Neural Networks (CNN) and deep learning algorithms, achieving a commendable success rate of 97% with a minimal loss of 0.1606 in the classification process.

Soil categorization techniques can be approached in two primary ways. The first method involves using image processing and computer vision to classify soil based on parameters such as texture, color, and particle size, employing traditional image processing algorithms. The second category employs advanced techniques like Convolutional Neural Networks (CNN) within deep learning and machine learning, delivering cutting-edge outcomes. The utilization of deep learning applications streamlines the process, reducing dependence on spatial-form designs and pre-processing methods [6].

In their research, Shravani et al. [7] aim to develop a model that accurately classifies soil instances and links soil types with crop data for improved predictions. This model considers both crop classifications and regional characteristics in soil prediction. It involves analyzing real-time soil data in two phases: training and testing. The researchers utilize two datasets—one for soil and another for crops and generate a list of suitable classes by comparing expected and actual classes. In their work, Rahman et al. [8] present a model for predicting soil series based on land type and suggesting suitable crops. They utilize several machine learning techniques, such as weighted K-Nearest Neighbor (K-NN), bagged trees, and Support Vector Machines (SVM) with a Gaussian kernel, for soil classification. The SVM-based approach they propose shows superior performance over various existing methods, as demonstrated by experimental data.

Soil, defined as the loose surface material covering a significant portion of the Earth [9], comprises organic matter and inorganic particles essential for sustaining life. It plays a critical role in providing structural support for agricultural plants and serves as a vital growth medium [10]. Additionally, soil acts as the main source of water and nutrients for plants. The diverse physical and chemical properties of soil are the result of processes such as weathering, leaching, and microbial activity, leading to the formation of various soil types. Each soil type has distinct characteristics that offer both advantages and limitations for agricultural production [11]. Identifying the specific soil type required for a project is crucial for promoting the healthy growth of plant life.

Soil classification involves categorizing soil into distinct groups or classes, each with potentially similar behaviour and characteristics that can be geo-referenced and mapped [12]. This classification is based on measuring and describing various characteristics found in representative soil profiles, providing insights into soil formation processes [13]. These characteristics can be used individually or collectively to create soil datasets or maps tailored to classify soils based on specific end-user requirements. Figure 1 illustrates the scale of different soil parameters, including Nitrogen (N), Phosphorus (P), Potassium (K), Temperature, Humidity, pH value of the soil, and Rainfall (mm), aiding in understanding the crop's requirements.

Understanding the fertility, moisture content, and overall suitability of soil for various crops hinges on these parameters. The research leverages these key factors to

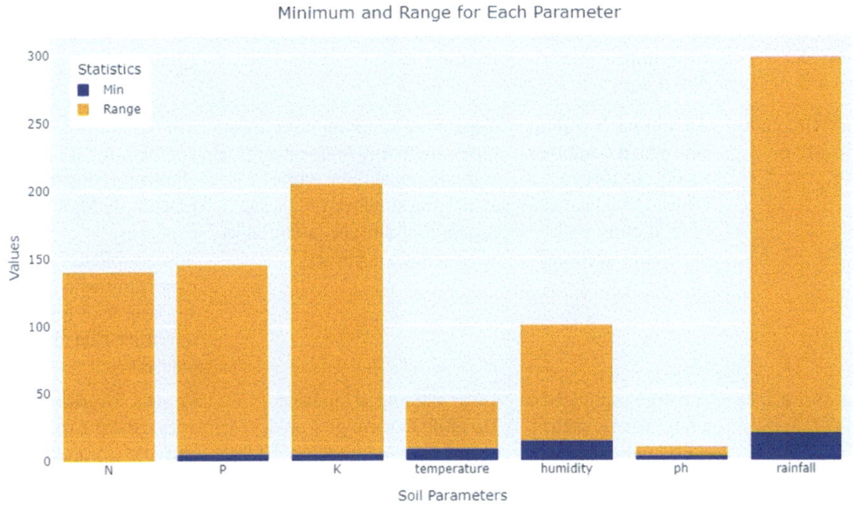

Fig. 1 Soil parameters

train machine learning algorithms, ensuring precise soil classification and subsequent accurate crop recommendations.

ML, a branch of AI and computer science (CS), is dedicated to leveraging data and algorithms to replicate the learning process observed in humans, gradually refining its accuracy [10]. Machine learning algorithms, which assess traditional algorithms, are broadly categorized as supervised, unsupervised, or reinforcement learning [14]. Among the various ML models are LR, Random Forest, Decision Tree, Artificial Neural Networks, Naïve Bayes, and more. Table 1 elaborates on the systematic analysis of ML models employed for various applications.

Supervised machine learning models, particularly in classification, involve the algorithm learning from given data inputs and then using this knowledge to classify new observations [15]. This method is used to determine the recognition of specific data based on a set of sample data with pre-assigned classes. The process of building a classifier consists of two stages. The training phase is critical, as it determines how parameters should be adjusted and consolidates various types of data into a unified format. Moving on to the testing phase, this set is applied to test data with known targets, and the results are then compared with the selected data from the training set. During testing, additional information is collected to evaluate the time required to produce results for each data point and assess the accuracy of data interpretation, determining whether it achieves a high level of accuracy. In Table 1, the following algorithms are discussed as the classification models which are used to predict the category or class to which a given input belongs.

Table 1 Systematic analysis of the ML models

Models	Algorithms description	Features	Limitations
GB-Gradient Boosting	An approach within ML which combines models to form a robust model capable of predicting with maximum accuracy	Capable of handling diverse data types, including both categorical and numerical data, while exhibiting superior predictive capabilities and the potential to achieve high accuracy	This can pose computational challenges due to potentially high expenses. Susceptible to overfitting when working with noisy or outlier-ridden datasets
Decision Tree (DT)	A model organized as a tree, where each node signifies a feature, and each branch represents a decision based on that feature	Clear and understandable. Versatile in managing both categorical and numerical data. Demonstrates competence in addressing missing values and outliers	Prone to overfitting with data and may struggle to capture complex relationships between features
Random Forest (RF)	An ensemble method that reduces overfitting and boosts overall accuracy by generating several decision trees and aggregating their predictions	Capable of managing high-dimensional data containing multiple features. Exhibits reduced susceptibility to overfitting compared to Decision Tree. Versatile in handling both numerical and categorical data	Can involve significant computational expenses. Performance may be compromised when confronted with imbalanced data
Gaussian Naïve Bayes (GNV)	A probabilistic model based on Bayes' theorem, assuming independence among features	Efficient and quick algorithm. Proficient in managing high-dimensional data. Capable of processing both categorical and numerical data	Relies on the assumption of feature independence, which may not align with real-world datasets Performance may be subpar when dealing with data containing correlated features
Multimodal Naïve Bayes (MNV)	An enhanced version of Gaussian Naïve Bayes is specifically designed to handle data characterized by multiple modes and clusters	This algorithm is efficient, capable of managing data with multiple clusters. It exhibits versatility in processing both categorical and numerical data	May exhibit suboptimal performance when dealing with data containing correlated features

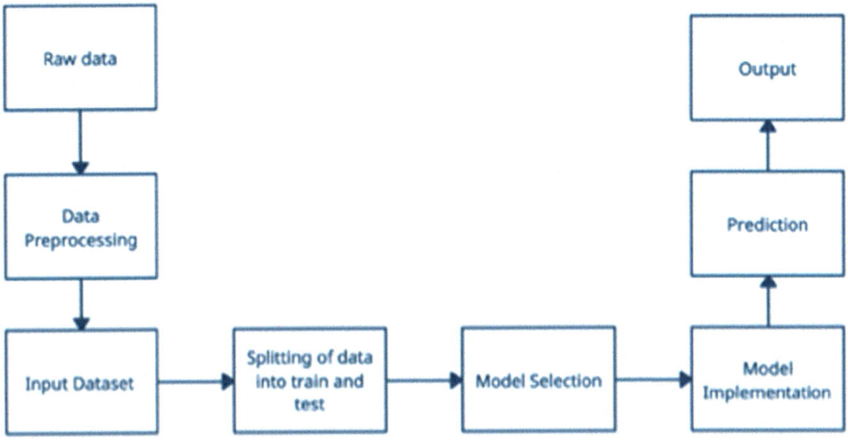

Fig. 2 Crop recommendation model

3 System Methodology

3.1 ML Model Framework

Figure 2 depicts the model framework, initiating with phases such as (1) Data Selection through Dataset, (2) Data pre-processing, (3) Data transformation using ML models, and (4) Feature selection. Following these steps, the selected data undergoes training and testing using classification algorithms.

3.2 System Architecture

The system architecture involves collecting soil data, pre-processing it, training and testing machine learning algorithms, evaluating their performance, and recommending the best crop for the specific soil type. System models can be implemented using software tools and technologies, such as Python. Figure 2 shows the system architecture for the proposed model.

3.2.1 Data Selection

The study incorporates two types of data, both of which will be merged. The dataset is the Crop prediction dataset obtained from the Kaggle website. The dataset was constructed by integrating information from soil-based crop datasets to categorize soil for 22 different crops. With a total of 22 crops, this dataset comprises 2200 instances, each with seven data fields and a single class label. The data fields include the soil nutrient ratios of Nitrogen (N), phosphorus (P), and Potassium (K), followed by temperature in degrees Celsius, humidity in percentage, and the soil's pH value.

The data is presented in numerical form as continuous values, making it suitable for classification methods to assess accuracy in each classification approach.

3.2.2 Data Pre-processing

In the data pre-processing phase, various techniques are applied to the crop prediction dataset before employing the chosen algorithms, including KNN, SVM, Naïve Bayes, and Random Forest. Common data processing methods like data cleaning [16], normalization, and reduction will be utilized. Data cleaning is usually employed to address any missing values in the crop recommendation dataset. However, the crop recommendation dataset showed no missing values, eliminating the necessity for data clean up.

Conversely, the real-world data from data.gov.in requires the pre-processing phase to enhance the dataset by addressing missing values. This step aims to improve the data quality by reducing the occurrence of missing values in the real-world dataset.

On the contrary, data reduction is a method wherein a sizable dataset is streamlined to expedite the results acquisition process. Ultimately, using fewer datasets tends to yield more accurate and precise data [17].

The identification of the particular parameter for each algorithmic technique applied during data training is termed parameter selection. The selection of each strategy depends on understanding how the manipulation of a particular parameter affects its performance. For k-NN, the parameter is the value of "K," as the selected parameter for tuning in each technique to enhance accuracy.

3.2.3 Parameter Selection

The paper investigates the essential process of identifying optimal values for the parameters used in machine learning models. This step is crucial, as the performance of machine learning algorithms often depends on the precise configuration of their parameters. The section likely provides a detailed exploration of how specific parameters for each algorithmic technique undergo examination and selection to enhance the accuracy and effectiveness of the model. Parameter selection involves a systematic experiment with various parameter values, ultimately choosing the combination that produces the most favorable outcomes in terms of model performance [18]. This meticulous process ensures that the machine learning model is finely tuned to achieve optimal predictive capabilities. Table 2 illustrates the evaluation parameters used to evaluate the performance of machine learning models.

The process of designing the crop recommendation system is outlined in a concise step-by-step procedure as follows:

Table 2 Systematic analysis of the ML models

Parameters	Description
Accuracy	Accuracy is a straightforward metric that gauges the frequency with which the classifier makes correct predictions $\text{Accuracy} = \frac{TP+TN}{TP+TN+FP+FN}$
Precision	The precision of a prediction indicates the percentage of accurately predicted positive cases among the total number of depicted positive cases. It becomes particularly relevant in situations where a higher emphasis is placed on minimizing False Positives compared to False Negatives $\text{Precision} = \frac{TP}{TP+FP}$
Recall	It represents the number of actual positive cases correctly predicted by our model. Recall is crucial in situations where reducing False Negatives is more important than reducing False Positives $\text{Recall} = \frac{TP}{TP+FN}$
F-score	It provides a comprehensive understanding of both Precision and Recall metrics. Its highest value occurs when Precision equals Recall $\text{F1-Score} = \frac{\text{Precision} \times \text{Recall}}{\text{Precision} + \text{Recall}}$

1. The soil dataset must be pre-processed before being integrated into a comma-separated values file.
2. Implement pre-processing techniques on the input dataset, including tasks such as handling missing values, encoding categorical data, and scaling values within the appropriate range.
3. Utilize the specified split ratio to partition the pre-processed dataset into training and testing datasets. The recommended method involves an 80:20 split ratio, where 80% of the dataset is used for training the ensemble model, and the remaining 20% is reserved for the test dataset.
4. Develop separate classifiers using the training dataset. Each unique base learner receives the training dataset and generates a classifier.
5. Evaluate the data by using the testing dataset with each classifier, obtaining specific class labels for analysis.
6. Utilize the Majority Voting Technique to aggregate the outputs from individual classifiers, resulting in the outcome of soil classification and crop recommendation [19].

4 Result and Discussion

This research proposes a methodology that uses machine learning algorithms to predict and classify crops according to soil nutrient levels, thereby helping farmers choose the best crops based on soil type. Six crop types and eleven distinct nutrient types are included in the dataset. An 80–20 pattern was used for training and testing the models. Models used to predict the crops in our system are discussed below.

Support Vector Machine (SVM): SVM method is an effective tool for tasks involving classification. It operates by locating the hyperplane in the feature space that best separates various classes. When there are more dimensions than samples, SVM performs well in managing high-dimensional data.

Decision Tree: Based on the most informative features, decision trees are simple yet powerful models that recursively separate the data into subsets. Their ability to handle both numerical and category data makes them straightforward to interpret.

K-Nearest Neighbors (KNN): KNN is an easy-to-understand technique that uses the majority class of its K nearest neighbors to classify fresh data points. It is a non-parametric method, meaning it does not make assumptions about the data distribution.

Naive Bayes: Based on the Bayes theorem, the Naive Bayes classifier uses probability. It operates under the simplifying the assumption of feature independence, which is often true in actual use. Naive Bayes is computationally efficient and effective for high-dimensional data.

Extreme Gradient Boosting (XG-Boost): XG-Boost is an ensemble learning method that predicts using a group of weak learners (decision trees). This model is known for its speed and performance and is commonly used in machine learning competitions.

Random Forest: Another ensemble learning system, Random Forest, generates predictions by assembling a set of decision trees. Each tree is trained using a different portion of the data, and predictions are subsequently merged to generate the final prediction. Random Forest works well with high-dimensional data and is robust to overfitting.

XG-Boost and Random Forest are often preferred for their ability to handle complex relationships and achieve high accuracy. However, in our system, we have tested different models to achieve maximum accuracy and tuned their parameters to find the best-performing model for a particular dataset [20].

Table 3 shows the comparison of different machine learning models used to predict crops according to the nutritional status of the soil.

Figures 3, 4, 5 and 6 illustrate the learning curves for accuracy (A), precision (P), recall (R), and F1-score (F1) based on an 80% training set and 20% testing set. The evaluation indicates that XG-Boost outperforms other machine learning models, achieving an accuracy of 98.64% (Fig. 7).

Table 3 ML models analysis using evaluation parameters

Parameters/algorithms	KNN	SVM	LR	SG	XG-Boost
Accuracy	96.82	96.14	94.55	96.59	98.64
Precision	96.74	96.33	94.39	96.57	98.49
Recall	97.04	96.29	94.63	96.97	98.76
F-score	96.64	95.99	94.40	96.70	98.59

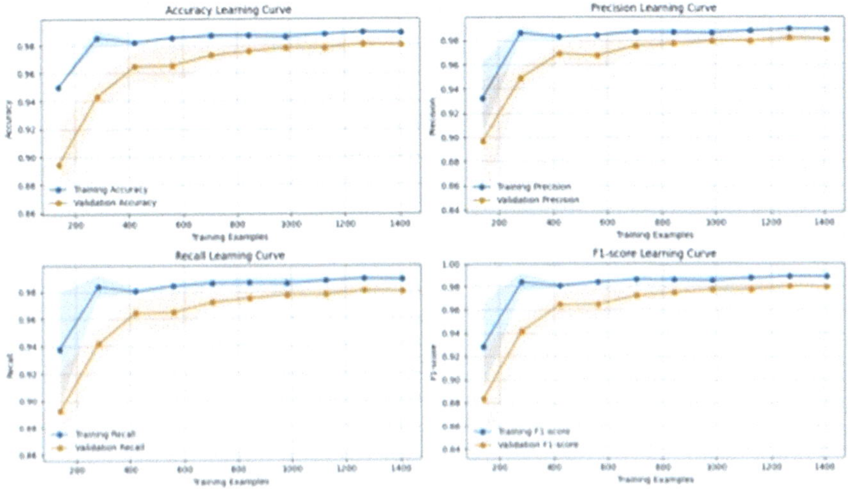

Fig. 3 KNN evaluation metrics

5 Conclusion

A predictive model is implemented for predicting optimal crop yield recommendations for specific soil types. This ML model's effectiveness is evaluated using KNN, Logistic Regression, SVM, Stochastic Gradient, and XG-Boost algorithms, to achieve the desired level of accuracy. The model's ultimate goal is to empower farmers with cutting-edge technologies, enabling them to cultivate multiple crops simultaneously, considering the variations in soil types across different fields. The model outlines specific parameters essential for cultivating particular crops on distinct soil types. The algorithm exhibits the highest accuracy, establishing it as the core algorithm for this model.

To summarize, this paper significantly contributes to the agricultural sector by introducing a data-driven methodology for soil classification and crop recommendations. The proposed approach holds promise for improving agricultural practices and maximizing crop yields. This technology has the potential to revolutionize agricultural practices, assisting farmers to make optimal decisions for crop selection and management. The authors suggest that future research should focus on enhancing

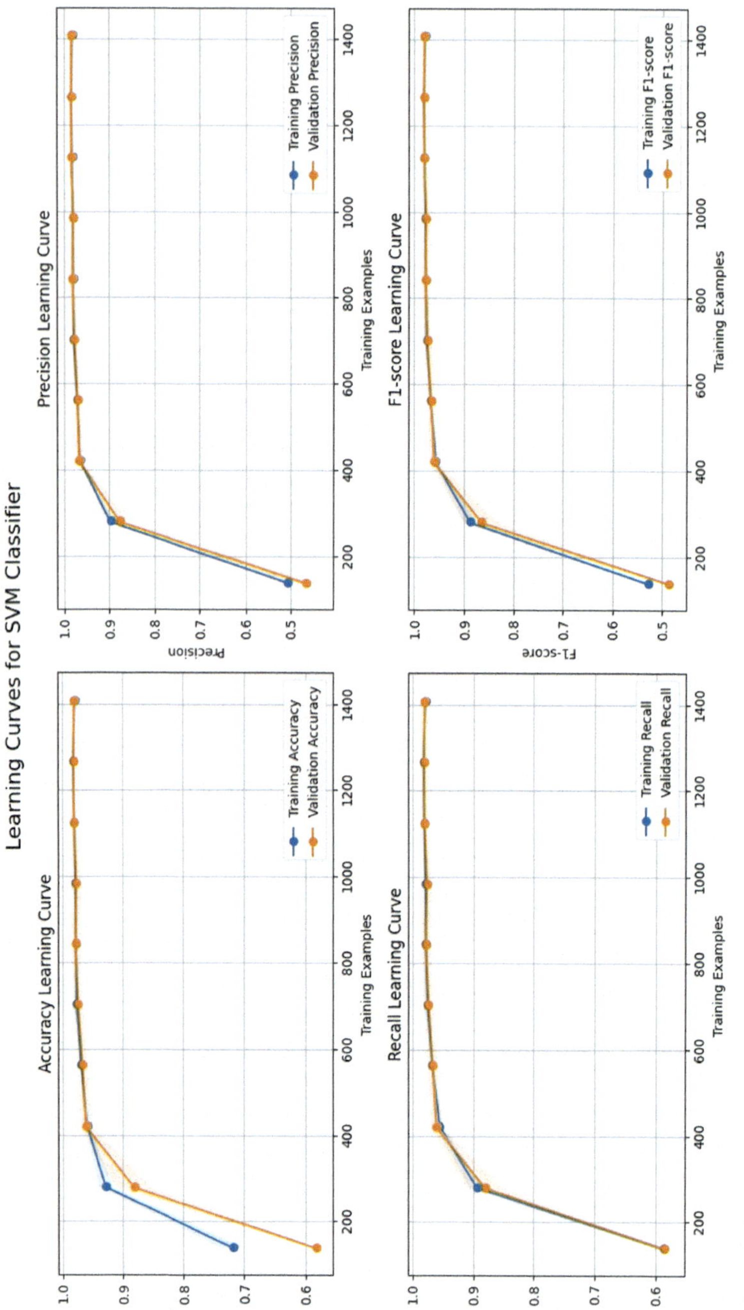

Fig. 4 SVM evaluation metrics

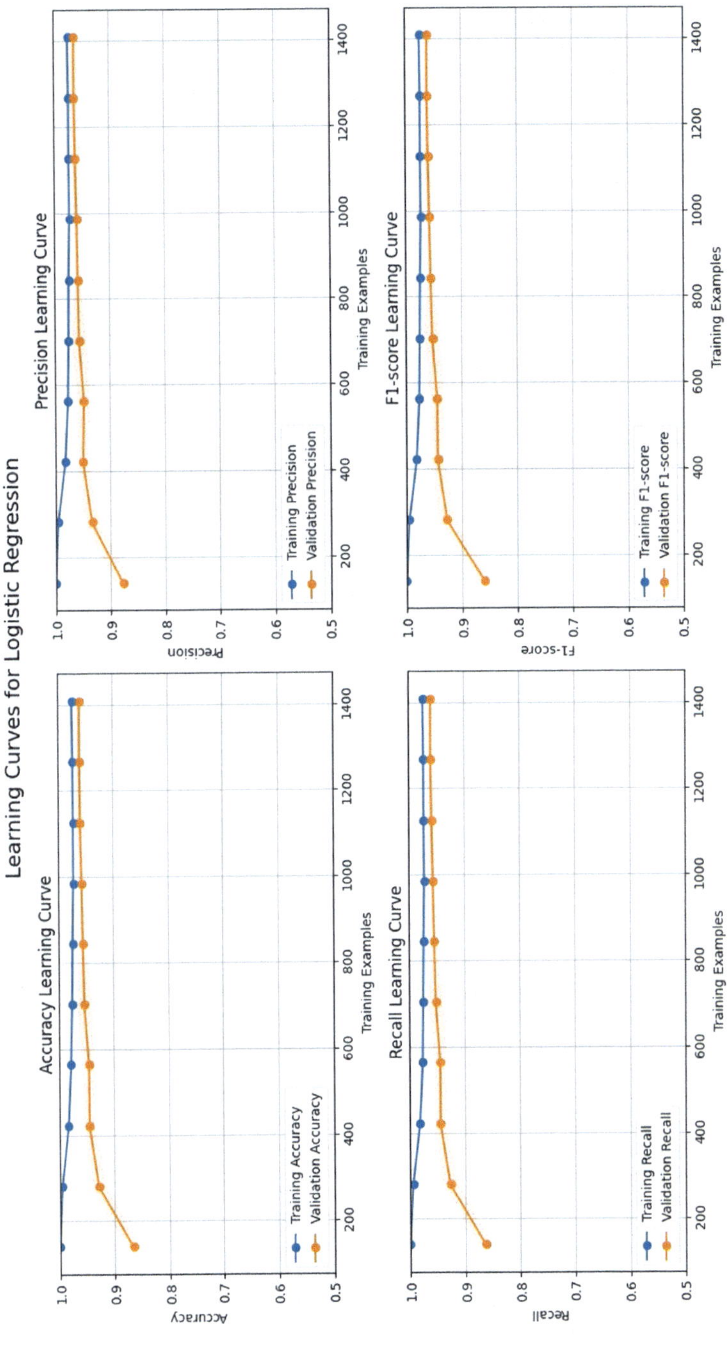

Fig. 5 Logistic regression evaluation metrics

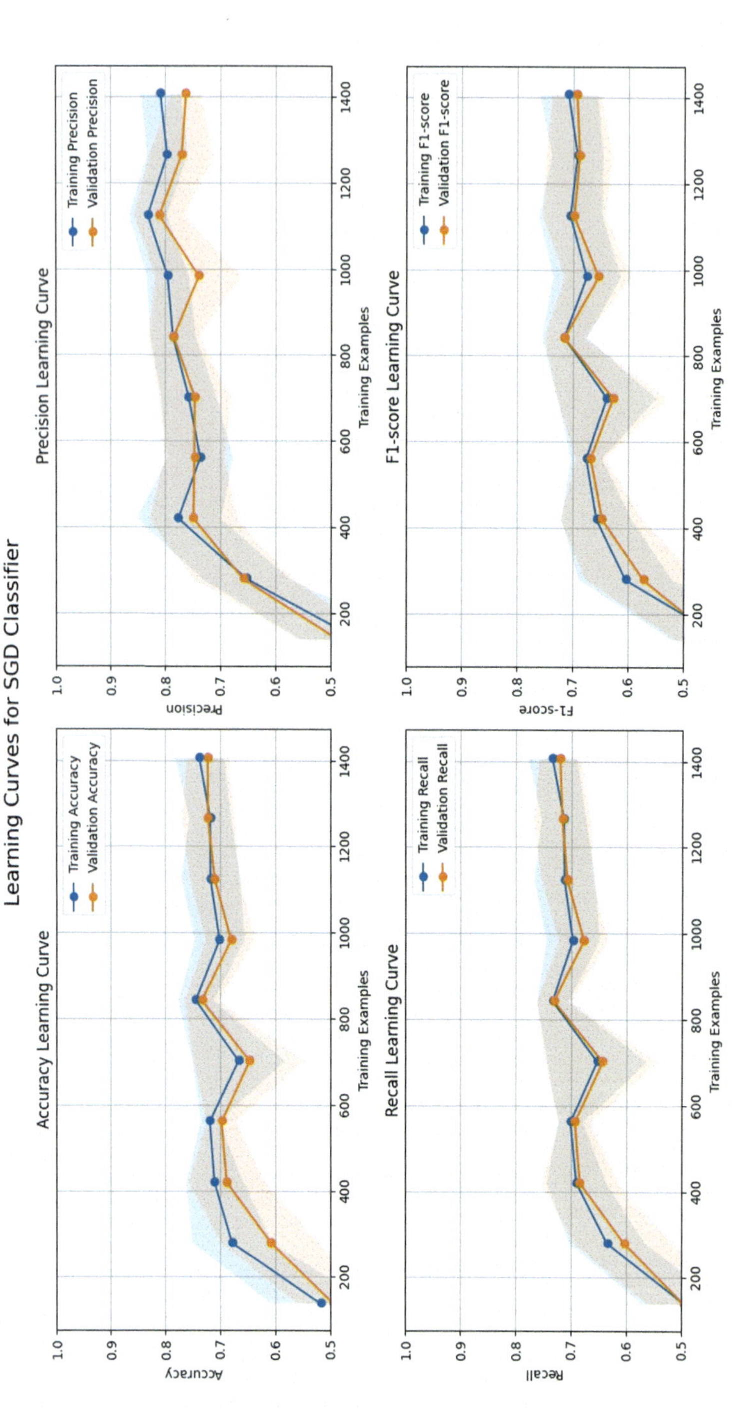

Fig. 6 Stochastic gradient evaluation metrics

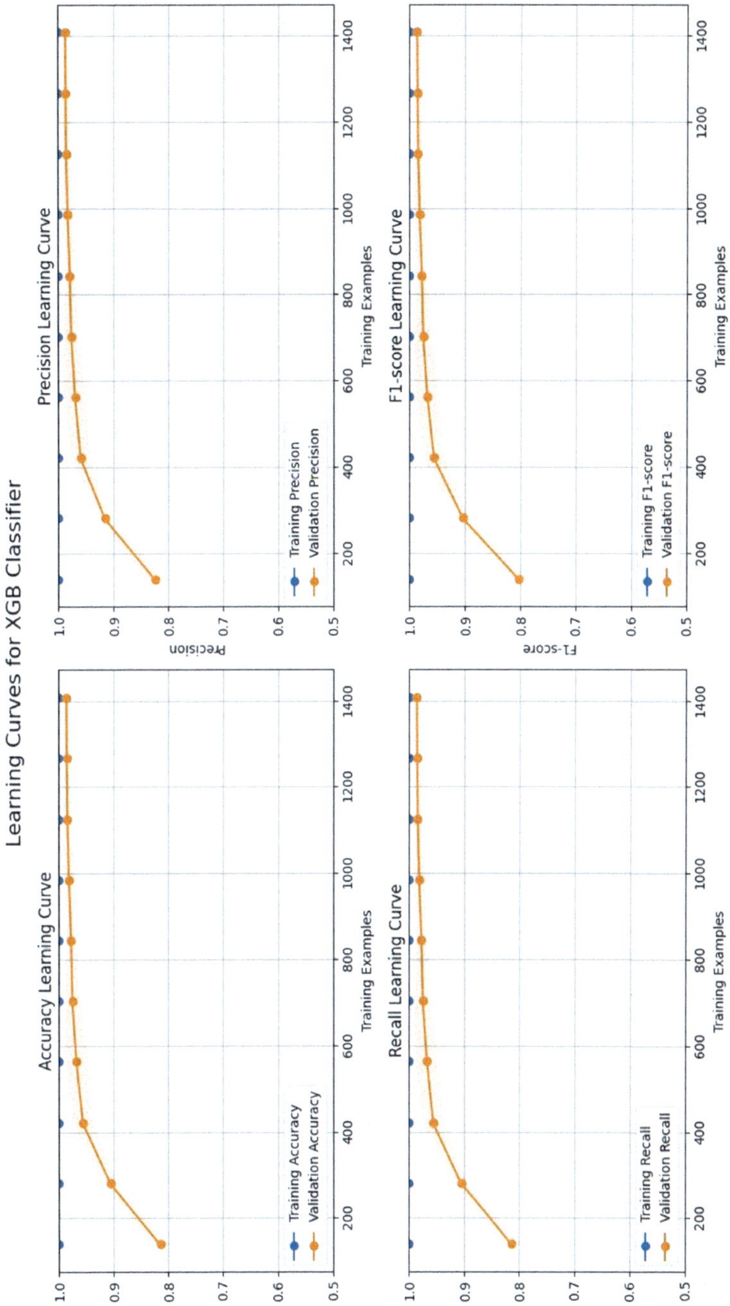

Fig. 7 XG-Boost evaluation metrics

the accuracy and efficiency of machine learning algorithms for soil classification and fertilizer recommendations based on crop disease detection using Convolutional Neural Network as well as exploring broader applications of this technology in agriculture.

References

1. Brady NC, Weil RR (2010) Elements of the nature and properties of soils. Pearson Prentice Hall
2. Lombardozzi DL, Lu Y, Lawrence PJ, Lawrence DM, Swenson S, Oleson KW, Wieder WR, Ainsworth EA (2020) Simulating agriculture in the community land model version 5. J Geophys Res Biogeosci 125(8). https://doi.org/10.1029/2019jg005529
3. Atkinson J, Davison L, Springman S (2000) Soil description and classification. Soil Classif. https://environment.uwe.ac.uk/geocal/SoilMech/classification/default.htm
4. Raj GS, Lijin Das S (2020) Survey on soil classification using different techniques. Int Res J Eng Technol (IRJET) 07(03), p-ISSN: 2395-0072
5. Hamzah MD, Ronaldo AD (2021) Effective soil type classification using convolutional neural network 3(1) (2021). ISSN: 2685-8711, E-ISSN: 27145263 https://doi.org/10.35842/ijicom
6. Tava PS, Shukla A, Bansa A A comprehensive review on soil classification using deep learning and computer vision. https://doi.org/10.1007/s11042-021-10544-5
7. Shravani V, Uday Kiran, Yashaswini J, Priyanka D (2020) Soil classification and crop suggestion using machine learning. Int Res J Eng Technol (IRJET) 07(06), p-ISSN: 2395-0072
8. Rahman SAZ, Chandra Mitra K, Mohidul Islam SM (2018) Soil classification using machine learning methods and crop suggestion based on soil series. In: 2018 21st international conference of computer and information technology (ICCIT), pp 1–4. https://doi.org/10.1109/ICCITECHN.2018.8631943
9. Bronick CJ, Lal R (2005) Soil structure and management: a review. Geoderma 124(1–2):3–22. https://doi.org/10.1016/j.geoderma.2004.03.005
10. Lin BB (2011) Resilience in agriculture through crop diversification: adaptive management for environmental change. Bioscience 61(3):183–193. https://doi.org/10.1525/bio.2011.61.3.4
11. Mitchell J, Soga K (2005) Fundamentals of soil behavior, 3rd edn. Wiley, Hoboken, NJ
12. Food and Agriculture Organization of the United Nations (2020) Soil classification. Soil classification | FAO SOILS PORTAL | Food and Agriculture Organization of the United Nations. Retrieved from http://www.fao.org/soils-portal/data-hub/soilclassification/en/
13. Butler BE, Butler B Soil classification for soil survey. Clezedon Press, Oxford
14. Prabhakaran S (2018) How naive Bayes algorithm work? (with example and full code): ML+ Machine Learning Plus. https://www.machinelearningplus.com/predictive-modeling/how-naive-bayes-algorithm-workswith-example-and-full-code/
15. Zhang Z (2016) Introduction to machine learning: K-Nearest Neighbors. Ann Transl Med 4(11):218–218. https://doi.org/10.21037/atm.2016.03.37
16. Bandara P, Weerasooriya T, Ruchirawya T, Nanayakkara W, Dimantha M, Pabasara M (2020) Crop recommendation system. Int J Comput Appl 975:8887

17. Pruthviraj, Akshatha GC, Shastry KA, Nagaraj, Nikhil (2022) Crop and fertilizer recommendation system based on soil classification. In: Recent advances in artificial intelligence and data engineering: select proceedings of AIDE 2020. Springer Singapore, pp 29–40
18. Pudumalar S, Ramanujam E, Rajashree RH, Kavya C, Kiruthika T, Nisha J (2017) Crop recommendation system for precision agriculture. In: 2016 eighth international conference on advanced computing (ICoAC). IEEE, pp 32–36
19. Gosai D, Raval C, Nayak R, Jayswal H, Patel A (2021) Crop recommendation system using machine learning. Int J Sci Res Comput Sci Eng Inf Technol 7(3):558–569
20. Motwani A, Patil P, Nagaria V, Verma S, Ghane S (2022) Soil analysis and crop recommendation using machine learning. In: 2022 international conference for advancement in technology (ICONAT). IEEE, pp 1–7

Employing Machine Learning Methods and RFM Model for Customer Clustering: Case Study of an Agricultural Retailer

Sara Sohrabi, Amirhossein Nosrati Malekjahan, and Ali Husseinzadeh Kashan

Abstract Customer segmentation is a tool to communicate effectively with customers. Companies cannot satisfy all potential buyers of their products or services. They should choose a group of customers that they can serve better than their competitors. To do this, customers must be segmented. Analyzing customer perceptions to distinguish between diverse customer groups and devising tailored strategies for each group poses a significant challenge for customer-centric organizations. Utilizing machine learning techniques for customer segmentation serves as a method to assess and categorize customers. In addition, customer clustering is crucial for any enterprise, especially agricultural goods retailers. In this article, we leverage transactional records from an online retail company, extracting pertinent features to construct RFM model indicators including Recency, Frequency, and Monetary from sales invoices. Following data preparation and processing, the optimal number of customer clusters is determined through both hierarchical methods and the k-means algorithm and also the silhouette score for model evaluation. Performance evaluation reveals a slightly better preference for the k-means algorithm. Finally, a correlation analysis between RFM indicators identified the optimal customer group, referred to as the "golden customers."

Keywords Customer clustering · RFM indicators · K-means algorithm · Machine learning

S. Sohrabi · A. Nosrati Malekjahan · A. Husseinzadeh Kashan (✉)
Faculty of Industrial and Systems Engineering, Tarbiat Modares University, Tehran, Iran
e-mail: A.Kashan@modares.ac.ir

A. J. Kulkarni and M. Hajiaghaei-Keshteli (eds.), *Advances in Intelligent Systems for Sustainable Agriculture*, Lecture Notes in Networks and Systems 1201,
https://doi.org/10.1007/978-981-97-9839-1_15

1 Introduction

In recent decades, the agricultural industry has faced significant changes. With the development of technology, economic shifts, and sustainable development, there is a need for intelligent marketing strategies for agricultural products. One key tool in this area is customer segmentation. Customer segmentation for agricultural products is a fundamental topic in marketing and sales that enables the optimization of strategies and enhances business efficiency [1]. Through precise analysis of customer characteristics and alignment with market needs, agricultural businesses can improve performance and attract new customers.

Segmentation is the process of dividing customers into smaller groups (clusters) based on common characteristics. In the field of agricultural products, customer segmentation can be related to farmers, producers, or specific markets for agricultural products [2]. Customer segmentation, by increasing accuracy in understanding target markets, facilitates the identification of customer needs and enhances customer satisfaction with the purchasing experience. By understanding the characteristics of customers in each segment, businesses can optimize their marketing strategies.

Recently, data mining and machine learning have strengthened customer segmentation and behavior prediction. Most clustering methods are based on recent purchase methods, purchase frequency, and purchase amount (RFM) [3–5]. In these methods, each customer is categorized into a group based on the predominant behavior in historical transaction data. For predicting behavior, Supervised Machine Learning (SL) methods such as Decision Trees [6], Regression [7], and Neural Networks [8] have been employed. In some cases, a combination of vari-ous data mining and machine learning methods has been utilized [9]. The RFM model, popular for its simplicity and ease of use by business managers, is a go-to method for calculating customer lifetime value.

First introduced by Hughes [10], the RFM model is a popular tool in customer value analysis [11], aiding in both lifetime value measurement and customer segmentation for behavior analysis [12]. RFM stands for Recency, Frequency, and Monetary. The RFM scores customers based on Recency (when they last bought) and Frequency (how often they buy). Higher scores indicate potentially valuable customers. The RFM model also considers the total amount a customer spends (Monetary) during the period [10]. The RFM model allows for identifying the highest probability of responding to a new offer among current customers.

Customer segmentation is a tool for establishing effective communication with customers [13]. Limited resources force companies to target specific customer segments. Segmentation helps identify the most profitable groups for tailored marketing strategies [14]. In the real world, most segmentations are done using various RFM methods. In [9], segmentation is introduced using the RFM/P method, where, in addition to the main variables R, F, and M, the product type (P) is al-so used for segmentation. In customer segmentation, differentiation is based on common characteristics and features to accurately analyze the behavior of each group and provide a plan to retain customers based on the characteristics of each group.

Market segmentation often relies on techniques like Artificial Neural Networks (ANNs) and K-means clustering. The K-means algorithm is simple and efficient, making it a popular choice for various applications. It is favored for its straightforward understanding, interpretation, and application [15]. For small values of k, K-means is considered a linear algorithm, and hierarchical clustering algorithms provide better-quality clusters compared to K-means. Studies suggest Hierarchical Agglomerative Clustering (HAC) may create better customer segments than K-means, potentially leading to more effective marketing strategies [16, 17].

K-means clustering is a popular tool for customer segmentation, enabling businesses to identify valuable customers and develop targeted marketing strategies [18, 19]. Studies like Cheng et al. [11] further demonstrate its effectiveness in customer value analysis. This study utilized the Silhouette method [20] to deter-mine the optimal number of clusters. vEmploying the RFM model and k-means clustering, industrial equipment manufacturing company customers were catego-rized into six segments based on RFM indicators. Segment-specific customer characteristics were analyzed through CLV assessment. The study concludes by recommending tailored promotion strategies for each customer segment [21].

Retailers can boost sales through customer segmentation [22]. This article explores two methods: by customer lifespan and transaction volume, and by pur-chase history. A case study has been conducted on the segmentation and profiling of customers based on their lifetime value using the RFM model for customer da-ta collected from a gas station in Istanbul, Turkey. Recommendations have been provided based on the valuable customer profiles categorized into two segments, VIP and Gold [11].

In today's fiercely competitive market, where marketing companies and retail stores are at their peak in attracting new customers and retaining existing ones, each company strives to consider a customer segmentation approach and rec-ommend suitable products for each customer to gain a competitive edge. In [16], a method for recommending products to customers is presented, utilizing the K-means clustering algorithm and the RFM model based on customers' purchase histories. This approach involves customer segmentation and then proceeds to product recommendation. Additionally, real-world data from an e-commerce company validated the system's performance. Past research extensively explored customer clustering using data mining and machine learning tools. Nowadays, ex-tracting valuable information from data poses a fundamental challenge in the modern business environment. This challenge compels industry managers to em-ploy data mining techniques to extract hidden and unknown information and patterns from abundant customer data, using them in customer relationship management.

The use of information technology, particularly the establishment of customer relationship management information systems and the application of data mining methods for systematic analysis of customer profiles in service and manufacturing industries, is recognized as a necessity. These initiatives empower companies to categorize customers into different clusters and intelligently implement marketing strategies for each customer group. Accurate information about customer priorities and preferences aids companies in selecting more effective marketing strategies, contributing to their success in the marketing domain. Based on this foundation, in this research, clustering techniques have been employed to identify different customer groups within an online retailer specializing in agricultural products. The examined data consists of transaction datasets covering 2 years, including all transactions recorded between December 1, 2021 and December 9, 2022. In the rest of this paper, in Sect. 2, the utilized data, methodology, and methods are introduced; Sect. 3 presents the obtained results; and finally Sect. 4 illustrates the conclusions.

2 Methods and Materials

2.1 *K-means Algorithm*

Due to its widespread application in the field of data analysis, k-means has gained significant popularity as a clustering approach [23]. The fundamental principle of the algorithm involves dividing the dataset into K clusters, each characterized by a centroid (or cluster center) [23]. This partitioning technique aims to elucidate the inherent structure and patterns within the data by grouping similar data points into clusters based on their proximity and similarity.

K-means exhibits commendable performance across various scenarios, particularly in identifying globular clusters [24]. For instance, it excels in categorizing consumers who tend to form clusters based on similar purchase preferences, especially in market segmentation contexts. By specifying the desired number of clusters (k), k-means efficiently organizes customers into distinct groups, enabling organizations to gain insights into the characteristics and needs of different client groups. It is noteworthy that the versatility of the k-means algorithm extends beyond consumer behavior analysis, encompassing diverse applications in data-driven decision-making processes. Its ability to discern patterns and group similar entities together renders it a valuable tool in enhancing organizational understanding and facilitating targeted strategies based on identified clusters. Consequently, k-means remains a robust and widely utilized method in data analysis and clustering applications.

2.2 *Hierarchical Clustering*

Hierarchical clustering stands out as a potent clustering technique due to its unique ability to create a hierarchy of clusters, eliminating the need to predefine the number of clusters. This distinctive feature allows for the grouping of data at various fine-grained levels, providing enhanced flexibility for in-depth data analysis. The technique excels in accommodating clusters of varying sizes and shapes, making it particularly adept at adapting to different data distributions [25].

In hierarchical clustering, the determination of similarity between clusters relies on the separation or similarity between individual data points. The algorithm initiates the clustering process with single data points and progressively merges them into larger clusters, resulting in the establishment of a hierarchical structure of clusters. This hierarchical structure can be visually represented as a tree diagram known as a "dendrogram," offering an intuitive means to comprehend the relationships and connections between different data points. The utilization of hierarchical clustering offers a dynamic and versatile approach to exploratory data analysis. Its inherent capacity to capture the inherent hierarchical structure within datasets makes it an invaluable tool for researchers and analysts seeking a comprehensive understanding of complex data relationships. The dendrogram serves as a visual aid, facilitating the interpretation of the clustering hierarchy and aiding in the identification of meaningful patterns and relationships within the data.

2.3 Methodology

In this research, the dataset has been collected from sales reports over 2 years of an online retail company specializing in agricultural products. The data mining and analysis process for knowledge extraction was carried out using the standard CRISP-DM methodology. This methodology comprises some major steps, which include understanding the business environment, data selection, data preparation, modeling, model evaluation, and evaluation of the results (Fig. 2). The steps of this method are described in Fig. 2. In this research, effective features for customer clustering based on the RFM model have been identified; in addition, we use two clustering techniques including the hierarchical and k-means algorithms to determine the optimal cluster number. In these two techniques, the average values of Recency, Frequency, and Monetary are calculated for each cluster, and also correlation coefficients are determined among the indicators then Gold, Silver, and Bronze customers are specified. Finally, the silhouette scores are calculated and the performance of the mentioned techniques is compared based on them and more analyses are presented. The overall procedure of the methodology is shown in Fig. 1.

The rationale behind focusing on k-means and hierarchical clustering techniques in our research stemmed from their widespread recognition and applicability in customer segmentation tasks, particularly within the context of RFM (Recency, Frequency, Monetary) analysis. These two methods are well established and extensively documented in the literature for their effectiveness in uncovering meaningful patterns within customer datasets.

Furthermore, including these specific techniques was informed by their distinct advantages and interpretability, which align with the objectives of our study. K-means clustering, for instance, is widely known for its simplicity and efficiency in partitioning data into clusters based on similarity, making it suitable for large-scale datasets and real-world applications. On the other hand, hierarchical clustering offers

Fig. 1 Methodology of the research

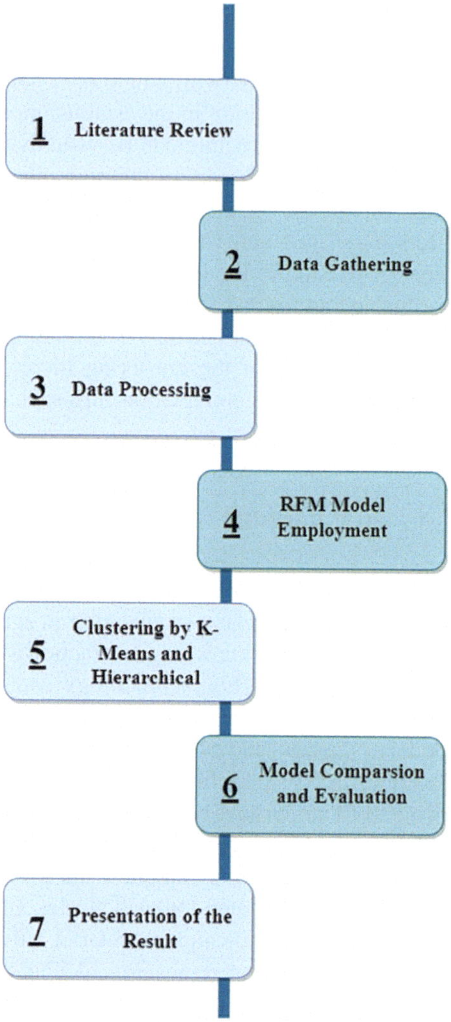

a hierarchical structure of clusters, providing insights into the hierarchical relationships among segments, which can be insightful for understanding customer behavior. While it is acknowledged that there exist numerous other machine learning techniques for clustering, we believe that the selection of k-means and hierarchical clustering was justified based on their relevance to our research context, their interpretability, and their established efficacy in customer segmentation tasks.

We acknowledge the importance of considering a diverse range of methodologies for customer segmentation beyond the machine learning techniques employed in this study. While present research primarily focuses on the application of k-means and hierarchical clustering alongside the RFM (Recency, Frequency, Monetary) model,

Fig. 2 Crisp DM process

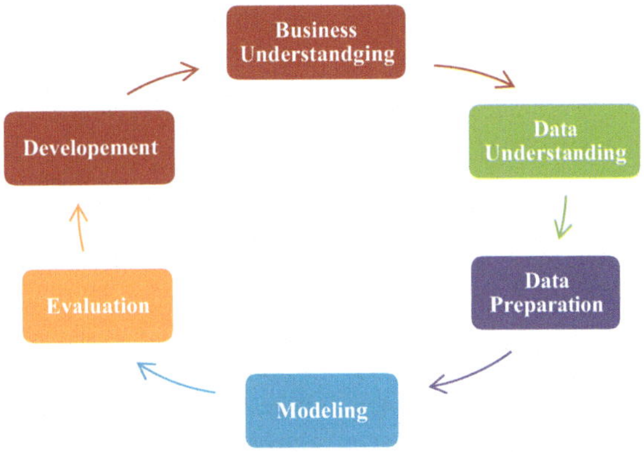

it is essential to recognize and compare alternative methods for customer segmentation. Traditional non-machine learning methods in customer clustering scope offer valuable insights and also considerable weaknesses which forced this research to employ machine learning for this end. Some of the mentioned methods and their weaknesses are presented below briefly:

- Demographic Segmentation:
  - Limitations: Assumes homogeneity within demographic groups and may overlook individual variations in behavior.
- Behavioral Segmentation:
  - Limitations: Requires comprehensive data collection and analysis to capture nuanced behavioral differences.
- Psychographic Segmentation:
  - Limitations: Requires detailed psychographic data and may be subject to interpretation bias.

While our study emphasizes the application of machine learning techniques for their ability to uncover complex patterns within data, we acknowledge the complementary role of traditional segmentation methods in understanding customer dynamics. Future research endeavors may benefit from a comprehensive comparison of these methodologies to inform strategic marketing decision-making and enhance customer relationship management strategies. By including the advantages of k-means and hierarchical clustering, this study provided a balanced perspective on the strengths of both traditional and machine learning-based segmentation approaches.

2.4 Data Collection and Preprocessing

This research focuses on analyzing the RFM (Recency, Frequency, Monetary) aspects of customer behavior using an international dataset from an online retailer specializing in agricultural products. The dataset initially comprises 541,909 records and 8 registered features. To ensure the quality of the data and enhance its suitability for analysis, a comprehensive data preprocessing approach is employed. The first step involves the removal of duplicate data to eliminate redundancy and ensure the accuracy of the dataset. Subsequently, feature selection is performed to identify the most relevant variables for the analysis, streamlining the dataset and enhancing computational efficiency. Lastly, the process includes estimating missing data and addressing any gaps or incomplete information in the dataset. By executing these data preprocessing steps, the research aims to refine the dataset, ensuring its accuracy, completeness, and relevance for the subsequent analysis of customer segments based on RFM parameters.

2.4.1 Data Collection

The features are defined as follows:

- **Invoice Number**: Each invoice is identified by a specific number, and a customer may have a wholesale purchase that includes one or more orders, recorded based on the time of customer purchase. This means that more than one invoice number may be recorded for each customer over time.
- **Stock Code**: The location in the product inventory where products are stored. According to the sales invoice, dispatch notes are issued for the items leaving the inventory.
- **Item Description**: This includes the product name and its specifications. It is defined and entered into the current system through the organization's coding pattern. After defining the product coding, all tracking and planning are done based on the product code.
- **Quantity Sold for Each Item**: This includes the number of units ordered for each item and recorded on each invoice.
- **Invoice Date**: The date and time related to the invoice for items purchased by customers.
- **Unit Price of the Item**: The price for each unit of the item is recorded on the invoice.
- **Customer ID**: To facilitate systematic examination of customers and to maintain customer confidentiality, organizations assign a customer code to the system. This code is used to identify the specific customer within the organization after their initial purchase.
- **Country**: This feature represents geographical regions and the distribution of sales. Analyzing this feature indicates the diversity of retail sales.

2.4.2 Data Preprocessing

The data preprocessing process is carried out as follows:

- **Removing Duplicate Data**: After this stage, 5268 duplicate data entries were removed, and the record count was adjusted to 536,641.
- **Feature Selection**: For RFM analysis, five necessary features, including customer ID, invoice number, invoice date, quantity of sold items, and the unit price of each item, were selected. The type of each of these features was determined, and the remaining features were removed due to their lack of necessity in RFM analysis.
- **Estimating Missing Data**: For the "Customer ID" feature, 135,037 records had missing values, accounting for 25.16% of the total data. Considering the relatively low percentage, it was decided to remove these records. If the percentage were higher, their removal might not have been feasible, and their values would need to be replaced. At this stage, the record count was reduced to 135,037, resulting in a final count of 401,604 records.

The data processing continued with the mentioned 5-feature dimensions, and the count was maintained at 401,604 records.

3 Results

3.1 RFM Indicators Calculation

In this step, the procedure of calculating values for each indicator in the RFM model is explained below:

- **Monetary**: This indicator is obtained using the sales invoice date table; it means when the customer made his last purchase (in terms of date and time). This index is concluded by multiplying the two characteristics of the number of goods sold and the unit price of the goods for each customer ID.
- **Frequency**: It is obtained using the invoice number and customer ID, and it means the number of purchases made by each customer in a certain period. This index is obtained separately by customers' IDs and their purchase invoice numbers, and then it is included in a table along with Monetary.
- **Recency**: This indicator is calculated using the number of goods purchased in the customer's invoice and the unit price of the product, and it means how much money the customer spends on the purchase. To calculate this index, we first converted the invoice date format into year, month, and day, and then separated the maximum value and based on the time vector, for each customer, the minimum time difference according to the new invoice time format with the maximum time The registration of the invoice has been obtained under the title of index (R).

In Table 1, the three explained indicators are determined for each customer ID.

Table 1 RFM indicators for each customer

Customer ID	Monetary	Frequency	Recency
0	12346	2	325
1	12347	182	1
.	.	.	.
4369	18282	13	7
4370	18283	721	3
4371	18287	70	42

3.2 Detect and Eliminating Irrelevant Data

At this stage, the distribution of data was drawn based on each of the RFM indicators, and then the first quartile and the third quartile were determined, and based on Eqs. (1) and (2), irrelevant data were removed, and finally the number of records was decreased from 4371 to 3628 including customers IDs, Recency, Frequency, and Monetary. The first one calculates the distance between the first and third quartiles and the second equation determines the irrelevant data.

$$IQR = Q_3 - Q_1 \tag{1}$$

$$outlier = RFM\,(\forall R\&F\&M \geq Q_1 - 1.5IQR)\&RFM\,(\forall R\&F\&M \geq Q_1 + 1.5IQR) \tag{2}$$

3.3 Data Normalization

In this research, data are normalized based on Max and Min functions using Eq. (3). This transforms data in the range of [0, 1].

$$\frac{x - \min(x)}{\max(x) - \min(x)} \tag{3}$$

3.4 Customer Clustering with K-means Algorithm

3.4.1 Model Construction

Now, according to the experts' opinion, the number of 1 to 10 clusters was examined to implement loops in the k-means algorithm in the Python tool with the elbow

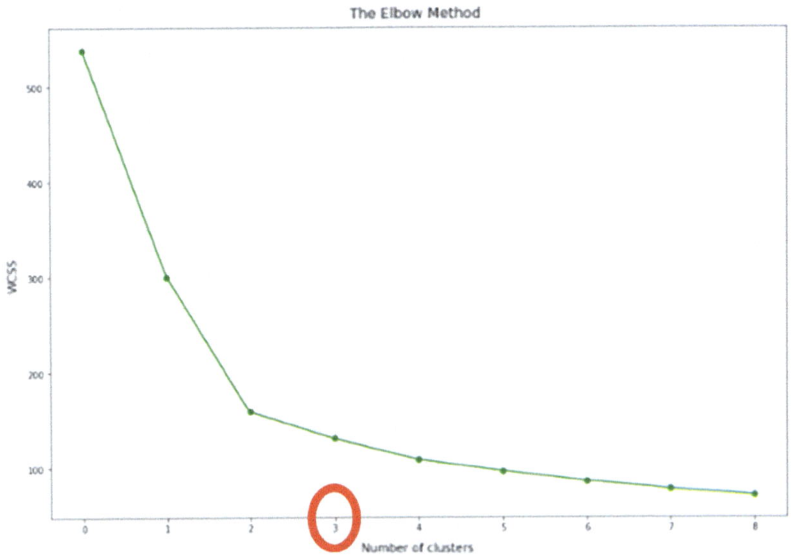

Fig. 3 Elbow method diagram

method, which is shown in Fig. 3; In this method, the optimal number of clusters is determined where the slope of changes becomes imperceptible; according to the mentioned figure, the optimal number of clusters is selected as 3. After that, we carried out the model considering three clusters.

3.4.2 Model Evaluation

Here silhouette score evaluation method is used and its value for three determined clusters is calculated as 0.4775 according to Eq. (4). In this equation, (i) represents the average distance of a point from a certain cluster with other points of that cluster, and $b(i)$ represents the minimum average distance of the point with other clusters.

$$S(i) = \frac{b(i) - a(i)}{\max(b(i), a(i))} \tag{4}$$

3.4.3 Model Employment

In this step, the clustering label from 0 to 2 was assigned to the customer ID. After that, based on each RFM indicator (Table 2) and also three determined clusters, the average value was calculated and used to obtain the correlation between RFM indicators.

Table 2 RFM indicators for each cluster

Cluster-ID	Monetary	Frequency	Recency
0	600.7551	31.4221	49.95
1	1692	112.8765	38.95
2	373.451	22.73	241.27

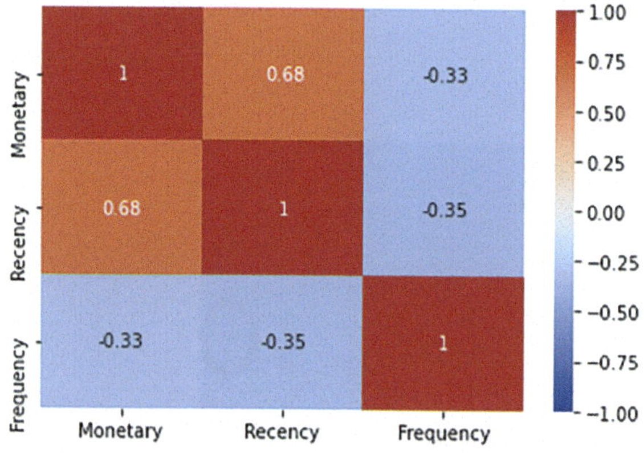

Fig. 4 Correlation diagram

Table 3 Customer evaluation based on the RFM model

Customer classification	Cluster-ID	Monetary	Frequency	Recency
Gold customers	1	Max	Max	Min
Silver customers	0			
Bronze customers	2	Min	Min	Max

According to the correlation between indicators shown in Fig. 4, the highest correlation was determined between Recency and Monetary, and the corresponding analysis is brought in Table 3. To concretely show the separation of clusters, its 3D diagram is also shown in Fig. 5.

3.5 *Hierarchical Customer Clustering*

3.5.1 Model Construction

First, the required libraries were called in the Python tool, and then the normalized data was used in the k-means method. In the following, using the Ward method and Euclidean distance, we draw the dendrogram diagram cumulatively (bottom-up approach) in Fig. 6, and the optimal number of clusters according to this figure is 3, similar to what we obtained in the k-means method.

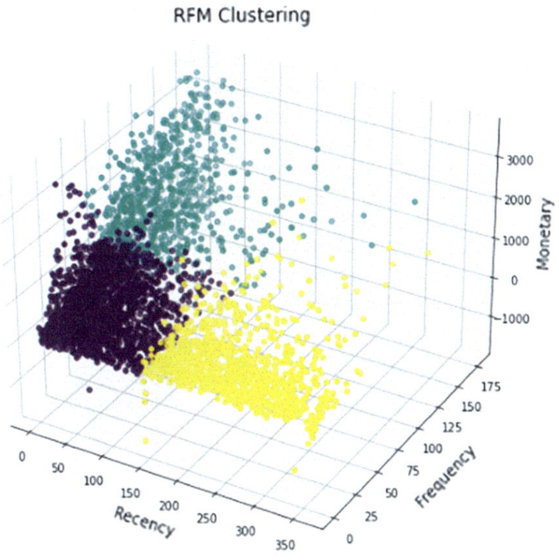

Fig. 5 3D clustering

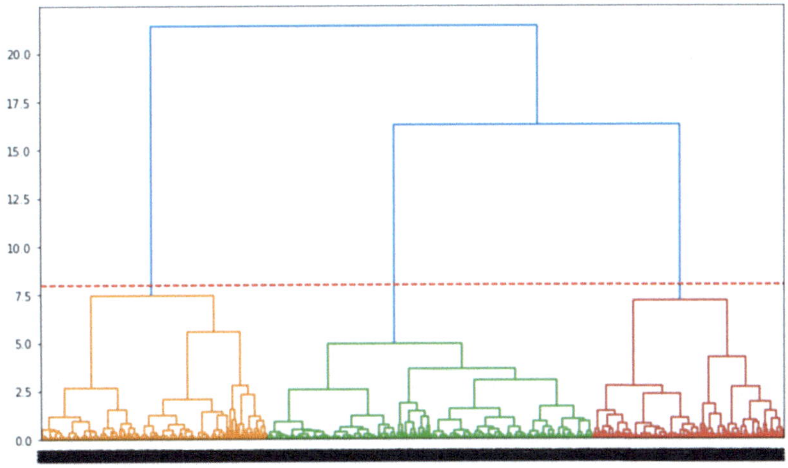

Fig. 6 Dendrogram diagram

3.5.2 Model Evaluation

In this clustering model, the silhouette score was calculated as 0.4321 which is less than the value obtained in the k-means method, and the k-means method performed better with a difference of 0.045.

Fig. 7 3D customer clustering diagram

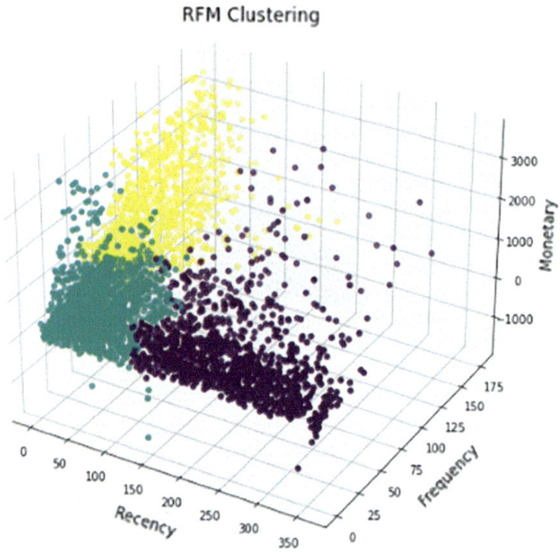

Table 4 Hierarchical clustering analysis

Cluster label	Monetary	Frequency	Recency
0	436.425	26.98	225.04
1	573.78	27.53	46.24
2	1546.36	104.19	30.93

3.5.3 Model Deployment

Here, the cluster label is assigned to the customer ID and the clustering is specified for each customer; then, the clustering diagram is shown in Fig. 7. According to the analysis of the clusters in the hierarchical method shown in Table 4, due to having the highest Monetary and Frequency and the lowest Recency, the best cluster of customers has been marked with blue color, which corresponds to the label of cluster 2. Also, the box diagram of all three groups of customers for three indicators, Recency, Frequency, and Monetary, in model RFM is shown in Figs. 8, 9, and 10, respectively.

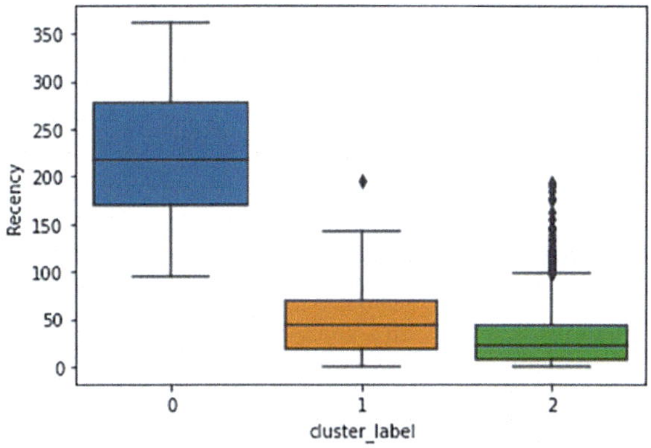

Fig. 8 Boxplot for recency

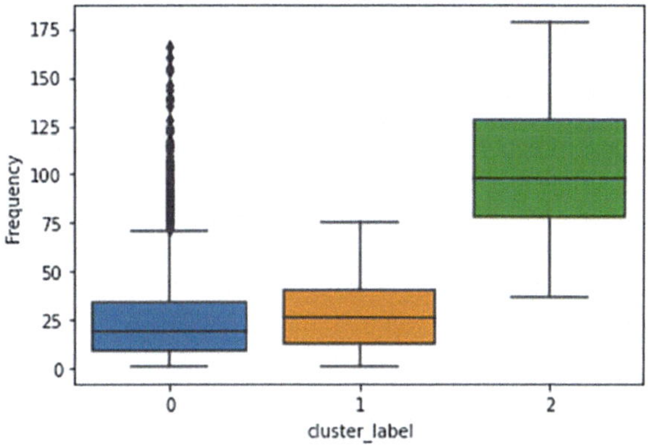

Fig. 9 Boxplot for frequency

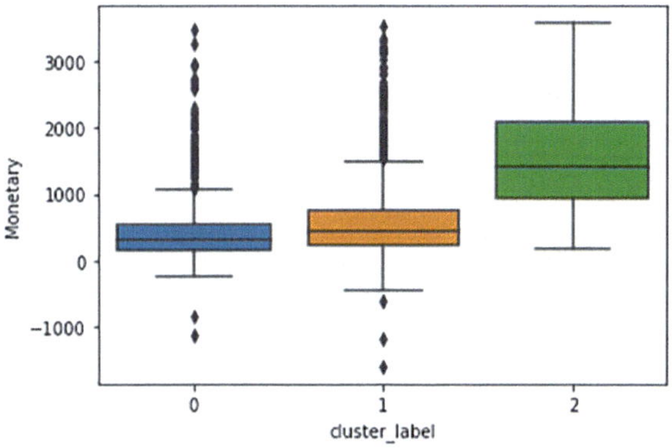

Fig. 10 Boxplot for monetary

4 Conclusion

In the realm of sustainable agriculture, the significance of retailing and customer clustering cannot be overstated. As we navigate the complex landscape of cultivating crops and providing essential equipment for the agricultural sector, understanding and adapting to the unique needs of our clientele becomes paramount. Retailing practices play a pivotal role in facilitating access to environmentally friendly and sustainable farming solutions, ensuring that farmers have the tools and resources needed to implement ecologically sound practices. Aligning with the proposed approach in this research and tailoring it to customer preferences not only enhances the overall efficiency of the supply chain, but it also contributes to the overarching goal of strengthening sustainable agriculture.

In this research, the most important set of customers, which we call the golden customers, have been identified. This affair is crucial for the crops and equipment retailers to identify this specific group of customers. RFM model indicators were determined based on the characteristics of customers' sales invoices, and based on this, clustering was done using two methods, hierarchical clustering and k-means; the optimal number of clusters was the same from both methods as 3. The evaluations showed that the k-means method is performing slightly better in this scope. Then, based on the analysis specifically the correlation between the RFM indices, the golden customers identified in both methods were selected as the best cluster for having the highest value in the Monetary and Frequency index and the lowest value in the Recency index. Accordingly, for this category of customers, different strategies can be applied, including discounts or payment terms, and more cooperation can be established in the future.

Through the strategic segmentation of customers based on Recency, Frequency, and Monetary value, our study aims to unravel patterns that empower retailers to provide targeted support to farmers. The RFM model, rooted in data-driven insights, not only aids in optimizing inventory management but also facilitates the identification of environmentally conscious farming practices preferred by distinct customer segments. Based on the results presented in this research, we intend to contribute valuable insights to the discourse on sustainable agriculture, emphasizing the pivotal role of customer-centric strategies in advancing the broader mission of environmentally friendly and resilient farming practices.

References

1. Gil JM, Gracia A, Sanchez M (2000) Market segmentation and willingness to pay for organic products in Spain. Int Food Agribus Manag Rev 3(2):207–226
2. Bui T, Nguyen P (2020) Market research for an agricultural company in South Karelia
3. Monalisa S, Juniarti Y, Saputra E, Muttakin F, Ahsyar TK (2023) Customer segmentation with RFM models and demographic variable using DBSCAN algorithm. TELKOMNIKA (Telecommunication Computing Electronics and Control) 21(4):742–749

4. Xian Z, Keikhosrokiani P, XinYing C, Li Z (2022) An RFM model using K-means clustering to improve customer segmentation and product recommendation. In: Handbook of research on consumer behavior change and data analytics in the socio-digital era. IGI Global, pp 124–145
5. Gustriansyah R, Suhandi N, Antony F (2020) Clustering optimization in RFM analysis based on k-means. Indones J Electr Eng Comput Sci 18(1):470–477
6. Križanić S (2020) Educational data mining using cluster analysis and decision tree technique: a case study. Int J Eng Bus Manag 12:1847979020908675
7. Seyedan M, Mafakheri F (2020) Predictive big data analytics for supply chain demand forecasting: methods, applications, and research opportunities. J Big Data 7(1):1–22
8. Li J, Pan S, Huang L (2019) A machine learning based method for customer behavior prediction. Tehnički vjesnik 26(6):1670–1676
9. Heldt R, Silveira CS, Luce FB (2021) Predicting customer value per product: from RFM to RFM/P. J Bus Res 127:444–453
10. Hughes AM (1994) Boosting response with RFM. Mark. 5:4–10
11. Cheng CH, Chen YS (2009) Classifying the segmentation of customer value via RFM model and RS theory. Expert Syst Appl 36(3):4176–4184
12. Chen D, Sain SL, Guo K (2012) Data mining for the online retail industry: a case study of RFM model-based customer segmentation using data mining. J Database Mark Cust Strategy Manag 19:197–208
13. Engineering MPI (2023) Retracted: an empirical study on customer segmentation by purchase behaviors using a RFM model and K-means algorithm. Math Probl Eng
14. Parikh Y, Abdelfattah E (2020) Clustering algorithms and RFM analysis performed on retail transactions. In: 2020 11th IEEE annual ubiquitous computing, electronics & mobile communication conference (UEMCON). IEEE, pp 0506–0511
15. Haghighatnia S, Abdolvand N, Rajaee Harandi S (2018) Evaluating discounts as a dimension of customer behavior analysis. J Mark Commun 24(4):321–336
16. Kamthania D, Pawa A, Madhavan SS (2018) Market segmentation analysis and visualization using K-mode clustering algorithm for E-commerce business. J Comput Inf Technol 26(1):57–68
17. Hung PD, Lien NTT, Ngoc ND (2019) Customer segmentation using hierarchical agglomerative clustering. In: Proceedings of the 2nd international conference on information science and systems, pp 33–37
18. Arunachalam D, Kumar N (2018) Benefit-based consumer segmentation and performance evaluation of clustering approaches: an evidence of data-driven decision-making. Expert Syst Appl 111:11–34
19. Mesforoush A, Tarokh MJ (2013) Customer profitability segmentation for SMEs case study: network equipment company. Int J Res Ind Eng 2(1):30–44
20. Anitha P, Patil MM (2022) RFM model for customer purchase behavior using K-Means algorithm. J King Saud Univ Comput Inf Sci 34(5):1785–1792
21. Wu HH, Chang EC, Lo CF (2009) Applying RFM model and K-means method in customer value analysis of an outfitter. In: Global perspective for competitive enterprise, economy and ecology: proceedings of the 16th ISPE international conference on concurrent engineering. Springer London, pp 665–672
22. Miguéis VL, Camanho AS, e Cunha JF (2012) Customer data mining for lifestyle segmentation. Expert Syst Appl 39(10):9359–9366
23. Teslenko D, Sorokina A, Smelyakov K, Filipov O (2023) Comparative analysis of the applicability of five clustering algorithms for market segmentation. In: 2023 IEEE open conference of electrical, electronic and information sciences (eStream). IEEE, pp 1–6
24. Kansal T, Bahuguna S, Singh V, Choudhury T (2018) Customer segmentation using K-means clustering. In: 2018 international conference on computational techniques, electronics and mechanical systems (CTEMS). IEEE, pp 135–139
25. Murtagh F, Contreras P (2017) Algorithms for hierarchical clustering: an overview, II. Wiley Interdisc Rev Data Min Knowl Discov 7(6):e1219

Enhanced Vegetation Cover Assessment Using Sentinel-2A: A Unified Perspective on NDVI, SAVI, and GNDVI

Pradnya Awate and Ajay Nagne

Abstract This study uses Sentinel-2A multispectral imagery to provide a thorough analysis of the vegetation cover in the Sillod area of the Aurangabad District, Maharashtra. In order to distinguish between different vegetation types, such as sparse, moderate, and dense vegetation, our research focuses on the computation of well-known vegetation indices, such as NDVI, SAVI, and GNDVI. By using a semiautomatic object-based classification method, we were able to identify features with excellent results. The accuracy assessment of the study yielded encouraging results. The NDVI demonstrated a remarkable 90.69% overall accuracy, with SAVI and GNDVI following closely behind at 83.58% and 93.23%, respectively. These results highlight the NDVI and GNDVI's potential usefulness for extracting vegetation in our study area. It illustrates how important it is to use Sentinel-2A imagery and sophisticated spectral indices for accurate vegetation analysis in Sillod, Maharashtra's changing landscape.

Keywords Vegetation indices · NDVI · SAVI · GNDVI

1 Introduction

Precision agriculture, also known as smart farming or precision farming, is a contemporary farming method that makes use of data, information, and technology to maximize a range of farming operations. Its objectives are to raise total farm efficiency, reduce resource waste, and maximize agricultural yields. The agricultural sector is facing several challenges, such as resource scarcity, population increase, and climate change. One important way to address these issues is through precision agriculture. In the last few decades, geospatial technologies such as Geographic Positioning System (GPS), Remote Sensing (RS), and Geographic Information System (GIS) have been used to produce precision agriculture [1].

P. Awate (✉) · A. Nagne
Dr. G. Y. Pathrikar College of CS and IT, MGM University, Aurangaba, India
e-mail: awate.pradnya@gmail.com

A. J. Kulkarni and M. Hajiaghaei-Keshteli (eds.), *Advances in Intelligent Systems for Sustainable Agriculture*, Lecture Notes in Networks and Systems 1201,
https://doi.org/10.1007/978-981-97-9839-1_16

Precision agriculture's data-driven approach and cutting-edge technologies make it extremely helpful for crop identification and analysis. This methodology improves the precision and effectiveness of crop identification and analysis by utilizing a range of technologies and approaches [2].

Crop analysis requires the use of vegetation indices since they provide vital information on the growth and health of crops. They give quantitative and objective measurements of the vegetation conditions using data from remote sensing. These indexes are essential to modern agriculture for a number of reasons. Utilizing spectral and spatial data, vegetation indices are essential tools for agricultural analysis because they shed light on the development and health of crops [3].

Vegetation indices make use of both spatial and spectral data to produce detailed maps of agricultural areas and to evaluate the quality of the vegetation [4]. Crop analysis requires the use of vegetation indices, such as the Normalized Difference Vegetation Index (NDVI), the Soil-Adjusted Vegetation Index (SAVI), and the Green Normalized Difference Vegetation Index (GNDVI). Plant health is measured by NDVI, soil brightness is corrected by SAVI, and vegetation greenness is emphasized by GNDVI. When combined, they offer a thorough understanding of agricultural conditions. By combining these indicators, crop health may be assessed in a more accurate and nuanced way, enabling more accurate decisions about management and resource optimization [5].

Accurately identifying the vegetation cover is essential for agriculture. These vegetation indices make it possible to distinguish between sparse, moderate, and dense vegetation; monitor crop health; and pinpoint stress factors [6]. Adjusting interventions to specific vegetation features enables the optimization of resource utilization, reduction of environmental impact, and maximization of agricultural yields [7]. Vegetation indices are, in short, critical tools in precision agriculture that help make well-informed decisions for the use of profitable and sustainable agricultural techniques [7].

Different vegetation covers may be identified and managed with the use of vegetation indices such as the NDVI, SAVI, and GNDVI. These indices take use of the unique ways that crops absorb and reflect light at different wavelengths. The Normalized Difference Vegetation Index (NDVI) is a standardized measure for assessing the health of vegetation. It calculates the variation in red and near-infrared light reflectance [8, 9]. SAVI and GNDVI improve accuracy in estimating plant cover by concentrating on the green spectrum and integrating soil brightness, respectively [10].

The research article is organized, starting with a thorough overview of related work in Sect. 2. Section 3 provides insights into the carefully chosen study area, while Sect. 4 elaborates on the research methodology. The study's conclusion is finally summarized in Sect. 5, which offers a thorough overview of the entire research process.

2 Related Work

Numerous research papers have illustrated the use of vegetation indices and remote sensing in crop analysis. For instance, the author of the study evaluated the effects of drought on India's wheat harvests using remote sensing and VIs. The Normalized Difference Vegetation Index (NDVI) had the highest association with yield, indicating that VIs were useful in identifying drought stress in wheat crops, according to the author [11].

In order to segment rice for automated classification of blast and brown spot diseases, the author's suggested study employed Enhanced Vegetation Indices (EVI), Green Normalized Difference Vegetation Index (GNDVI), Normalized Difference Vegetation Index (NDVI), and Soil-Adjusted Vegetation Index (SAVI). To extract the affected areas from the VI images, Otsu's approach was used. Then, the infected areas' five distinct texture features—homogeneity, correlation, contrast, energy, and entropy—are calculated [12].

The author of the related work classified LISS IV imagery using decision tree methods using 11 vegetation indices (NDVI, DVI, GEMI, GNDVI, MSAVI2, NDWI, NG, NR, NNIR, OSAVI, and VI green). The alternative method involves integrating these vegetation indices with texture features (mean, variance, kurtosis, and skewness). A comparison of these two approaches has been conducted. The findings show that it is possible to construct categorized maps for Indian satellite IRS-P6, LISS IV sensor imagery with 8.33% greater accuracy by including textural features with vegetation indices [13].

The author of another research remotely analyzed the density of ground cover vegetation for olive groves using 11 vegetation indexes. Whereas, at intervals of 10% to 25%, Inverse Ratio Vegetation Index (IRVI) is most sensitive to achieving a satisfactory level of ground cover vegetation. The author presents an improved understanding of how vegetation indicators behave in response to varying densities of soil cover [14].

Another study created models for estimating rice Leaf Area Index (LAI) using various multispectral bands, vegetation indices, and their combinations using Extreme Gradient Boosting model (XGBoost), then assessed the accuracy of those models against Random Forest (RF) and Semi-empirical Model (SEM) models. The outcomes showed, according to the author, that different approaches produced diverse outcomes for multispectral bands and VIs used in LAI estimate. The NIR was the best band in both the RF and XGBoost models, while the GNDVI had the highest LAI estimate accuracy in the SEM. In the meanwhile, NDRE and LCI outperformed other VIs in the XGBoost and RF models, respectively. Nevertheless, combining several VIs could only substantially improve the accuracy of the LAI calculation. In the meanwhile, the accuracy in both RF and XGBoost may be enhanced by the bands plus VIs combinations [15].

In the next sections, we will look at the technique, datasets, and outcomes of the vegetation indices NDVI, SAVI, and GNDVI in determining vegetation area.

3 Study Area

The selected study area is Sillod which is an informative area for agriculture and is situated in the Aurangabad District of Maharashtra, India. As per Fig. 1, the Survey of India website's shapefile data (https://onlinemaps.surveyofindia.gov.in/) was used to build the research area map. This area has a long history of agriculture, mostly focused on the production of sugarcane, cotton, and soybeans. The area is well known for producing high-quality cotton, and one of the main industries there is cotton cultivation. Pulses like lentils and soybeans grow well in Sillod's rich soils in addition to cotton. The availability of irrigation and contemporary infrastructure in Sillod is facilitated by its close proximity to the Godavari River and the Aurangabad Industrial Zone.

For this study, Sentinel-2A satellite data (S2A_MSIL2A) is used. This image was retrieved from Copernicus Open Access Hub (https://scihub.copernicus.eu/) on the date 2022–12-30 T05:32:31.024Z. Thirteen spectral bands are sampled by the SENTINEL-2 MultiSpectral Instrument (MSI): four bands at 10 m, six bands at 20 m, and three bands at 60 m in spatial resolution. With a swath width of 290 km and a spatial resolution of 10 m (four visible and near-infrared bands), 20 m (six red edge and shortwave infrared bands), and 60 m (three atmospheric correction bands), MSI covers 13 spectral bands (443–2190 nm). However, as the three atmospheric bands

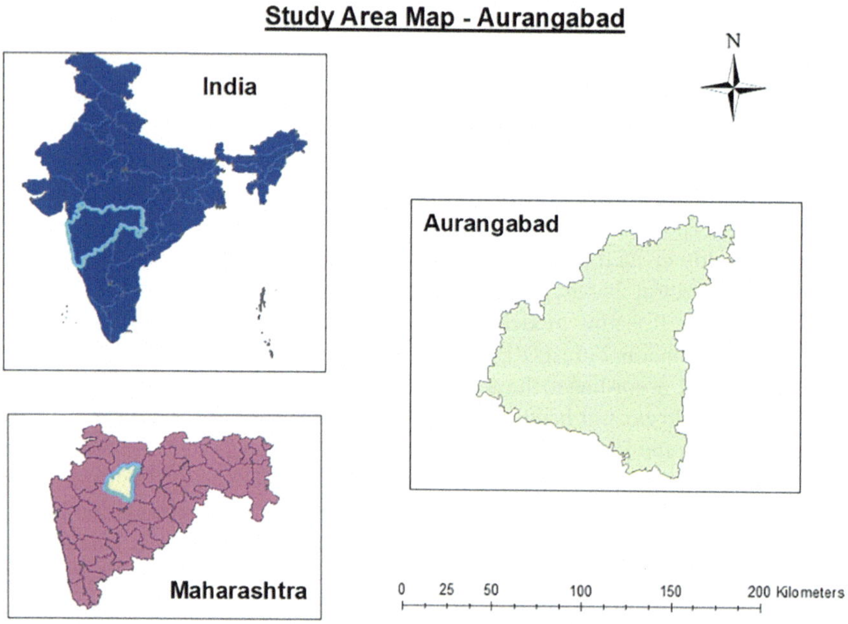

Fig. 1 Study area map

are primarily employed for cloud filtering and atmospheric corrections, they were not used in this study [16].

4 Methodology

The first stage in the methodology part, after collecting Sentinel-2A imagery, is data preparation. Important processes like layer stacking and region clipping are performed at this step to help create more specialized and manageable datasets. After that, the computation of vital vegetation indices—like the NDVI, SAVI, and GNDVI—is carried out in order to enable a thorough evaluation of the health and vegetative cover of the research region. Additionally, each vegetation index image is processed using density slicing procedures, which improve the capacity to identify and emphasize certain regions of interest. Subsequently, a semiautomatic classification procedure is implemented to efficiently classify land cover categories, supported by ground truth data gathered from Google Earth images. To confirm the classification findings accuracy and dependability, accuracy assessment is then carried out. The study then moves on to the results and discussion phase, where the conclusions and their consequences are discussed, evaluated critically, and contrasted with the existing literature of current knowledge. A concluding piece that highlights the main conclusions and highlights the importance of the study's findings comes next. Figure 2 shows pictorial representation of methodology.

4.1 Preprocessing

According to the technical manual supplied by Copernicus, the European Space Agency, the Level-1C product for Sentinel-2 images offers orthorectified Top-of-Atmosphere (TOA) reflectance together with sub-pixel multispectral registration. The kit includes masks for land and ocean as well as clouds. The Level-2A product offers sub-pixel multispectral registration together with orthorectified atmospherically adjusted surface reflectance. So this step includes two major techniques: layer stacking and region clipping.

4.1.1 Layer Stacking

Layer stacking is an important image processing approach used in our research, with a particular emphasis on the layer stacking of Sentinel-2A's 10 m resolution bands, which include the red, green, blue, and near-infrared bands as shown in Fig. 3. Through this method, a single, multi-band composite image that provides a more thorough perspective of the Earth's surface is created from the separate spectral bands [17]. This method was used in our study to produce a comprehensive and

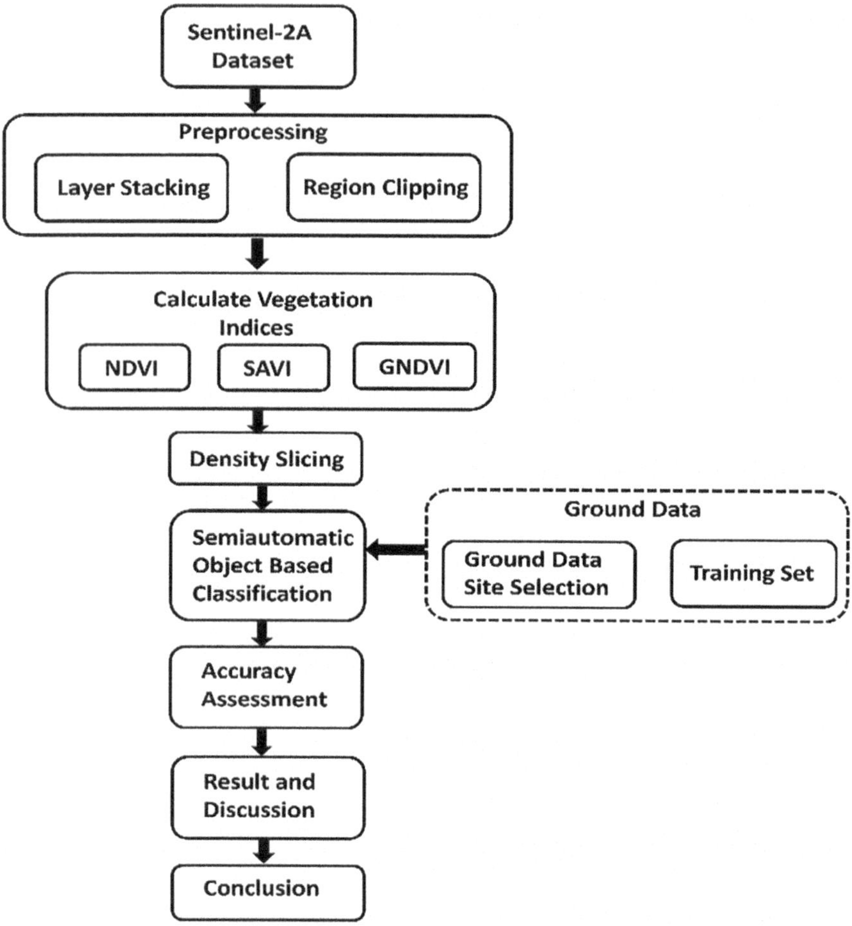

Fig. 2 Prescribed methodology

in-depth map of the Earth's surface. We were able to determine the relative value of different forms of land cover within our study region by studying this composite image and its significance for our research. Furthermore, it offered insights into vegetation dynamics, advancing our comprehension of environmental circumstances and changes. Table 1 shows list of Sentinel-2 bands [17, 18].

4.1.2 Region Clipping

After the layer stacking procedure of Sentinel-2 images was completed in our study, a critical operation known as area clipping was undertaken. In this process, particular regions or areas of interest are extracted from the composite image. We sought to

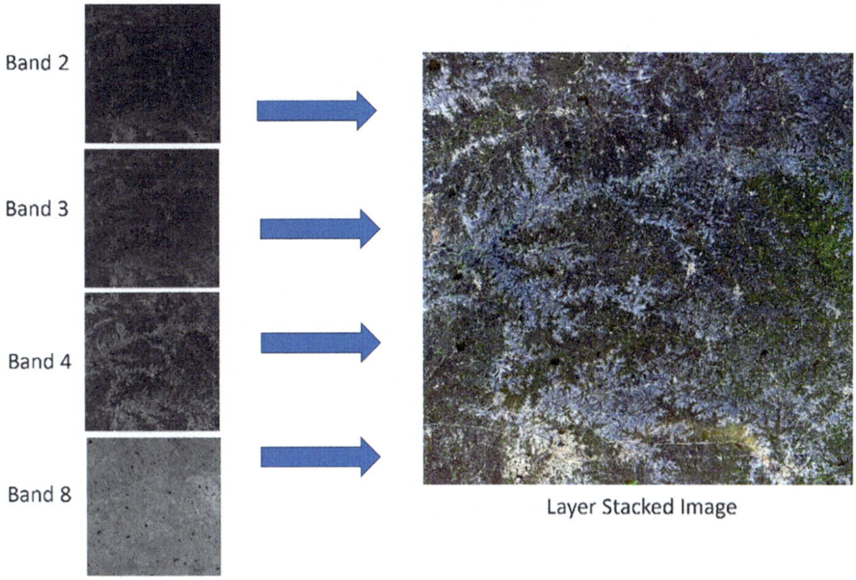

Fig. 3 Sentinel-2A image (Date: 2022–12-30 T05:32:31.024Z) layer stacking

Table 1 List of sentinel-2 spectral bands

Band number and name	Wavelength (nm)	Resolution (m)
Band 1 (Coastal/Aerosol)	443	60
Band 2 (Blue)	490	10
Band 3 (Green)	560	10
Band 4 (Red)	665	10
Band 5 (Vegetation Red Edge)	705	20
Band 6 (Vegetation Red Edge)	740	20
Band 7 (Vegetation Red Edge)	783	20
Band 8 (Near-Infrared)	842	10
Band 8A (Narrow Near-Infrared)	865	20
Band 9 (Water Vapor)	945	60
Band 10 (Cirrus)	1375	60
Band 11 (Short-Wave Infrared)	1610	20
Band 12 (Short-Wave Infrared)	2190	20

improve the accuracy of our research by narrowing the scope of our investigation to certain places inside the study site by isolating these targeted regions. Region clipping minimizes the amount of data, makes better use of computing resources, and simplifies further analysis [19]. When working with large-scale remote sensing datasets, this approach is extremely helpful as it allows us to focus on the precise

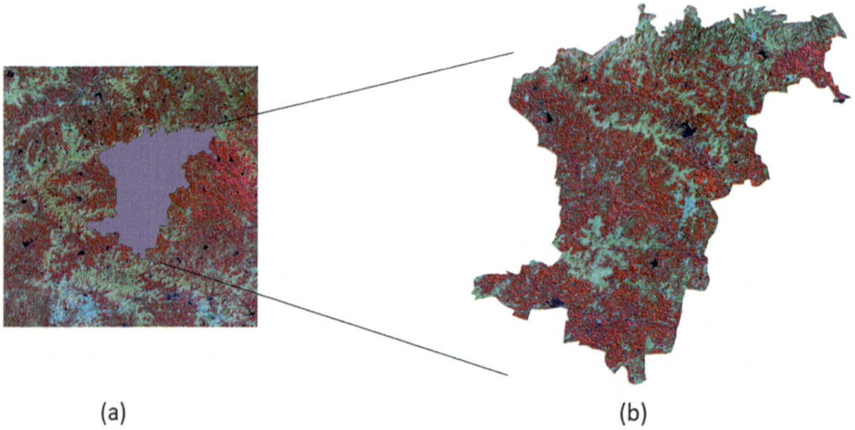

Fig. 4 **a** Before clipping image and **b** after clipping ROI

geographic areas that are pertinent to research goals [20]. Figures 3b and 4a shows Sentinel-2A image before clipping and after clipping.

4.2 Calculate Vegetation Indices

Normalized Difference Vegetation Index (NDVI).

The Normalized Difference Vegetation Index (NDVI), which can provide quantitative insights into the photosynthetic activity and health of plants, is extremely important for crop analysis. By comparing the reflectance of the red and near-infrared (NIR) spectral bands, the NDVI is computed, providing a numerical depiction of the biomass and chlorophyll content of the crop. The image was categorized by the NDVI into values ranging from -1 to $+1$; a negative number indicates no vegetation, while a positive value indicates vegetation. Equation (1) illustrates the standard NDVI [21]:

$$\mathrm{NDVI} = \frac{\mathrm{NIR} - \mathrm{RED}}{\mathrm{NIR} + \mathrm{RED}} \tag{1}$$

Soil-Adjusted Vegetation Index (SAVI).

For crop analysis, the Soil-Adjusted Vegetation Index (SAVI) is essential, especially in situations where changes in soil brightness have a major effect on vegetation indices. In the red spectral region, SAVI reduces the impact of soil reflection by adding a correction factor for soil brightness. This adjustment reduces noise in the data connected to soil, which improves the accuracy of vegetation estimates. The Soil-Adjusted Vegetation Index (SAVI) was developed with the intention of lessening the impact of soil brightness on vegetation evaluations. Equation (2) is the SAVI formula [2]:

$$\mathrm{SAVI} = \frac{(1 + \mathrm{L})(\mathrm{NIR} - \mathrm{RED})}{\mathrm{NIR} + \mathrm{RED} + \mathrm{L}} \tag{2}$$

where L, which usually ranges from 0 to 1, is the soil-brightness correction factor. By modifying this parameter, SAVI may be optimized to reduce the effect of soil brightness on the vegetation index. Because of its adaptability, SAVI is especially useful in environmental and agricultural applications, where different soil types can have an impact on how accurately plant evaluations are made [22].

Green Normalized Difference Vegetation Index (GNDVI).

One of the most important technical tools for crop analysis is the Green Normalized Difference Vegetation Index (GNDVI), which measures the health of the vegetation with an emphasis on green biomass. The green and near-infrared reflectance bands in vegetation are used by GNDVI to measure the quantity of green biomass and chlorophyll present. The accurate assessment of crop vigor and the timely identification of stress factors like disease, nutrient shortages, or water scarcity are made possible by this technological accuracy [23]. Equation (3) is the GNDVI formula:

$$\mathrm{GNDVI} = \frac{\mathrm{NIR} - \mathrm{GREEN}}{\mathrm{NIR} + \mathrm{GREEN}} \tag{3}$$

The normal range of GNDVI readings is –1 to 1, with greater positive values indicating healthy, green vegetation. Precision agriculture relies heavily on this index since it measures vegetation health and greenness in an efficient and comprehensive manner [24].

4.3 Density Slicing

Density slicing is an image processing method that uses certain pixel intensity or spectral values to separate and isolate areas of interest within a digital image. It involves dividing pixel values into discrete ranges, or "slices," each of which represents a specific spectral characteristic or intensity level. The final image then highlights these slices for better analysis and examining. Density slicing is very useful in remote sensing applications because it may isolate certain characteristics or regions of interest in an image, making identification and measurement more accurate. With multi-layered or complicated datasets, this method is especially useful since it enables analysts to extract relevant information from a larger dataset by isolating key traits [25, 26].

4.4 Ground Data

In the context of Sentinel-2 imagery analysis, with an emphasis on the Sillod area, choosing a suitable training dataset and site for training samples is an essential step in guaranteeing the accuracy of remote sensing applications. A carefully chosen site is essential for Sillod. It should represent the whole variety of land cover classes encountered in the region, including crops, natural vegetation, urban areas, and water bodies. In order to effectively represent the different land cover categories, the training dataset should have spectral signatures and metadata. Accurate and representative training sample collection helps classification models work better, making it possible to distinguish between different types of land cover in Sentinel-2 data. A key component of efficient land cover categorization and information extraction is making sure that the training data match the precise analysis objectives [27].

4.5 Semiautomatic Object-Based Classification

Following the computation of NDVI, SAVI, and GNDVI in our work, we performed a vital preprocessing step known as density slicing. In order to create a basis for additional study, this method involved segmenting NDVI imagery into several classes and creating boundaries around areas with particular NDVI values. The same procedure is used for GNDVI and SAVI.

Next, we classified these divided areas using a semiautomatic object-based classification method. Our classification technique strikes a compromise between automation and human interaction, improving the precision and dependability of our mapping of land cover. Figure 5 shows classified image of NDVI, SAVI, and GNDVI. It makes it easier to conduct a thorough evaluation of our research region, which leads to deeper understandings of the dynamics of land cover, crop health, and plant distribution. Our ability to monitor the environment, analyze land use, and practice precision agriculture is strengthened by this approach [28, 29].

4.6 Accuracy Assessment

An essential step following classification is accuracy evaluation, in which the accuracy and dependability of the outcomes of the classification are methodically assessed. To measure the degree of agreement, the categorized data are compared to ground truth or reference data. The accuracy evaluation carried out following classification in our study produced significant findings. Our overall NDVI accuracy of 90.6860% demonstrates the categorization model's robustness in accurately capturing the various land cover types in our research area. GNDVI showed an astounding overall accuracy of 93.2317%, while SAVI showed an overall accuracy

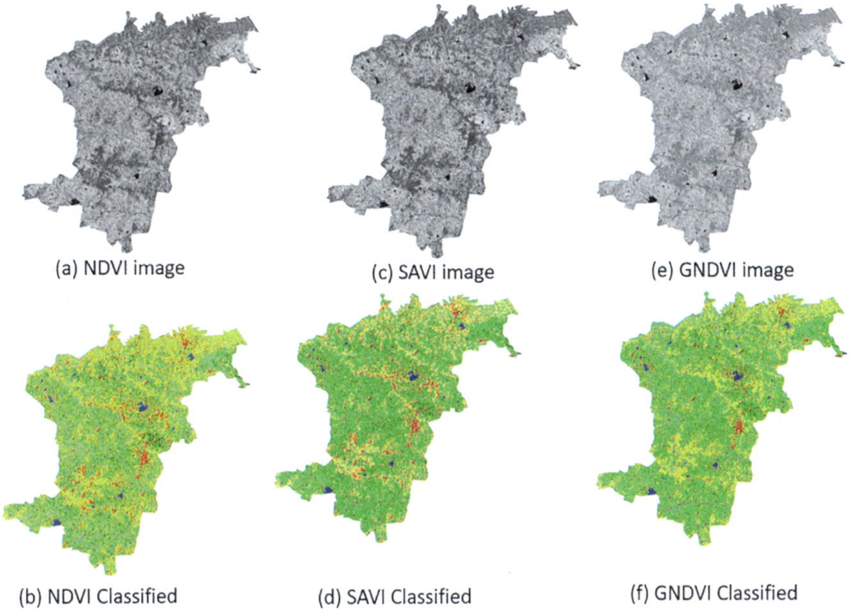

Fig. 5 NDVI, SAVI, and GNDVI classified image

Table 2 Overall accuracy for NDVI, SAVI, and GNDVI

Vegetation index	Overall accuracy (%)
NDVI	90.69
SAVI	83.58
GNDVI	93.23

of 83.5823%. These outcomes, which are displayed in the accompanying Table 2, highlight the accuracy and dependability of our classification strategy.

4.7 Result and Discussion

The three key indices for feature extraction in our work are NDVI, SAVI, and GNDVI, each of which provides distinct insights. In order to distinguish between dense, moderate, and sparse vegetation, the NDVI assesses the health of the vegetation. SAVI refines discriminating between barren land and built-up areas due to its soil-adjusted nature. GNDVI improves the identification of water bodies by highlighting green biomass. When combined, these indices allow for the accurate classification of land cover types, making it easier to identify areas with different densities of vegetation, water bodies, and built-up areas. Their technological precision enhances

Table 3 NDVI image result

Index class	Percentage (%)	Area (sq. km)
Water	1.13	17.1255
Built up	6.45	97.4721
Barren land	43.37	655.2686
Sparse vegetation	29.78	449.9065
Moderate vegetation	11.06	167.1332
Dense vegetation	8.21	124.0791

our understanding of varied landscapes by enabling precise land cover analysis, which is essential for applications in environmental assessment, urban planning, and agriculture.

The distribution of land cover categories serves as an example of the informative results we were able to obtain in our accuracy assessment for the NDVI classification. As Table 3 illustrates, 6.45% of the study area was made up of built-up areas, with water bodies making up 1.13% of the total area. Barren lands predominated, making up a significant 43.37% of the area, demonstrating how common they are. Although at a lesser density, the 29.78% covered by sparse vegetation indicates a significant presence of vegetation. There was a balance between dense vegetation (representing 8.21% of the research area) and intermediate vegetation (11.06%). The results of this accuracy evaluation offer insightful information about the composition of land cover and the accuracy of the categorization, supporting a variety of applications in environmental monitoring, land management, and agriculture.

We have obtained informative statistics that offer a thorough understanding of the land cover categories in our study area through our SAVI-based accuracy assessment. The water bodies, which make up a relatively small proportion of 1.13%, have a unique spectral signature. Built-up areas, which make up 6.45% of the total, are places with infrastructure and structures created by humans. Barren lands represent places with little to no vegetation, making up a significant 28.39% of the total. Lower vegetative density zones are highlighted by sparse vegetation, which accounts for 14.98% of the total. With 29.77% of the vegetation being moderate, there is a noticeable presence of healthy vegetation; the remaining 19.27% of the vegetation is dense, indicating places with a rich and dense vegetative cover. Table 4 shows SAVI results.

We have uncovered important data from our accuracy evaluation using GNDVI that provide insight into the various land cover categories in our study area. Water bodies, which make up a negligible 1.15% of the total, have unique spectral properties. The 4.41% of the area that is built up represents built-up areas and urban growth. Barren lands, which are defined as large areas with little vegetation, make up 39.36% of the total. At 24.63%, areas with modest vegetative presence are indicated by sparse vegetation. At 18.83%, moderate vegetation predominates, indicating areas with a robust and healthy plant population. Dense vegetation, which makes up 11.62% of the land, denotes regions with a high concentration of luxuriant flora as shown in

Table 4 SAVI image result

Index class	Percentage (%)	Area (sq. km)
Water	1.13	17.1262
Built up	6.45	97.4822
Barren land	28.39	428.9213
Sparse vegetation	14.98	226.3443
Moderate vegetation	29.77	449.8918
Dense vegetation	19.27	291.2192

Table 5 GNDVI image result

Index class	Percentage (%)	Area (sq. km)
Water	1.15	17.4003
Built up	4.41	66.6444
Barren land	39.36	594.6915
Sparse vegetation	24.63	372.1627
Moderate vegetation	18.83	284.5187
Dense vegetation	11.62	175.5674

Table 5. The data produced from the GNDVI provide a thorough understanding of the dynamics of land cover in our research area, which is highly advantageous for applications in environmental assessment, land management, and agriculture.

5 Conclusion

Evaluating agricultural vegetation cover is essential to sustainable land management. Sparse vegetation may be a sign of erosion, overgrazing, or soil deterioration, which calls for actions like soil conservation or replanting. The resilience of ecosystems is enhanced by moderate cover, which promotes ideal crop growth, soil fertility, and biodiversity. Dense vegetation cover provides valuable insight for erosion as well as habitat for wildlife. Observing changes over time provides information on the resilience of ecosystems to climate change. For long-term food security and environmental well-being, integrating vegetation evaluation helps with resource allocation, risk reduction, and the promotion of sustainable agricultural techniques.

Our work delves into an extensive technique for Sentinel-2 imagery-based vegetation analysis. Sentinel-2A L2A multispectral data that we carefully chose and analyzed needed very little preprocessing such as layer stacking and region clipping. We made use of the capabilities of important vegetation indices inside this framework, such as NDVI, SAVI, and GNDVI, each of which has a unique function in feature identification.

Density slicing and semiautomated object-based categorization, which precisely defined land cover types, are the fundamental components of our methodology. The results from the accuracy assessment showed that NDVI had a remarkable overall accuracy of 90.69%, followed closely by SAVI and GNDVI at 83.58% and 93.23%, respectively. As a result, our analysis indicates that the NDVI and GNDVI are effective methods for extracting vegetation cover in our study area. This work lays the groundwork for strong applications in land use analysis, agriculture, and environmental monitoring, providing a basis for resource management and well-informed decision-making.

References

1. Suh S-H, Kim D-Y, Jhang J-E, Byamukama E, Hatfield G, Shin S-Y (2017) Identification of the White-Mold affected soybean fields by using multispectral imageries, spatial autocorrelation and support vector machine. In: International Conference on Research in Adaptive and Convergent Systems (pp. 104–109). Krakow, Poland: Association for Computing Machinery
2. Awate P, Nagne A (2023) Recent trends in crop assessment and identification based on spatial and spectral features using geospatial technology. Ind Eng J 16(3):142–157
3. Yeom J, Jung J, Chang A, Ashapure A, Maeda M, Maeda A, Landivar J (2019) Comparison of vegetation indices derived from UAV data for differentiation of tillage effects in agriculture. Remote Sen 11(13):1548
4. Marino S, Alvino A (2019) Detection of spatial and temporal variability of wheat cultivars by high-resolution vegetation indices. Agronomy 9(5):226
5. Kaya Y, Polat N (2023) A linear approach for wheat yield prediction by using different spectral vegetation indices. Int J Eng Geosci 8(1):52–62
6. Fakhri SA, Sayadi S, Latifi H, Khare S (2019) An optimized enhanced vegetation index for sparse tree cover mapping across a mountainous region. In: 2019 IEEE International Workshop on Metrology for Agriculture and Forestry (MetroAgriFor) (pp 146–151). Portici, Italy: IEEE
7. Rahman M, Islam MS, Chowdhury TA (2018) Change of vegetation cover at Rohingya refugee occupied areas in cox's bazar district of Bangladesh: Evidence from remotely sensed data. J Environ Sci Nat Resour 11(1–2):9–16
8. Gao L, Wang X, Johnson BA, Tian Q, Wang Y, Varrelst J, Gu X (2020) Remote sensing algorithms for estimation of fractional vegetation cover using pure vegetation index values: A review. ISPRS J Photogramm Remote Sens 159:364–377
9. Richards DR, Belcher RN (2020) Global changes in urban vegetation cover. Remote Sens 12(1):23
10. Khunrattanasiri W (2023) Application of remote sensing vegetation indices for forest cover assessments. In: S N Singapore, Concepts and Applications of Remote Sensing in Forestry (pp 153–166). Singapore: Springer, Singapore
11. Ijaz M, Zafar Q, Khan AA, Hassan SS (2022) Assessing drought and its impacts on wheat yield using remotely sensed observations in rainfed Potohar region of Pakistan. Environ Dev Sustain 25:3699–3721
12. Phadikar S, Goswami J (2016) Vegetation indices based segmentation for automatic classification of brown spot and blast diseases of rice. In: 2016 3rd International Conference on Recent Advances in Information Technology (RAIT) (pp 284–289). Dhanbad, India: IEEE
13. Verma AK, Garg PK, Hari Prasad KS, Dadhwal VK (2016) Classification of Liss Iv imagery using decision tree MethODS. Int Arch Photogramm Remote Sens Spat Inf Sci 41:1061–1066
14. Lima-Cueto FJ, Blanco-Sepúlveda R, Gómez-Moreno ML, Galacho-Jiménez FB (2019) Using vegetation indices and a UAV imaging platform to quantify the density of vegetation ground cover in olive groves (Olea Europaea L.) in Southern Spain. Remote Sens 11(21):2564

15. Liu S, Zeng W, Wu L, Lei G, Chen H, Gaiser T, Srivastava AK (2021) Simulating the leaf area index of rice from multispectral images. Remote Sens 13(18):3663
16. Kobayashi N, Tani H, Wang X, Sonobe R (2020) Crop classification using spectral indices derived from Sentinel-2A imagery. J Inf Telecommun 4(1):67–90
17. Nivedita Priyadarshini K, Kumar M, Rahaman SA, Nitheshnirmal S (2018) A comparative study of advanced land use/land cover classification algorithms using Sentinel-2 data. In: The International Archives of the Photogrammetry, Remote Sensing and Spatial Information Sciences, 665–670
18. Park H, Choi J, Park N, Choi S (2017) Sharpening the VNIR and SWIR Bands of Sentinel-2A imagery through modified selected and synthesized band schemes. Remote Sens 9(10):1080
19. Nagne AD, Dhumal RK, Vibhute AD, Gaikwad S, Kale K, Mehrotra S (2017) Land use land cover change detection by different supervised classifiers on LISS-III temporal datasets. In: 2017 1st International Conference on Intelligent Systems and Information Management (ICISIM) (pp 68–71). Aurangabad, India: IEEE
20. Irons JR, Dwyer JL, Barsi JA (2012) The next Landsat satellite: The landsat data continuity mission. Remote Sens Environ 122:11–21
21. Gaikwad SV, Vibhute AD, Kale KV, Dhumal RK, Nagne AD, Mehrotra, S C, Surase RR (2018) Drought severity identification and classification of the land pattern using landsat 8 data based on spectral indices and maximum likelihood algorithm. In: S. Springer, Microelectronics, Electromagnetics and Telecommunications (vol 521, pp. 517–524). Springer, Singapore
22. Nagy A, Szabó A, Adeniyi OD, Tamás J (2021) Wheat yield forecasting for the tisza river catchment using landsat 8 NDVI and SAVI time series and reported crop statistics. Agronomy 11(4):652
23. Zhou J, Khot LR, Bahlol HY, Boydston R, Miklas PN (2016) Evaluation of ground, proximal and aerial remote sensing technologies for crop stress monitoring. IFAC-PapersOnLine 49(16):22–26
24. Zhou J, Pavek MJ, Shelton SC, Holden ZJ, Sankaran S (2016) Aerial multispectral imaging for crop hail damage assessment in potato. Comput Electron Agric 127:406–412
25. Knight AW, Tindall DR, Wilson BA (2009) A multitemporal multiple density slice method for wetland mapping across the state of Queensland Australia. Int J Remote Sens 30(13):3365–3392
26. Odeh J, Ahmad HF, Othman M, Johari R (2004) Image retrieval system based on density slicing of colour histogram of images subareas and colour pair segmentation. Int Arab J Inf Technol 1:196–202
27. Van Niel TG, McVicar TR, Datt B (2005) On the relationship between training sample size and data dimensionality: Monte Carlo analysis of broadband multi-temporal classification. Remote Sens Environ 98(4):468–480
28. Lv Z, Shi W, Zhou X, Benediktsson JA (2017) Semi-automatic system for land cover change detection using bi-temporal remote sensing images. Remote Sens 9(11):1112
29. Dong J, Huang Y, Zhuang D, Zhu Y, Xu X, Ren H (2012) A simple semi-automatic approach for land cover classification from multispectral remote sensing imagery. PLoS ONE 7(9):e45889
30. Awate P, Nagne A (2023) Recent trends in crop identification and analysis by using remote sensing and gis. Online J Distance Educ E-Learn 11(1), 275–283.

AgroGenie: A Smart Approach to Agriculture Using Machine Learning

Abhiroop Sarkar, Kinshuk Ganguly, Snehal Ghosh, Debayan Ghosh, Subhajit Saha, Piyali Datta, and Deepsubhra Guha Roy

Abstract Precision agriculture is an instrumental approach for a nation whose economy is greatly influenced by agricultural yield. Crop Recommendation is a very valuable tool in the field of agriculture. This research involves the utilization of a machine learning-based approach to analyze the values of several parameters influencing crop growth. It then accepts a novel set of input parameters and recommends the most probable crops that can be cultivated. Adequate preprocessing methods have been applied to ensure the optimum removal of partiality from data. The ML models have been compared for their performance using a well-accepted validation measure. Then crops with similar features have been clustered to provide a more flexible farming option for the farmer.

Keywords Machine learning in agriculture · Agriculture 4.0 · Crop recommendation · Precision agriculture · Smart agriculture

1 Introduction

Applications of machine learning are found in many fields, including urbanization, education, health, and the military. These applications have a big influence on analysis and decision-making [1]. Utilizing 60% of the land resources in India, agriculture

Supported by IEM Grant in Aid Project from Institute of Engineering and Management, Kolkata, India.

A. Sarkar (✉) · K. Ganguly · S. Ghosh · D. Ghosh · S. Saha
Institute of Engineering and Management, University of Engineering and Management, Kolkata, India
e-mail: abhiroopsarkar2004@gmail.com

P. Datta
IEM Centre of Excellence for Data Science, Department of CSE (AIML), Institute of Engineering and Management, University of Engineering and Management, Kolkata, India

D. G. Roy
IEM Centre of Excellence for Cloud Computing and IoT, Department of CSE (AIML), Institute of Engineering and Management, University of Engineering and Management, Kolkata, India

A. J. Kulkarni and M. Hajiaghaei-Keshteli (eds.), *Advances in Intelligent Systems for Sustainable Agriculture*, Lecture Notes in Networks and Systems 1201,
https://doi.org/10.1007/978-981-97-9839-1_17

is a significant industry that provides food for the nation's 1.3 billion inhabitants. Reduced yield and soil pollution are caused by variables including depending just on the farmer's experience, not knowing enough about the nutrients in the soil, and lacking precision when predicting crops and yields [2]. These days, it's typical practice to forecast and analyze crop development using machine learning approaches [3, 4]. IoT and ML technologies are integrated with smart farming to maximize output and reduce resource use [5]. Acquiring accurate data and applying machine learning algorithms to analyze it are necessary for transforming agriculture. ML models built on large-scale farm data enhance crop growth and output analysis and prediction with high accuracy [6]. Crop productivity is affected by agroclimatic conditions, which makes data collecting difficult. The Indian Meteorological Department provides vital meteorological data for predicting major agricultural impacts. Various forecasting methods employed worldwide, including studies carried out in Pakistan on the correlation between pesticide use and productivity, illustrate how data mining may be leveraged to enhance agricultural operations [7]. Notwithstanding challenges like sensor costs and data preparation, the potential benefits of applying ML in agriculture are beginning to become apparent. Utilizing machine learning to its greatest potential in this field is the aim of continuing research [8].

To tackle various obstacles, a system was created that offers answers by suggesting suitable crops based on weather conditions and soil composition, such as temperature, humidity, pH, and rainfall [9, 10]. Furthermore, data mining techniques have been applied in research on droughts, floods, and forecasting extreme rainfall occurrences, indicating their relevance in various agricultural settings [11, 12]. Soil classification [13], weather simulations [14], intelligent agriculture land grading systems, pollution forecasts [15], support vector machines [16], and decision tree algorithms [17] are a few uses for spatial data mining.

This project aims to help farmers recommend the most suitable crop(s) to farm on their land from the list of 22 crops present in the dataset [18] based on some parameters: Nitrogen, Phosphorus, and Potassium content of soil, pH level of soil, Yearly rainfall, average humidity and temperature of the region. Beneficiaries and Usage of our system:

- Farmers This project aims to help farmers get the best yield for their soil condition.
- Residents of drought-prone areas: Farmers in India generally default to water-intensive crops like rice or wheat. This project should help reduce farmers' over-reliance on rice and wheat, thus helping save groundwater resources.
- Environmentalists: A large part of India's electricity mix is generated by burning coal in thermal power plants. Due to large power subsidies for farmers, this often results in them running their pumps for long periods to pump out the ever-reducing groundwater to irrigate their water-intensive crops like paddy. Since we also take rainfall into account, we can recommend crops that require less water, thereby reducing electricity demand and consequently, the pollution levels in the country.
- Impoverished people: A large number of people in India and around the world cannot afford to buy food at market rates. Hence, they are often dependent on the government to access subsidized food grains. Our project aims to optimize and

maximize food production in the country and abroad, which should bring down crop prices worldwide. People everywhere should have access to affordable food.

Help farmers improve their yield, and farm more efficiently, which may indirectly lower the food prices in our country.

Agriculture although beneficial for the existence of human beings faces challenges due to the rising population and climate change. Thus, agriculture 4.0, based on industry 4.0, is crucial for developing agriculture. In 2021, researchers found that the number of publications increased by 36.5% between 2011 and 2020 [19]. This reflected an increase in the emergence of advanced technologies. They also utilized NLP techniques to perform frequency analysis and find out the top 50 bi-grams and tri-grams. Internet of Things and machine learning were the top 2 bigrams. Decision support systems, support vector machines, machine learning techniques, machine learning algorithms, and wireless sensor networks have the highest frequencies, which aligns with the previous findings. They identified five major Agriculture 4.0 technologies: sensor and robotics, Internet of Things, cloud computing, data analytics, and decision support systems.

2 Literature Survey

Being a very crucial and growing problem domain, innumerable research has been continuing in leading agriculture towards its smarter form which fundamentally focuses on making agriculture efficient and uncomplicated for farmers. The very first step is to suggest suitable crops based on the independent parameters of the field into consideration. Kumar et al. [20] used Support Vector Machine, Decision Tree, and Logistic Regression for crop selection on a dataset consisting of 52611 data values from different 15 crops, where the SVM classification model offers a better accuracy of 89.66% compared to other algorithms. Pudumalar, in [21], deploys Naive Bayes and KNN classifier algorithms for crop yield prediction. Artificial neural networks and multi-layer perceptron were used for the prediction of food production and import achieving high efficiency and accuracy in crop yield prediction.

Ruchirawya, Thewahettige [22] used Arduino microcontrollers to collect data while several ML techniques including Naïve Bayes, Support Vector Machine, and K-Means Clustering were employed afterward. They also included Natural Language Processing for sentiment analysis within their model. An accuracy of over 92% was achieved, with the potential to reach 95% accuracy through continuous learning from farmer feedback. Udutalapally et al. [23] implemented a CNN model for disease prediction in plants, utilizing a health maintenance system with an edge device layer for real-time sensor data analysis. They attained a 99% accuracy with the proposed system and demonstrated prototype resilience to rust and weather variations for three months. Madhuri Shripathi Rao et al. [24] compared three supervised machine learning models: KNN, Decision Tree, and Random Forest to predict the most suitable crop for a particular land and climate on the same dataset that is used in this paper.

Table 1 List of the machine learning models deployed in the state-of-the-art works. **LR** Logistic Regression, **SVM** Support Vector Machine, **DT** Decision Tree, **kNN** k Nearest Neighbour, **NB** Naive Bayes, **RF** Random Forest, **CNN** Convoluted Neural Network, **NLP** Natural Language Processing, **ANN** Artificial Neural Network

Techniques	LR	SVM	DT	KNN	NB	RF	CNN	NLP	ANN
Kumar et al.	✓	✓	✓						
Pudumalar				✓	✓				✓
Ruchirawya, Thewahettige		✓		✓	✓			✓	
Udutalapally et al.							✓		
M. Shripathi Rao et al.		✓	✓			✓			

They concluded that Random Forest Classifier both in Entropy and Gini Criterion had the highest accuracy of 99.32%, Decision Tree Classifier had accuracy scores of 97.04% and 97.95% for Entropy and Gini Criterion respectively, and KNN had the lowest accuracy score of 97.04% (Table 1).

3 Methodology

3.1 Initial Approach

Collection and preprocessing of training dataset In this investigation, the dataset, integral to our project, was acquired from Kaggle [18]. Comprising 2200 instances distributed across 7 features and 22 crop labels, each class is represented by 100 instances. The 7 features represent significant parameters that contribute to crop growth. They include soil nutrient content (Nitrogen, Phosphorus, and Potassium), Average Annual Temperature, Average Humidity percentage, Annual Rainfall in cm, and pH level of the soil (Fig. 1).

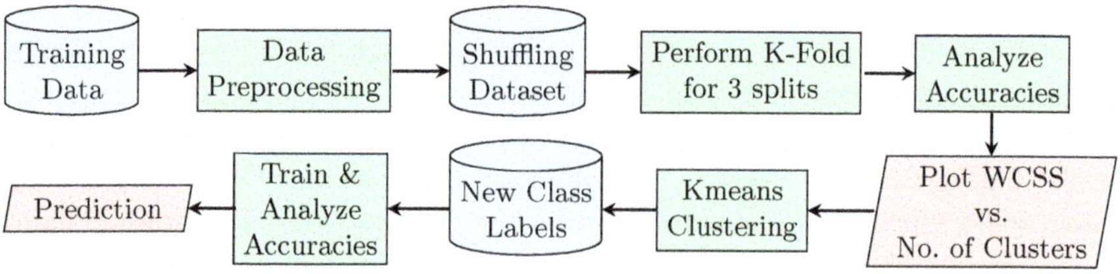

Fig. 1 Flowchart of the approach

The analysis revealed an absence of null values and duplicate instances within the dataset. To gain insights into the interrelationships among features, a correlation matrix and its visual counterpart in the form of a heatmap were computed. It was concluded that a negligible correlation exists between features.

A distinctive characteristic of the original dataset is its sequential arrangement, where all classes and their corresponding instances are ordered consecutively. To prevent any inadvertent biases that could be imprinted upon the model during training, the dataset was shuffled. This ensures the integrity and objectivity of the model training process, shielding against unwarranted presumptions that may arise due to the sequential nature of the original dataset.

Additionally, scaling operations such as Min-Max scaling and standard scaling have been performed. This ensures better performance of clustering algorithms decreasing biases arising due to abnormal data values in the dataset (Figs. 2, 3 and 4).

Selection and Training of ML Model The research objective aimed at effecting classification among 22 distinct crop categories, leveraging a suite of classifier models, including Logistic Regression, Support Vector Machine, Decision Tree, K nearest neighbor, and Random Forest Classifier. It was imperative to address the non-mutually exclusive nature of features and their values, as the feature-pH exhibited susceptibility to variations in soil nitrogen, phosphorus, and potassium content. Consequently, the Naive Bayes Classifier was excluded from consideration.

The training and testing data were subjected to min-max scaling and standardized scaling before cross-validation. The evaluation of the final set of machine learning classification models utilized the K-fold cross-validation method, employing a 3-fold configuration, and the Random Forest Classifier was found to outperform the rest. The

To facilitate comprehensive analysis, learning curves for each model were generated, contributing to refined visualization and interpretation of the model's behavior. This methodological approach ensures the robustness and scientific rigor of the research findings within the given dataset (Table 2).

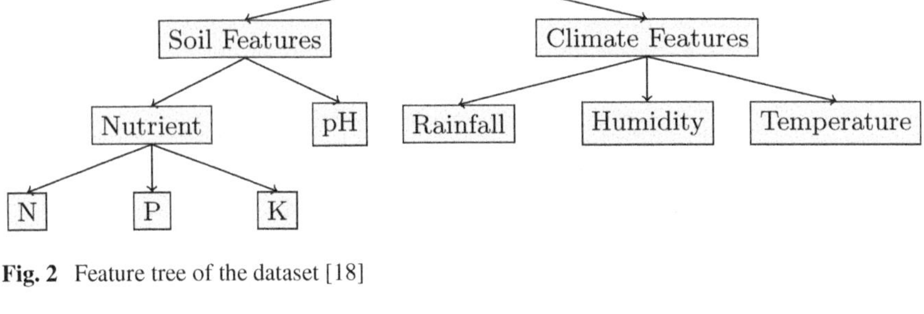

Fig. 2 Feature tree of the dataset [18]

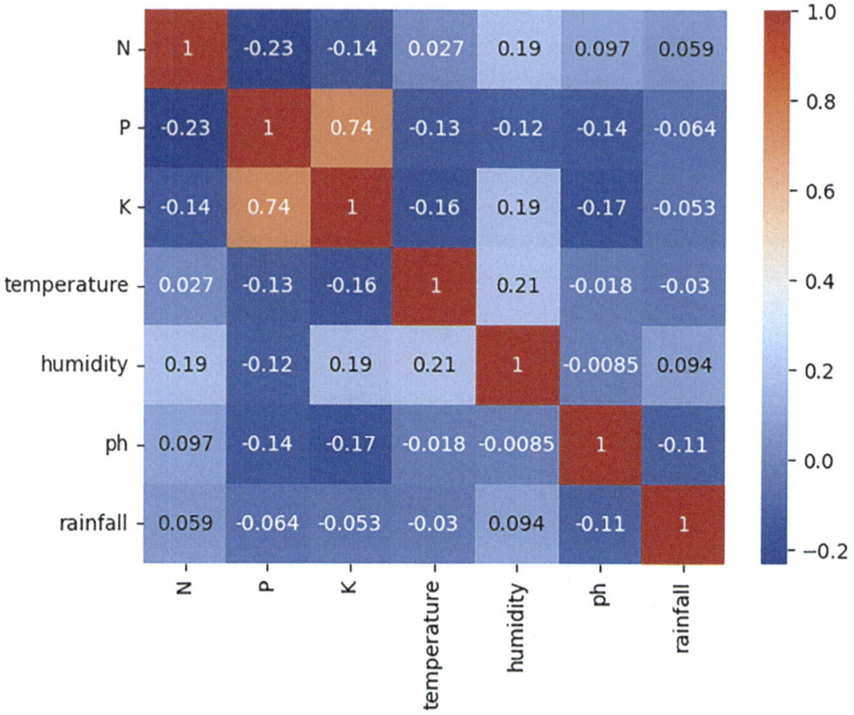

Fig. 3 Heatmap illustrating the correlation between the features [18]

3.2 Improved Approach

Motivation behind clustering The initial dataset contained feature vectors representing individual crop labels as class values. However, suggesting a single crop for farmers to cultivate may not always be convenient, as they may already have the resources and experience for a specific crop. Abruptly switching to a new crop might not be feasible. Therefore, a more practical approach is to offer farmers a range of crops with similar feature values. This enables them to choose a suitable crop based on their existing resources and capabilities. To achieve this, clustering is applied to the data using the K-means algorithm, creating clusters or groups of crops with similar features as new class labels. This is thus an instance of clustering followed by classification on newly generated clusters as class labels [25].

Selecting an appropriate number of clusters and training models on new clusters as class labels To determine the optimal number of clusters for the dataset, an elbow curve was constructed. The X-axis of the curve represents the number of clusters, while the corresponding Within Cluster Sum of Squares (WCSS) distance is plotted on the Y-axis. Upon examination, the elbow curve from Fig. 5 exhibited a noticeable bend within the range of 8–11 clusters. After identifying the potential

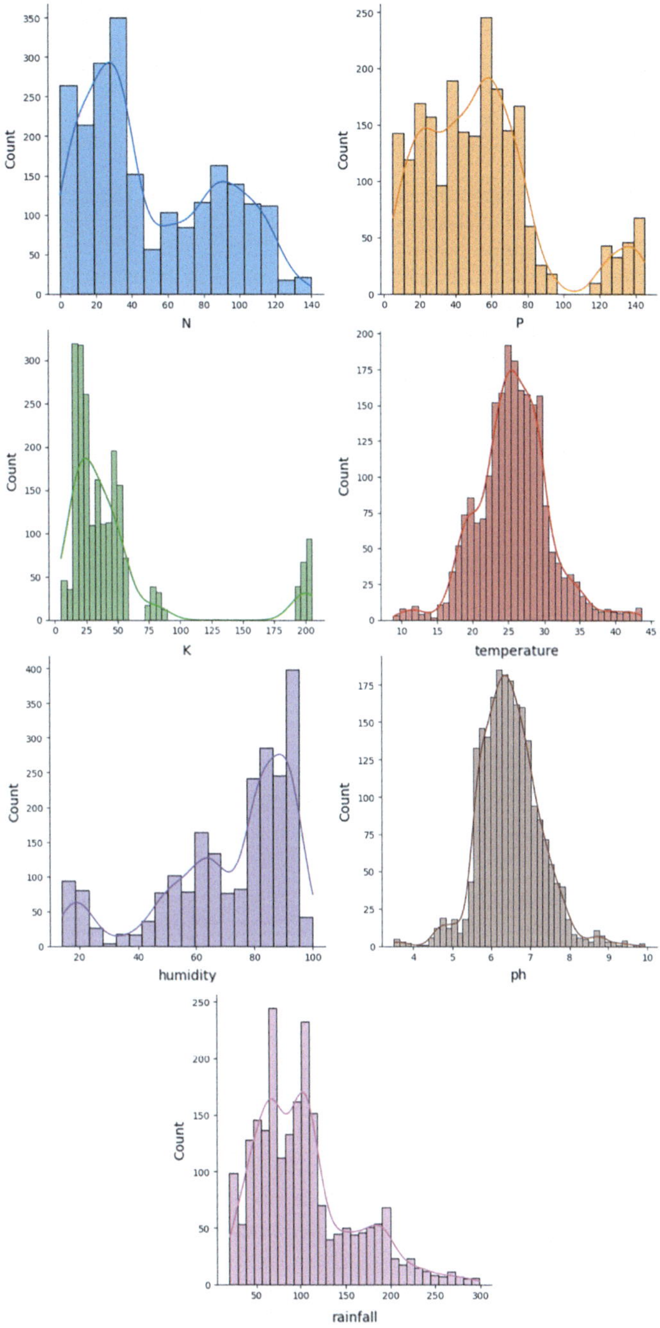

Fig. 4 Distplots for 7 features- N, P, K, temperature, humidity, pH and rainfall [18]

Algorithm 1 *Clustering and new label designation*

1. Start
2. Segregate the feature values from the dataset to serve as input for the unsupervised clustering algorithm
2. Compute and Store the **WCSS (Within-Cluster Sum of Squares)** for different cluster values 1 $\rightarrow$ 22 inside a list
5. Obtain the **Elbow Curve** to show the relation between WCSS and no of clusters and identify the potential range for the number of clusters needed
6. **Standardize numeric features** $\longrightarrow$ $standardscalar$
7. **Perform clustering** for each of the potential number of clusters and **check the corresponding accuracies** of the different classifier models by using the clusters as new class labels
9. Initialize a Dictionary **Cluster** $\longleftarrow$ **Original Crop labels** with their **respective Frequency**
10. During prediction we only predict the **class** labels belonging to the cluster with **majority frequency**

Table 2 Accuracies of model when trained for individual class labels

Models	Average accuracy
Logistic regression	98.4545
Support vector machine	99.3637
Random forest classifier	99.602
K nearest neighbor	99.409
Decision tree	99.482

number of clusters from the elbow curve, six classifier models were trained using the newly formed clusters for each candidate cluster number. The accuracies were then determined using the K-fold cross-validation technique with 3 folds. Following this evaluation, the number of clusters was finalized at 10, determined by selecting the cluster configuration that yielded the highest accuracy. The Random Forest Classifier was subsequently chosen as the preferred model for further analysis.

Mapping newly formed clusters to the existing crop labels for final prediction Upon the formation of the ten clusters, a dictionary was created. In this dictionary, each key denoted the cluster index, and the associated contents were the crop labels from the original dataset whose instances had been included in that particular cluster. Alongside the crop labels, the frequency of each crop label's appearance in the specific cluster was also recorded. This approach aimed to prevent the influence of outliers by maintaining a count of how often instances of a particular crop label appeared in a given cluster. The trained model when provided with an input vector predicts a cluster number. The crop labels mapped to that cluster number are then displayed as outputs only if the crop labels are the majority labels(greater than 50 out of 100 total instances) in the given cluster (Table 3).

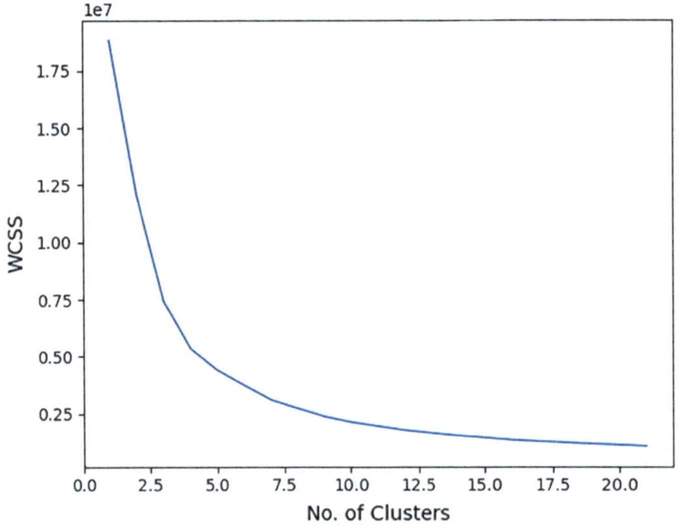

Fig. 5 Elbow curve—within cluster sum of squares (Y-axis) versus number of clusters (X-axis)

Table 3 Comparative analysis of accuracy of different models about the number of clusters formed

Models	Average accuracy models as per number of clusters		
	9 clusters	10 clusters	11 clusters
Logistic regression	94.5000	94.500	94.5005
Support vector machine	92.7277	92.2276	92.7277
Random forest classifier	97.3636	97.9545	97.4546
K nearest neighbor	93.0455	92.2732	93.0455
Decision tree	96.1816	96.7724	95.9545

4 Results and Analysis

For 10 clusters, which have the highest accuracy, the clusters are indexed from 0 to 9 and the respective frequencies for each crop are as follows in Table 4.

Table 3 illustrates the number of clusters and corresponding accuracy values of the different classifier models. It is evident that for 10 clusters, the Random Forest classifier shows a maximum accuracy of 97.9545%. Table 5 illustrates the classification report for the Random Forest Classifier model while being trained and tested on 10 clusters which are treated as class labels. Metrics used: Precision(Prec): Precision refers to a classifier's ability to identify the number of positive predictions that are relatively correct. It is calculated as the ratio of true positives to the sum of true and false positives for each class. Recall(Rec): Recall is the capability of a classifier to

Table 4 Cluster information and predominant crops in **Bold**

Cluster	Predominant crops	Frequency
0	Lentil, Chickpea, Pigeonpeas, **Kidneybeans**	4,5,32,100
1	Papaya, Coffee, **Maize, Cotton, Banana**	8, 38, 99, 100, 100
2	Pigeonpeas, Papaya, **Coffee, Jute, Rice**	1, 18, 62, 100, 100
3	**Apple, Grapes**	100, 100
4	**Watermelon, Muskmelon**	100
5	Mothbeans, **Pigeonpeas, Mango**	42, 55, 89
6	Maize, Orange, Pigeonpeas, Mango, Papaya, **Mothbeans, Lentil, Blackgram, Mungkbean**	1, 2, 9, 11, 13, 58, 96, 100, 100
7	Pigeonpeas, Orange, **Papaya**	3, 18, 58
8	Papaya, **Orange, Coconut, Pomegranat**e	3, 80, 100, 100
9	**Chickpea**	95

Table 5 Merged classification reports for folds 1, 2, and 3 for random forest

Folds	Fold 1				Fold 2				Fold 3			
Class	Prec	Rec	F1-S	Supp.	Prec	Rec	F1-S	Supp.	Prec	Rec	F1-S	Supp.
0	0.98	0.96	0.97	52	0.97	0.90	0.94	42	0.98	0.98	0.98	47
1	0.96	0.97	0.97	109	0.99	1.00	1.00	116	0.98	0.99	0.98	120
2	0.98	0.98	0.98	106	1.00	0.99	0.99	100	0.96	0.96	0.96	75
3	1.00	1.00	1.00	57	1.00	1.00	1.00	67	1.00	1.00	1.00	76
4	1.00	1.00	1.00	62	1.00	1.00	1.00	64	1.00	1.00	1.00	74
5	0.93	1.00	0.96	65	0.94	0.94	0.94	53	0.97	0.99	0.98	68
6	0.99	0.95	0.97	130	0.94	0.96	0.95	137	0.98	0.98	0.98	123
7	0.96	0.93	0.95	29	0.92	0.88	0.90	26	0.92	0.92	0.92	24
8	1.00	1.00	1.00	89	0.99	1.00	1.00	102	1.00	0.97	0.98	92
9	0.97	1.00	0.99	35	0.96	1.00	0.98	26	1.00	1.00	1.00	34
Accuracy			0.98	734			0.98	733			0.98	733
Macro Avg	0.98	0.98	0.98	734	0.97	0.97	0.97	733	0.98	0.98	0.98	733
Wt. Avg	0.98	0.98	0.98	734	0.98	0.98	0.98	733	0.98	0.98	0.98	733

discover all positive cases from the confusion matrix. It is calculated as the ratio of true positives to the sum of true positives and false negatives for each class. F1 score (F1-s): The F1 score is a weighted harmonic mean of precision and recall, with 0.0 being the worst and 1.0 being the best. Since precision and recall are used in the computation, F1 scores are often lower than accuracy measurements. Accuracy: The number of correct predictions divided by the total number of predictions is called accuracy (Table 6).

Table 7 shows two examples of prediction using our trained ML model with some arbitrary values as input.

Table 6 Accuracy scores for random forest models

Model	Accuracy (%)
1	97.9564
2	97.6807
3	98.2265
Average	**97.9545**

Table 7 Crops prediction based on input values

N	P	K	Temperature	Humidity	pH	Rainfall	Prediction
61	37	40	27	92	5.5	148	Rice, jute, coffee
99	18	47	31	85	6.2	27	Muskmelon, watermelon

5 Conclusion

In conclusion, this work employed K-means clustering, an unsupervised machine learning technique, on a dataset comprising 2200 entries with 7 features and 22 crop labels. The primary objective was to recommend suitable crops for input data of different climate and soil-based parameters to maximize yield and avoid wastage of efforts, time, and resources. It was also aimed at listing crops with similar features under a single cluster to allow farmers to make an informed choice as on the availability of their resources and their experience. This helps to ensure that farmers can engage in cultivation at their convenience. Subsequently, K-fold cross-validation was employed to assess the performance of various classifier models using the generated clusters as target class labels. Among the classifiers investigated, the Random Forest classifier exhibited the highest accuracy, reaching 97.9545% with the utilization of 10 clusters.

In light of the results obtained, this research provides valuable insights into optimizing classification accuracy in agricultural datasets. The identification of an optimal number of clusters and the subsequent application of K-fold cross-validation contribute to the betterment of machine learning methods for crop classification. As precision agriculture increasingly relies on data-driven decision-making, the findings presented in this study hold significance for improved accuracy in crops. The outcomes show the practical application of unsupervised learning techniques in enhancing the efficiency of precision agriculture practices.

6 Future Scope

The Nitrogen, Phosphorus, and Potassium content of the soil of the agricultural land could be measured with the help of an NPK sensor, while the pH level of the soil could be measured with the help of a pH meter. These sensors would be connected to a microcontroller ESP module powered with Wi-Fi-enabled communication ability. It can transmit and receive data from a server where the values recorded by sensors are stored. The data related to climate, i.e. Temperature, Rainfall, and Humidity can be collected from Openweather API for prediction. The model's functionality can be utilized using an interactive user interface as a web application.

Acknowledgements We extend our sincere appreciation to the IEM Centre of Excellence for Data Science and the IEM Centre of Excellence for Cloud Computing and IoT for their invaluable support and resources through the IEM Grant in Aid project led by Dr. Piyali Datta and Dr. Deepsubhra Guha Roy. Their assistance has been instrumental in facilitating our research endeavors. Additionally, we express our sincere gratitude to the IEDC-CSE of the Institute of Engineering and Management, Kolkata, for generously providing the workspace essential for the completion of this work.

References

1. Markov T, Zhang C, Agarwal S, Eloundou T, Lee T, Adler S, Jiang A, Weng L (2022) New and improved content moderation tooling
2. Zingade D, Buchade O, Mehta N, Ghodekar S, Mehta C (2017) Crop prediction system using machine learning. Int J Adv Eng Res Dev Spec Issue Recent Trends Data Eng 4(5):1–6 (2017)
3. Li L, Wang B, Feng P, Li Liu D, He Q, Zhang Y, Wang Y, Li S, Lu X, Yue C et al (2022) Developing machine learning models with multi-source environmental data to predict wheat yield in china. Comput Electron Agric 194:106790
4. Van Klompenburg T, Kassahun A, Catal C (2020) Crop yield prediction using machine learning: a systematic literature review. Comput Electron Agric 177:105709
5. Kuradusenge M, Hitimana E, Hanyurwimfura D, Rukundo P, Mtonga K, Mukasine A, Uwitonze C, Ngabonziza J, Uwamahoro A (2023) Crop yield prediction using machine learning models: case of irish potato and maize. Agriculture 13(1):225
6. Shaikh TA, Rasool T, Lone FR (2022) Towards leveraging the role of machine learning and artificial intelligence in precision agriculture and smart farming. Comput Electron Agric 198:107119
7. Abdullah A, Brobst S, Pervaiz I (2004) Learning dynamics of pesticide abuse through data mining. pp 151–156
8. Rashid M, Bari B, Yusup Y, Kamaruddin M, Khan N (2021) A comprehensive review of crop yield prediction using machine learning approaches with special emphasis on palm oil yield prediction. IEEE Access 9. https://doi.org/10.1109/ACCESS.2021.3075159
9. Samui P, Mandla VR, Krishna A, Teja T (2011) Prediction of rainfall using support vector machine and relevance vector machine. Earth Sci India 4
10. Sharma H, Kumar S et al (2016) A survey on decision tree algorithms of classification in data mining. Int J Sci Res (IJSR) 5(4):2094–2097
11. Dhanya C, Nagesh Kumar D (2009) Data mining for evolution of association rules for droughts and floods in india using climate inputs. J Geophys Res Atmosph 114(D2)
12. Kannan M, Prabhakaran S, Ramachandran P (2010) Rainfall forecasting using data mining technique. Int J Eng Technol 2:397–401

13. Verheyen K, Adriaens D, Hermy M, Deckers S (2001) High-resolution continuous soil classification using morphological soil profile descriptions. Geoderma 101(3–4):31–48
14. Rajagopalan B, Lall U (1999) A k-nearest-neighbor simulator for daily precipitation and other weather variables. Water Resour Res 35(10):3089–3101
15. Jorquera H, Perez R, Cipriano A, Acuna G (2001) Short term forecasting of air pollution episodes. In: Zannetti P (eds) environmental modeling 4
16. Tripathi S, Srinivas V, Nanjundiah RS (2006) Downscaling of precipitation for climate change scenarios: a support vector machine approach. J Hydrol 330(3–4):621–640
17. Jia Z (2009) An expert system based on spatial data mining used decision tree for agriculture land grading. In: 2009 Second international conference on intelligent computation technology and automation, vol 4, pp 142–145. IEEE
18. Crop recommendation dataset. url https://www.kaggle.com/datasets/atharvaingle/crop-recommendation-dataset/data (2020)
19. Araújo SO, Peres RS, Barata J, Lidon F, Ramalho JC (2021) Characterising the agriculture 4.0 landscape-emerging trends, challenges and opportunities. Agronomy 11(4). https://doi.org/10.3390/agronomy11040667, https://www.mdpi.com/2073-4395/11/4/667
20. Kumar A, Sarkar S, Pradhan C (2019) Recommendation system for crop identification and pest control technique in agriculture, pp 0185–0189. https://doi.org/10.1109/ICCSP.2019.8698099
21. Pudumalar S, Ramanujam E, Rajashree RH, Kavya C, Kiruthika T, Nisha J (2017) Crop recommendation system for precision agriculture. In: 2016 Eighth international conference on advanced computing (ICoAC), pp 32–36. https://doi.org/10.1109/ICoAC.2017.7951740
22. Ruchirawya T (2020) Crop recommendation system, vol 175
23. Udutalapally V, Mohanty S, Pallagani V, Khandelwal V (2020) Scrop: a novel device for sustainable automatic disease prediction, crop selection, and irrigation in internet-of-agro-things for smart agriculture. IEEE Sensors J PP. https://doi.org/10.1109/JSEN.2020.3032438
24. Rao MS, Singh A, Reddy NS, Acharya DU (2022) Crop prediction using machine learning. J Phys Conf Series 2161(1):012033. https://doi.org/10.1088/1742-6596/2161/1/012033
25. Yaswanth Kumar Alapati KS (2016) Combining clustering with classification: A technique to improve classification accuracy. Int J Comput Sci Eng (IJCSE) 5(6)

Implementation of DCNN Framework—Auto Identification and Categorization of Various Stress in Paddy Crop and Resource Management System

Vilas Gaikar, Caroleena Rane, Narendra Mustare, and Rajiv Nayan

Abstract To effectively mitigate the potential loss of agricultural production, it is imperative to promptly identify stressors in paddy crops during the booting development stage and implement appropriate measures to address them. This proactive approach aims to minimize both qualitative and quantitative impacts on agricultural output. The visual nature of symptoms has historically required the involvement of human specialists in the detection and classification of stress in traditional paddy crops. This study presents a novel deep convolution neural network (DCNN) framework designed to automate the identification and categorization of various biotic and abiotic stress factors affecting rice crops based on field photographs. The research employs the VGG-16 Convolutional Neural Network (CNN) model, which has undergone prior training, to autonomously categorize images of rice crops experiencing stress during the trunk growing phase. The disciplined models exhibit an common truth of 90.79% when evaluated on the concealed dataset. This outcome provides evidence for the practical viability of employing bottomless knowledge techniques on a dataset comprising 20,000 field photos. These photos depict five distinct varieties of rice crops and are annotated with ten different stress categories, including a category representing healthy or normal conditions. The findings presented in this study have the potential to enhance agricultural and resource management automation

---

V. Gaikar (✉)
Department of Economics, Smt. CHM. College, University of Mumbai, Mumbai, India
e-mail: gaikar_vilas@rediffmail.com

C. Rane
Department of Chemistry, Indrayani Education Society's, Arts, Comm. & Science College, University of Mumbai, Mumbai, India

N. Mustare
Department of EIE, CVR College of Engineering, Ibrahimpatan, Hyderabad, India
e-mail: narendra@cvr.ac.in

R. Nayan
Ramanujan College, University of Delhi, Delhi, India

A. J. Kulkarni and M. Hajiaghaei-Keshteli (eds.), *Advances in Intelligent Systems for Sustainable Agriculture*, Lecture Notes in Networks and Systems 1201,
https://doi.org/10.1007/978-981-97-9839-1_18

systems as well as mobile applications. The aforementioned approach is characterized by subjectivity and a susceptibility to errors, thereby potentially leading to the formulation of unsuitable assessments in the context of stress management.

Keywords Agricultural production · Paddy crops · VGG-16 · CNN · Field photos

1 Introduction

India's agricultural sector relies heavily on the cultivation of the paddy crop, which is grown on an area of 33.92 million hectares and yields 11.50 million tons per year [1–3]. Despite these remarkable totals, rising populations in Asia have put a strain on the region's rice harvest. Modifying elements of society and economy [4, 5]. The main rice-growing regions of the efficiency of manufacturing in the Asia area is barely 20% of the global total. Destruction brought on by drought, illness and pests [6]. To reduce the agricultural yield gap, one alternative is to adopt a more integrated strategy that incorporates the best crop improvement and resource management strategies already used by farmers, such as irrigation regime, nutrient management, and other agronomic direction aspects rice yield [7, 8]. As rice crops grow in a broad variety of habitats with climates, varying temperatures, and soil–water conditions, they are subjected to a number of stressors, both biotic and abiotic, that may have a negative impact on crop production and selection [9, 10]. The stressors are often seen as a major obstacle to long-term paddy crop success. The average global crop output is cut by more than 40 percent owing to environmental challenges such over-irrigation or drought, nutrient deficiencies, pathogens, toxicity due to over-feeding of fertilizers, and excessive salt [11, 12]. Through natural selection, the plans have developed specialized systems to cope with extreme adversity. Plants react differently to different stresses based on factors such as the kind of stress, its intensity, the age of the plant when it is exposed to the stress, and the plant's natural tolerance to stress. The negative impacts of stressors on crops may be mitigated via timely monitoring and detection of these strains, provision of suitable agricultural inputs, and prompt morphological diagnostics. Symptoms of stress in paddy crops manifest visually as changes in leaf color, shape, and texture [13, 14]. Microsymptoms are difficult to detect and measure by human eye perception alone. Figure 1 displays field examples of both healthy and damaged paddy crops [15, 16].

Two visual score signs utilized in classifying and determining stress categories are leaf injury and leaf color change [17]. Stress types are identified by a qualitative evaluation of these symptoms performed by qualified individuals. The process of visually inspecting each leaf is subjective, time-consuming, potentially harmful, and boring. Plant Nitrogen (N) nutrient insufficiency may be identified with the use of two simple decision instruments: a Leaf Color Chart (LCC) and a Chlorophyll Meter (SPAD). The LCC analyzes the leaf chlorophyll content of a rice crop to determine the plant's nitrogen status [18, 19]. Since LCC relies on human observation of leaf color, its reliability suffers when the illumination changes. Using a Leaf Color Chart

Fig. 1 Image **a** and **b** Field photos of normal and stressed paddy crops

(LCC) and chlorophyll, as shown in Fig. 2, one may calculate the amount of nitrogen (N) required by paddy plants. Potential farmers and researchers would benefit from an objective and speedy stress diagnosis system so that they may intervene and mitigate issues as soon as possible, hence increasing crop yields via the use of appropriate crop management practices [20, 21]. Using image process and machine learning mortal had developed an amount of methods identifying and categorizing plant stressors [22].

By using VGG model, we can improve the overall performance of image classification/categorization. For VGG model, preferred per-trained models are available for convenient transfer learning. VGG is simple to implement and interpret [23].

The SPAD meter and the Leaf Color Chart (LCC) are two typical methods for determining the nitrogen status of rice crops. In order to efficiently monitor and control nitrogen fertilizer in paddy fields, farmers, agronomists, and researchers employ these technologies. Why don't we look into both of them? [24].

Fig. 2 Present paddy crop nitrogen mandate tools, SPAD and LCC

1.1 Soil–Plant Analysis and Growth (SPAD) Meter

The SPAD meter is a portable instrument for determining the chlorophyll concentration in plant leaves, such as those of paddy crops. Photosynthesis relies on chlorophyll, and a plant's chlorophyll concentration may be a good indicator of its general health.

1.2 The Nuts and Bolts

The SPAD meter utilizes a measurement of light transmission to ascertain leaf health. To determine the nitrogen status a certain wavelength of light is passed onto the leaf and the amount of light reflected through the leaf's surface is recorded using SPAD meter. The recorded data is used to measure the nitrogen content as the amount of light transmitted is influenced by the chlorophyll content in the leaf [25].

1.3 Usage

To establish the crop's nitrogen fertilization rates, the SPAD readings are collected by farmers and academics from different leaves in a variety of fields using SPAD meters and are compared with calibration charts or thresholds [26].

1.4 Leaf Color Code (LCC)

A Leaf Color Chart (LCC) is used to tell whether plant needs any additional nitrogen. LCC is a simple visual tool which indicated the approximate nitrogen content in paddy and other crops [27].

1.5 The Nuts and Bolts

The LCC is a collection of color charts or strips displaying a spectrum of green tones. Farmers or agronomists use the color strips to determine the closest match between the paddy leaves and the paddy soil [28].

1.6 Usage

The LCC is used by farmers to determine the quality of their paddy by comparing the color of randomly selected leaves to corresponding LCC charts. The crop's nitrogen status may be deduced from the color match. A nitrogen deficit may be indicated if the leaf color is lighter on the LCC, whereas adequate nitrogen levels may be indicated if the leaf color is darker on the LCC. Optimizing nitrogen fertilization may enhance crop health, boost yields, and decrease environmental impact; the SPAD meter and LCC are both useful methods for measuring the nitrogen status of rice crops. However, correct use and interpretation need adhering to local rules and expert suggestions. It's also important to remain current with the newest changes in agricultural operations, since additional technology or upgrades may have surfaced since my previous update [29].

The study examines the relationship between the initial soil nitrogen levels and the demand for crops. A research investigation is required to examine the application of crop-demand-based nitrogen management by evaluating the development of soil plant analysis development (SPAD) values. Additionally, it is necessary to establish a correlation between SPAD values and leaf color chart (LCC) readings for a particular crop variety and season. There are two nondestructive diagnostic tools available for determining the nitrogen (N) status of leaves: the chlorophyll meter (SPAD 502) and the leaf color chart (LCC). The leaf color chart, developed by the International Rice Research Institute (IRRI) in Manila, Philippines, is a cost-effective and straightforward tool used to identify the optimal timing for nitrogen application in rice cultivation. This allows for the synchronization of nitrogen application with the crop's demand. The utilization of nitrogen (N) can be effectively enhanced through the utilization of an ideal and simplified tool, regardless of the amount of nitrogen applied.

2 Methodology

In order to address the aforementioned situation, a deep learning methodology has been implemented to identify and categorize instances of stress in paddy crops. The proposed methodology comprises two primary stages, specifically, the process of preparing and utilizing deep learning techniques and image dataset for classification purposes. Figure 3 depicts a block diagram that visually represents the various stages encompassed within the proposed methodology [30].

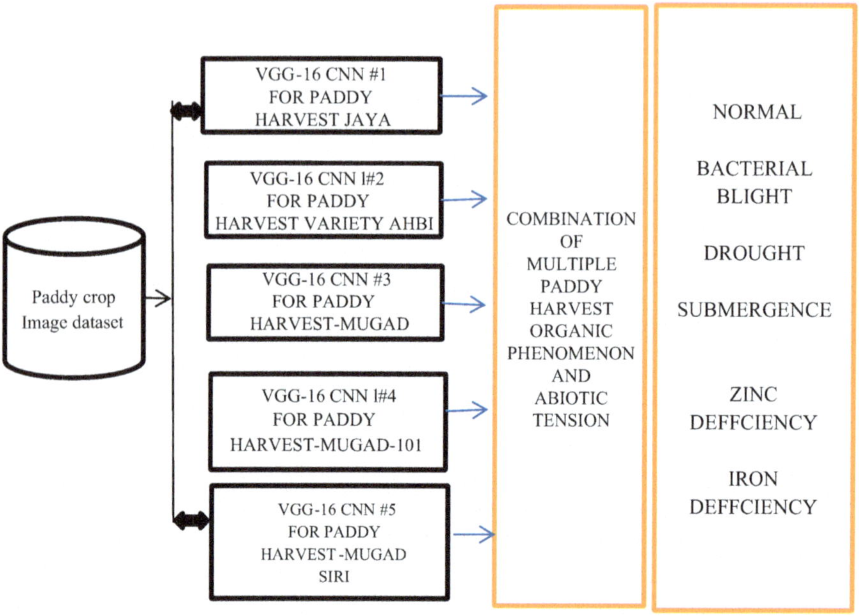

Fig. 3 Proposed method

2.1 *Study Area/Site Description*

The Coordinate Rice Improvement Project (AICRIP) in Mugad, University of Agricultural Sciences, India, is home to country's experimental paddy fields. The average annual precipitation in the studied region is between 364.5 and 685.7 mm, placing it in the dry zone. Large regions have a high-organic-matter, black-clay soil, whereas a smaller area has a high-organic-matter, sandy-loam soil. The unique stress of paddy crops in 60 experimental farmlands was evaluated.

2.2 *Preparing Crop Samples*

Experimental grain samples are chosen from five approved and popular paddy varieties in collaboration with the University of Agricultural Sciences in Dharwad, Karnataka State, India: Mugad Suganda, Jaya, Abhilasha, Mugad 101, and Mugad Siri. There is no genetic or physical contamination in any of the grain samples. According to protocol, the gathered rice grains were planted singly in a raised bed nursery. The plants were cultivated in a controlled environment with consistent nutrient levels, and all required measures were taken to ensure a consistent population of plants of each type. All 11 forms of stress were applied to these rice crop varieties, and observations were made under settings that promoted normal development and manifestation of all the crop strain signs. Field experiments were carried out in the Kharif season [31].

2.3 Image Capture

This research looks at the effects of 11 types of stress on 5 distinct paddy crop varieties. Approximately, 6000 photos of various types of rice crops have been collected. Within this collection, 500 photos represent each stress category. The collection has 500 photos from the field for each rice crop type, including both thriving and weakened fields. Experts have verified the stress identifiers. The observed pressures varied in intensity from mild to severe; for example, bacterial blight and fungal blast. An expert has labeled each picture in this dataset with a label indicating the stress class to which the image belongs. The photos were taken in daylight, with the help of the consistent and powerful natural lighting present during solar noon, when the sun is directly above. The 24 megapixel Nikon D3300 digital single-lens reflex camera was used to take the pictures. The camera's settings were changed to 1/500 of a second (for the shutter speed) and f/32 (for the aperture). The camera was fixed onto a tripod platform to provide a solid base and to allow for effortless movement. The lens of the camera never deviated more than around 45 degrees from the object's axis. The item was placed at a distance of 2.5 m from the crops to be photographed. To lessen the impact of shadows, we used white foam board as a reflecting surface, which helped to direct and boost light from the environment.

Thirty thousand annotated photos of paddy fields were collected to use as the basis for the experimental dataset. All kinds of crops, from strong and healthy examples to those showing various signs of stress, are shown here. The captured pictures, originally at 1910 by 1070 pixels, are physically harvested down to a uniform size of 120 pixels during dataset preparation process. The cultivation method is carried out in collaboration with agricultural specialists and rice scientists. Figure 4 shows photos from the field taken under a variety of stresses experienced by a Jaya paddy crop.

2.4 Capture of Picture

The effect of 11 forms of stress on 5 representative paddy crop varieties is analyzed. Five thousand photos are collected from each paddy crop variety, with 500 photographs of each stress class. In addition to photographs of strained rice fields, the collection also contains 400 images of healthy fields for each paddy crop type. The specialists validated the stress identities. Low to medium levels of stressors were prevalent, including bacterial blight and fungal blast. A stress class has been expertly labeled for each picture in this collection. The photographs were taken in the field mistreatment with a Nikon D300 Digital SLR camera, which has a document of

Layer type		Output layer	param #
Input-2	Input layer	(None, 150,150,3)	0
block_conv1	(Conv2d)	(None, 150,150,64)	1792
block_conv2	(Conv2d)	(None, 75,75,64)	36928
block1_pool	(maxpooling2D)	(None, 75,75,64)	0
block2_conv1	(Conv2d)	(None, 75,75,128)	73856
block2_conv2	(Conv2d)	(None, 75,75,128)	147584
block2_pool	(maxpooling2D)	(None, 37,37,128)	0
block3_pool	(Conv2d)	(None, 37,37,256)	295168
block3_conv2	(Conv2d)	(None, 37,37,256)	590080
block3_conv3	(Conv2d)	(None, 37,37,256)	590080
block3_pool	(maxpooling2D)	(none, 18,18,512)	0
block4_conv1	(Conv2d)	(none, 18,18,512)	1180160
block4_conv2	(Conv2d)	(none, 18,18,512)	2359807
block4_conv3	(Conv2d)	(none, 18,18,512)	2359807
block4_pool	(maxpooling2D)	(none, 9,9,512)	0
block5_conv1	(Conv2d)	(none, 9,9,512)	2359807
block5_conv2	(Conv2d)	(none, 9,9,512)	2359807
block5_conv3	(Conv2d)	(none, 9,9,512)	2359807
block5_pool	(maxpooling2D)	(none,4,4, 512)	0

Fig. 4 VGG-16 CNN model layers' diagram

24 megapixels, approaching solar noon, time with the most consistent lighting at the top of the atmosphere. The camera's ISO, aperture and shutter speed, settings configured to values of 300, 1/400, and f/22, respectively. The camera was positioned on a tripod platform to ensure both rigid stability and convenient portability. A consistent lens-to-object axis angle of approximately 45 degrees was maintained throughout the experiment. The objects were situated at a distance of 2.4 m from the crops in the photographs. In order to obviate the necessity of artificial lighting, a white foam board is employed as a reflector to redirect and step up earthy light toward the areas requiring illumination. The comprehensive compilation of paddy field photographs encompasses a vast assemblage of more than 20,000 meticulously annotated images, depicting both healthy and distressed crops. The acquired photographs, initially sized at 1920 by 1080 pixels, are manually curated to a designated dimension of 120 by 120 pixels in collaboration with agricultural experts and rice experts, as a component of

Fig. 5 Stressed Jaya paddy crop images. **a** Bacterial blight, **b** Fungal blast, **c** Drought, submergence, **d** Nitrogen, **e** Phosphorus, **f** Potassium, **g** Boron, **h** Zinc, **i** Iron deficiencies, and **j** Chemical harm

the dataset preparation procedure. Figure 5 presents a collection of field photographs showcasing the Jaya paddy harvest under various accent circonstance.

3 Experimental Findings and Debate

Five independent VGG-16 CNN models, paddy crop variety depicted in Figs. 6 and 7, make up the experimental classification model. There have been 12 divisions established for each paddy crop type. In order to boost the CNN's efficiency, in the models, the first 10,000-image field image dataset has been increased from 12,000 annotated photos to 30,000 by use of a wide range of photo-enhancement methods including shear transform, brightness disturbances, and random rescaling. Turning vertically,

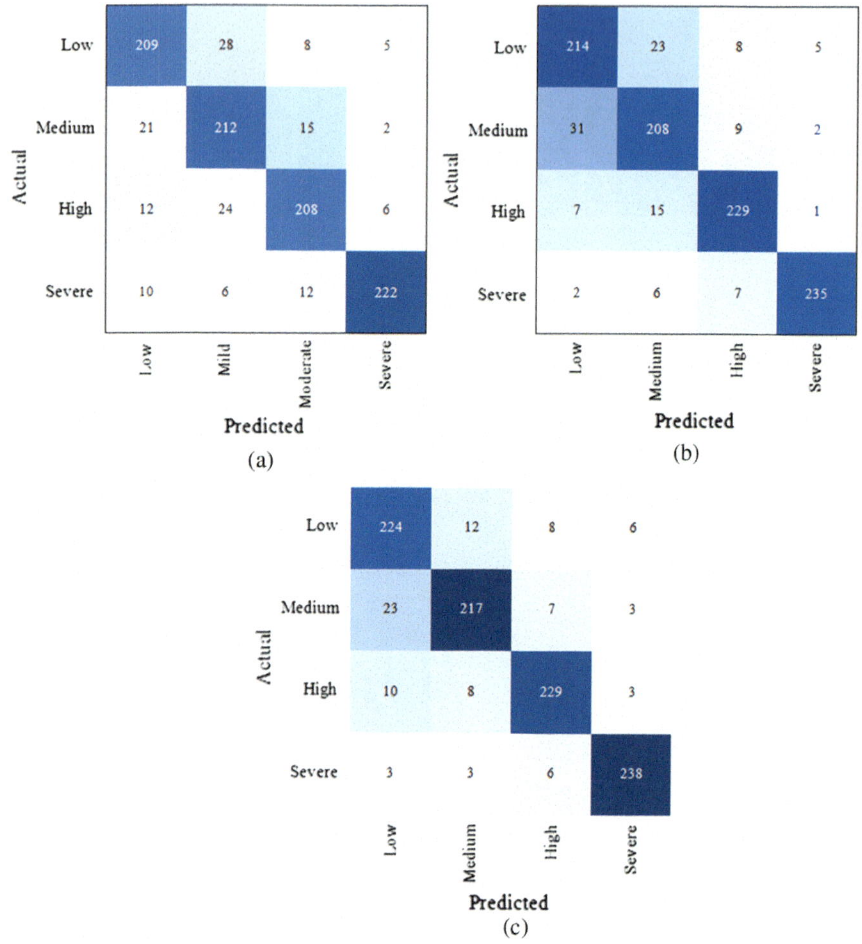

Fig. 6 The utilization of pre-trained VGG-16 CNN model for the classification of paddy crop stress images has resulted in the generation of confusion matrices. **a** Jaya, **b** Abhilasha, and **c** Mugad Suganda

turning horizontally, and skewing, every augmented picture displays striking dissimilarity to the source photos. This improved 60 percent of the photos will be used for training 40% trial and error to rank the precision of each unique CNN model. Examined portions' (10%) dataset was used as the test bed for this analysis.

An Intel Core i5-4210U desktop PC was used in the research, and it had an NVIDIA GeForce 1050 GPU, 8 GB of RAM, and a Discrete Graphics Array controller. Images of several types of paddy crops were used for model development and assessment. The Oxford group's pre-trained weights were used in each successive layer to ensure convergence. Confusion matrices are created for each model and shown in Fig. 6 to graphically represent the approximate predictions. Precision,

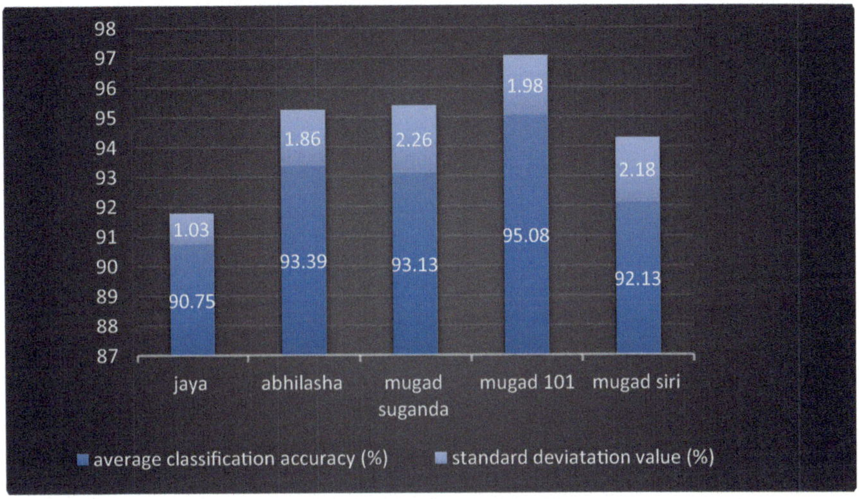

Fig. 7 VGG-16-based paddy harvest assortment

recall, and mean classification accuracy are computed for every stress class using the confusion matrices as the performance measures. The average correctness of stress classification is done for many types of paddy crops. Figure 6 depicts the outliers. Table 1 lists each recall, stress class's precision, and mean. The assessment of classification accuracy is ascertained through the examination and analysis of confusion matrices (Fig. 7).

On average, the stress classification accuracies of the paddy crop varieties Mugad 101 and Jaya were determined to be 90.08% and 80.75%, respectively. The technical feasibility of the deep learning approach is demonstrated by the implementation of a dataset consisting of 30,000 field images. The images provided illustrate 5 distinct varieties of paddy crops, which have been categorized into 12 distinct stress categories. The trained models exhibited an average accuracy of 82.89% in the classification of tenseness when evaluated on a separate dataset that was not utilized during the training process.

3.1 BPNN Versus VGG-16 CNN Categorization Results

The categorization effect of the current VGG-16 CNN model is validated by the newly proposed Backpropagation Neural Network (BPNN) categorization sort. The MPEG-7 standard encompasses two distinct descriptors for color representation, namely, the Dominant Color Descriptor (DCD) and the Color Layout Descriptor (CLD). These descriptors are employed in the training process of the Backpropagation Neural Network (BPNN) classifier to effectively convey the color arrangement and fluctuations within field images of paddy crops subjected to various forms of

Table 1 Paddy crop accent categorization

Paddy crop variety	Stress class	Precision	Mean stress classification accuracy (%)
Jaya	Normal	0.91	90.74
	Bacterial blight	0.92	
	Drought	0.89	
	Zinc deficiency	0.86	
		0.93	
Abhilasha	Normal	0.93	92.38
	Bacterial blight	0.94	
	Fungal burst	0.95	
	Submergence	0.91	
		0.90	
Mugad Suganda	Normal Bacterial blight Fungal blast	0.90	98.1
		0.91	
	Drought Submergence	0.93	
		0.94	
	Nitrogen deficiency	0.90	
Mugad 101	Normal Bacterial blight Fungal blast	0.91	95.08
		0.92	
		0.90	
	Drought Submergence Nitrogen deficiency	0.94	
		0.91	
		0.94	
Mugad siri	Normal	0.98	92.12
	Bacterial blight	0.95	
	Drought	0.94	
	Submergence	0.94	
	Nitrogen deficiency	0.95	

stress. BPNN versus VGG-16 used for classification jobs. Check for more latest deep learning research and developments. Table 2 presents stress classification data for all 12 classes across the 5 paddy crop types, facilitating a comparison between the Back-propagation Neural Network (BPNN) and VGG-16 Convolutional Neural Network (CNN) classification models. Table 2 illustrates that the VGG-16 CNN model exhibits a relatively superior performance in comparison to the BPNN classifier.

Table 2 VGG-16 CNN verses BPNN classification efficiency comparison

Paddy yield variety	VGG–16 CNN	BPNN
Jaya	91.74	89.00
Abhilasha	90.37	89.00
Mugad Suganda	95.07	87.91
Mugad 101	92.12	89.22
Mugad siri	92.88	89.11

4 Conclusion

The objective of this study is to develop an automated system for the identification and categorization of stress in paddy crops based on on-field images. The attainment of this objective is accomplished by employing deep learning methodologies. This research technique has the potential to be applied to a wide variety of abiotic pressures and biotic variables, including all five studied paddy crop types. For this classification job, we used the VGG-16 model, which is widely regarded as the gold standard for pre-training deep neural networks. By employing a dataset consisting of more than 5000 images of the Mugad 101 paddy crop variety, the VGG-16 model has attained a peak average stress classification accuracy of 95.08%. The findings derived from this investigation are promising, as they encompass a larger range of stress categories and crop varieties associated with paddy cultivation in comparison to prior scholarly undertakings. In upcoming research, a comparative assessment will be conducted to evaluate the efficacy of the VGG-16 model against other prominent models such as ResNet, GoogLeNet, Inception-v3, and LeNet. This project may face notable challenges due to the considerable variability observed in an outdoor environment, and impacts of environmental factors, such as elevated temperatures and soil composition. This model may establish a universality, for implementation in a wide range of field crops, such as wheat, maize, barley, soybean, and others. Further investigations may also predict the difference between potential yield and the yield achieved under stressful conditions.

References

1. Rosov KA, Mallin MA, Cahoon LB (2020) Waste nutrients from US animal feeding operations: Regulations are inconsistent across states and inadequately assess nutrient export risk. J Environ Manage 269:110738
2. Liu Z, Xiaozong S, Lihua J, Haitao L, Yu X, Xinhao G, Fuli Z, et al (2012) Strategies for managing soil nitrogen to prevent nitrate-N leaching in intensive agriculture system. Soil Health Land Use Management InTech China 133–154
3. Wang E, Robertson MJ, Hammer GL, Carberry PS, Holzworth D, Meinke H, Chapman SC, Hargreaves JNG, Huth NI, McLean G (2002) Development of a generic crop model template in the cropping system model APSIM. Eur J Agron 18(1–2):121–140

4. Lemaire G, Ciampitti I (2020) Crop mass and N status as prerequisite covariables for unraveling nitrogen use efficiency across genotype-by-environment-by-management scenarios: a review. Plants 9(10):1309
5. Hirel B, Le Gouis J, Ney B, Gallais A (2007) The challenge of improving nitrogen use efficiency in crop plants: towards a more central role for genetic variability and quantitative genetics within integrated approaches. J Exp Bot 58(9):2369–2387
6. Dabney SM, Delgado JA, Meisinger JJ, Schomberg HH, Liebig MA, Kaspar T, Mitchell J, Reeves W (2010) Using cover crops and cropping systems for nitrogen management. Adv Nitrogen Manag Water Qual 66:231–282
7. Zheng W, Liu Z, Zhang M, Shi Y, Zhu Q, Sun Y, Zhou H, Li C, Yang Y, Geng J (2017) Improving crop yields, nitrogen use efficiencies, and profits by using mixtures of coated controlled-released and uncoated urea in a wheat-maize system. Field Crop Res 205:106–115
8. Jin Z, Archontoulis SV, Lobell DB (2019) How much will precision nitrogen management pay off? An evaluation based on simulating thousands of corn fields over the US Corn-Belt. Field Crop Res 240:12–22
9. Singh C, Rao MSS, Mahaboobjohn YM, Kotaiah B, Kumar TR (2022) Applied machine tool data condition to predictive smart maintenance by using artificial intelligence. In: VE Balas, GR Sinha, B Agarwal, TK Sharma, P Dadheech, M Mahrishi (Eds) Emerging Technologies in Computer Engineering: Cognitive Computing and Intelligent IoT. ICETCE 2022. Communications in Computer and Information Science, vol 1591. Springer, Cham. https://doi.org/10.1007/978-3-031-07012-9_49
10. Agostini F, Tei F, Silgram M, Farneselli M, Benincasa P, Aller MF (2010) Decreasing nitrate leaching in vegetable crops with better N management. Genet Eng, Biofertilisation, Soil Qual Org Farming 147–200
11. Srivastava AK, Malhotra SK (2017) Nutrient use efficiency in perennial fruit crops—A review. J Plant Nutr 40(13):1928–1953
12. Yared A, Vara Prasad PV, Foster C, Wright Y, Young S, Bradley P, Stamm M, Ciampitti IA (2018) Major management factors determining spring and winter canola yield in North America. Crop Sci 58(1):1–16
13. Chiter Mal P, Singh AK, Lal Jat S, Dey A, Sankar Nayak H, Mandal BN, Saharawat YS, Jat ML, Yadav OP (2020) Soil quality and carbon sequestration under conservation agriculture with balanced nutrition in intensive cereal-based system. Soil Tillage Res 202:104653
14. Bai-Jian L, Ruo-Chen L, Ke-Chun L, Oladele OP, Zhi-Yu X., Lal R, Zhao X, Hai-Lin Z (2023) Management-induced changes in soil organic carbon and related crop yield dynamics in China's cropland. Glob Chang Biol 29(13):3575–3590
15. Zhang K, Yang D, Greenwood DJ, Rahn CR, Thorup-Kristensen K (2009) Development and critical evaluation of a generic 2-D agro-hydrological model (SMCR_N) for the responses of crop yield and nitrogen composition to nitrogen fertilizer. Agr Ecosyst Environ 132(1–2):160–172
16. Nezomba H, Mtambanengwe F, Tittonell P, Mapfumo P (2015) Point of no return? Rehabilitating degraded soils for increased crop productivity on smallholder farms in eastern Zimbabwe. Geoderma 239:143–155
17. Hochman Z, Horan H (2018) Causes of wheat yield gaps and opportunities to advance the water-limited yield frontier in Australia. Field Crop Res 228:20–30
18. Reimer AP., Denny RCH, Stuart D (2018) The impact of federal and state conservation programs on farmer nitrogen management. Environ Manag 62:694–708
19. Bindraban PS, Dimkpa CO, Pandey R (2020) Exploring phosphorus fertilizers and fertilization strategies for improved human and environmental health. Biol Fertil Soils 56(3):299–317
20. Murali Krishna S, Ray A, Padhy N (2023) IoT-enabled soil nutrient analysis and crop recommendation model for precision agriculture. Computers 12(3):61
21. Lars B, Kirchmann H, Djodjic F, Kyllmar K, Ulén B, Liu J, Andersson H, et al (2015) Turnover and losses of phosphorus in Swedish agricultural soils: Long-term changes, leaching trends, and mitigation measures. J Environ Qual 44(2):512–523

22. Patro P, Azhagumurugan R, Sathya R, Kumar K, Kumar TR, Babu MVS (2021) A hybrid approach estimates the real-time health state of a bearing by accelerated degradation tests, Machine learning. In: 2021 Second International Conference on Smart Technologies in Computing, Electrical and Electronics (ICSTCEE), Bengaluru, India, pp 1–9, https://doi.org/10.1109/ICSTCEE54422.2021.9708591
23. Pan FF, Tang J, Chen BH (2022) Cover crop effects on soil n retention and supply in fertilizer-intensive cropping systems (A review). Eurasian Soil Sci 55(9):1278–1294
24. Christian Bugge H, Hussey K, Holm PE (2011) Exploiting soil-management strategies for climate mitigation in the European Union: maximizing "win–win" solutions across policy regimes. Ecol Soc 16(4)
25. Muhammad I, Umar S, Mahmooduzzafar (2019) Nano-fertilization to enhance nutrient use efficiency and productivity of crop plants. Nanomater Plant Potential 473–505
26. Kim D-G, Giltrap D, Sapkota TB (2023) Understanding response of yield-scaled N2O emissions to nitrogen input: Data synthesis and introducing new concepts of background yield-scaled N2O emissions and N2O emission-yield curve. Field Crop Res 290:108737
27. Chen Q, Li X, Horlacher D, Liebig H-P (2004) Effects of different nitrogen rates on open-field vegetable growth and nitrogen utilization in the north China plain. Commun Soil Sci Plant Anal 35(11–12):1725–1740
28. Akasairi O, Harsanyi E, Széles A, János Holb I, Szabó S, Rátonyi T, Mohammed S (2023) A bibliographic review of climate change and fertilization as the main drivers of maize yield: implications for food security. Agric Food Secur 12(1):1–18
29. Alain M, Willigen PD, Heincn M, Morel C, Schneider A, Pellerin S (2008) A two-dimensional simulation model of phosphorus uptake including crop growth and P-response. Ecol Model 210(4):453–464
30. Łangowski Ł, Goñi O, Ikuyinminu E, Feeney E, O'Connell S (2022) Investigation of the direct effect of a precision Ascophyllum nodosum biostimulant on nitrogen use efficiency in wheat seedlings. Plant Physiol Biochem 179:44–57
31. Feng W, Li X, Dong H, Qin Y, Sun M, Shao J, Zheng C, Li P (2022) Fruits-based critical nitrogen dilution curve for diagnosing nitrogen status in cotton. Front Plant Sci 13:801968

Intelligent Monitoring and Crop Yield Systems-II

Optimized CNN Model to Develop a Decision-Making System for Spraying on Cotton Crop

Supriya Sudhir, Kalyani S. Kumar, and P. Santosh Reddy

Abstract Dynamic crop morphology capture aids real-time precision in field spraying decisions. Ultrasonic radar and costly LiDAR can't maximize their high precision. They may slow down decision speed because the current spraying amount control needs to include more information about branches and leaves, which highly relies on the statistical features of regional morphology. Since most agricultural tools in China are medium in size and relatively inexpensive, it makes sense to create a unique identification system that combines deep learning with cheap monocular vision. A crop's growth cycle was a helpful gauge for direct spraying because of the knowledge gained. For this reason, let's examine cotton as an example: This study aims to develop a single-camera acquisition system and propose an algorithm based on the Convolutional Neural Network (CNN) to identify different growth stages of cotton plants. An optimal CNN model structure is chosen among nine possible method structures using an optimization procedure based on the matrix of confusion and recognition efficiency, and its dependability is established by repeatedly switching the training and test sets by the k-fold test concept. This CNN model has an accuracy of 94.27%, an F1-score of 96.39%, a recall of 95.76%, a precision of 95.31%, and a recognition speed of 72.46 ms per picture. Also, the convolution neural network, the model suggested in this study, performs better than those of VGG16 and Alex Net. Finally, spraying deposition tests were conducted over three distinct cotton growth stages to validate the accuracy of the developed identification method and the viability of the sprayed decision-making process based on CNN. The findings of the studies confirm that the ideal spray parameters were applied during distinct growth times correspondingly, leading to an increase of up to 63.24% in the

S. Sudhir · K. S. Kumar · P. S. Reddy (✉)
Department of Computer Science & Engineering, BNM Institute of Technology, Bangalore, India
e-mail: santoshreddy@bnmit.in

S. Sudhir
e-mail: supriyasudhir@bnmit.in

K. S. Kumar
e-mail: kalyanisk@bnmit.in

A. J. Kulkarni and M. Hajiaghaei-Keshteli (eds.), *Advances in Intelligent Systems for Sustainable Agriculture*, Lecture Notes in Networks and Systems 1201,
https://doi.org/10.1007/978-981-97-9839-1_19

optimum index over the operations that did not discriminate growth periods, yielding an overall increase of 43.29%.

Keywords Spraying operations · Ultrasonic radar · Crops growth cycle · CNN algorithm · F1-score

1 Introduction

Cotton is essential since it may be used for fiber and oilseeds. It's used to make 36% of all fiber worldwide. China's importance in the international cotton industry makes cotton a crucial commodity [1, 2]. There is now a significant challenge in the agricultural industry regarding developing precision agriculture techniques. Target perception technology has therefore become essential to the precision spraying methods of crop protection [3–5]. Cotton is a crop that requires a lot of pesticides since its development cycles are so lengthy, and it is susceptible to many different types of pests and illnesses [6]. Researching how cotton is perceived by its intended audience is crucial for improving spray effectiveness. Agriculture has extensively used objective perception methods based on radar, laser, and ultrasound [7, 8]. Using ultrasonic sound envelopes and point quadrate analysis, the author verified the canopy density in real-time in the fruit orchard and vineyard [9]. While laser radar and ultrasonic-based target perception technology help assess the volume and form of big plants during precision spraying, it is challenging to implement in the goal recognition field of cotton in China. While China produces a lot of cotton, the country's fields are tiny and dispersed [10, 11]. The use of heavy agricultural gear and detecting equipment presents significant challenges [12]. Farmers are sensitive to the expense of planting due to this tiny and decentralized method of cotton cultivation; hence they tend to choose small, inexpensive agricultural machines [13]. The challenge is accepting costly detection tools like lasers, radar, and ultrasonic waves [14]. Furthermore, electrostatic spray and air-assisted spray equipment dominate China's domestic spray market, none of which has attained high spray accuracy [15]. Due to too many branches and leaf details, laser radar and ultrasonic cannot fully exploit their high precision and might slow decision time [16]. Exploring high-accuracy, real time, and low-cost perception methods for directing sprayed cotton based on current spray equipment has become a significant task in China [17].

Crop security in agriculture has traditionally relied heavily on the growth phase as a point of reference. Farmers and scientists have classified cotton into four distinct stages based on its appearance: the bud stage, the stage for seedling, the boll stage, and the opening boll harvesting stage. Precision spraying has variable ideal operation hours depending on the stage of crop development [18, 19]. Various stages of development are needed for the use of various insecticides [20, 21]. Defoliant should be administered at the boll opening harvest stage, whereas cotton aphid control should occur during the seedling stage and bollworm control during the bud stage. Furthermore, this paper's spray deposition studies on cotton at various development stages demonstrated that the spraying parameters differed when the most substantial deposition effect was produced [22, 23]. The timing of applications, operational periods, and spraying conditions may all be informed by knowledge of the cotton growth cycle [24, 25]. However, the development of cotton status can only be estimated by estimating the time for planting since the growth period is influenced by temperature, soil moisture, and sunlight length. Hence, the primary objective of this study is to design a cost-effective, real-time onboard system for accurately identifying the growth stages of cotton plants [26, 27]. This system will utilize a Convolutional Neural Network (CNN) to enable precise application in cotton fields [28].

The following are some of the contributions made by this paper: It was first intended to spray precisely during the growing stage. The first time a convolutional neural network has been utilized to detect cotton's development stages. Optimizations are made to a CNN model to improve its accuracy, stability, and speed. The deposition spray experimentation validated the importance of growth-period awareness as a guiding principle. The materials and techniques used to create a system for recognizing cotton's growth phases are detailed in Sect. 2. It covers the data collecting, calibration, and processing steps, the convolution neural network's guiding principle, the model optimization procedure, and the tools utilized in the project's creation. After training and evaluating many different CNN topologies, the outcomes are presented and analyzed in Sect. 3. This part compares the performance of VGG16 and other CNN models analyze the reliability of the optimized structure, visualize the CNN method, and validate the accuracy of spraying based on the identification of growth periods. The paper's conclusion may be found in Sect. 4.

2 Methodology

2.1 Collection of Data

The experimental field for cotton provides us with the experimental pictures needed to enable the creation and validation of CNN models. Since the quality of a picture is impacted by the uncontrollable lighting conditions in outside agricultural situations, numerous cameras and smartphones are utilized to acquire photographs from the experiment at a randomized period to strengthen the recognition method's resilience. Furthermore, the identification accuracy may suffer if the cotton is shot from a non-standard angle. Since the identification system will eventually be mounted on a sprayer, the camera will be positioned to point directly at the center of each cotton row, roughly 150 cm, with a horizontal shooting angle of 50 degrees. Specifically, Fig. 1 depicts the gunshot location. In all, 1445 photos were taken throughout each growth stage as part of the experiments.

All 1,445 cotton photographs were scored by three experts, who then classified them into four distinct stages of development. They have agreed on a dividing line to reduce the influence of personal preferences. Figure 2 depicts some sample images

Fig. 1 Gun shoot location for cotton

Fig. 2 Sample images for cotton developmental stages

from each of these four cotton developmental stages. Seedlings, buds, bolls, and bolls ready for harvest are shown left to right. Since the length of cotton's many development phases affects the quantity of data acquired throughout each phase, they notice less information available during the seedling and bud phases compared to the boll and boll phases. Table 1 displays the aggregate data amount for all harvest seasons. Finally, in this work, utilize 75% of the labeled photos for model building (training) and 25% for model validation. Both the training and testing datasets included cotton at every stage of development, from seedling through boll opening.

2.2 *Processing of Data*

Instead of immediately using the tagged cotton photos, the data processing approach shown in Fig. 3 is applied to them. This method involves scaling, data enhancement, and format conversion. The following describes the steps a Python application takes to perform the described procedure and analyze the data automatically. Since this model is intended for use in cotton pesticide application equipment, the original

Table 1 Aggregate data amount for all harvest seasons

	Original Pictures	Original picture training sets	Data-enhanced training set	Test set
Seedling stage	210	149	1481	64
Bud Stage	239	168	1671	72
Boll stage	462	324	3231	139
Bolls ready for harvest stage	534	375	3741	160
Total	1445	1016	10124	435

picture size was reduced to 100 × 100 to increase processing speed without sacrificing detail and to save computer resources. Since this study's data set is relatively small, data augmentation is used to remedy the problem of over-fitting and improve the generalizability of the models. Flipping the images horizontally, adjusting the brightness, blurring the images, introducing noise, and adjusting the color gradient are the five data augmentation techniques used for the training sets. Table 1 displays the enlarged data set. Finally, the information set is translated into recorded formats via Python script to simplify network input using the Tensor Flow framework (managed by Google) for CNN model training in this article.

2.3 CNN Design Principles and Architecture

CNNs are a subset of neural networks with unique architecture based on a series of layers. Combining forward feature extraction with backward parameter modification, CNN can perform picture identification and classification. The forward extraction of features procedure typically consists of three distinct levels.

Convolution Layer: The convolution layer is a weighted summation process. In the input picture, the template is in motion. The point in the goal picture is obtained by multiplying and summing the components inside the defining domains of the input image and the template. A template is a filter that specifies the characteristics to be sought by the convolution.

Pooling Layer: The network's output at a particular point is substituted using the function for pooling, which considers the global statistical features of outputs nearby. Average-pooling, Max-pooling, and pooling L2 are all examples of popular pooling processes. In this article, choose max-pooling as the pooling component of the neural system because it can better restrict the variation of the estimated mean brought on by variations in the convolution layer parameters while keeping texture information intact.

Entirely connected layer: Every node in the entirely connected layer is associated to each node in the layer above it, allowing it to classify the features retrieved in convolution and pooling layers. Applying a gradient descent method, a backward

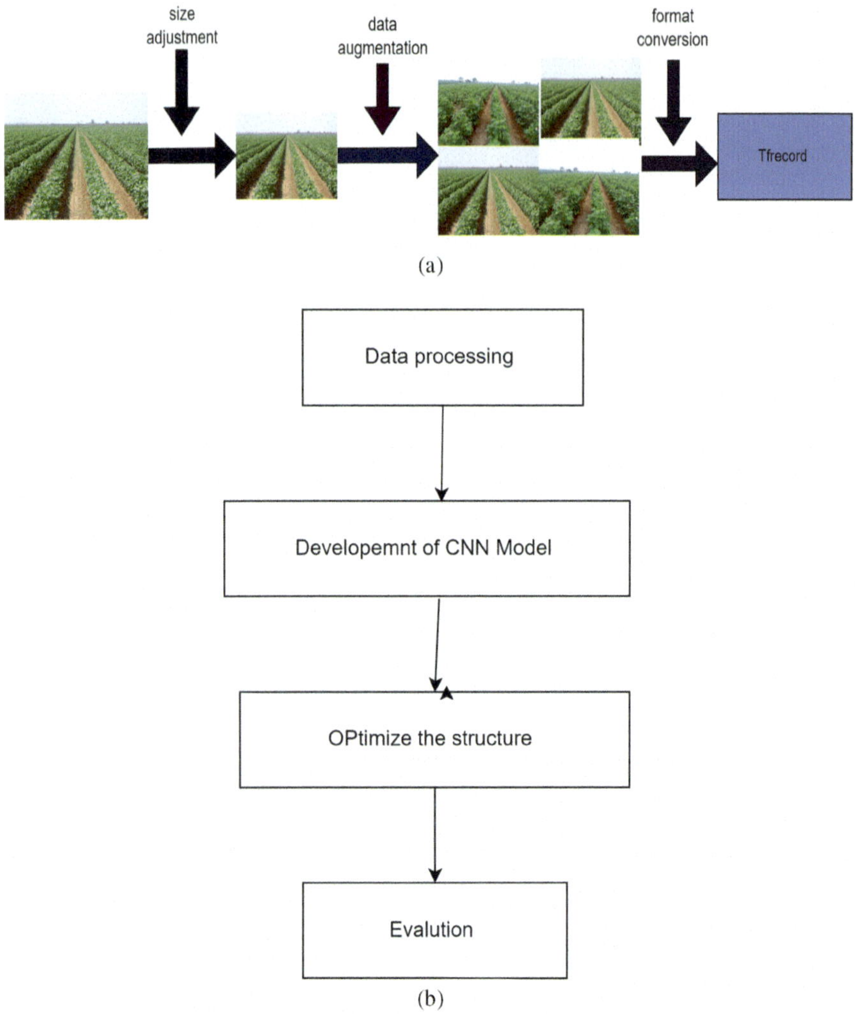

Fig. 3 **a** Data processing approach and **b** Workflow

parameter modification process in CNN continually adjusts the biases and weights to minimize the difference between the expected result and the actual value, increasing the training impact.

2.4 *Optimization of the Model Structure*

A lightweight CNN model prototype was developed, considering real-time and budgetary restrictions. There are 16 convolution kernels in each convolution layer, with sizes of 10×10 and 5×5. All three layers used for pooling are of size 2×2, the maximum allowed by the max-pooling layer algorithm. Two completely connected layers receive data flattened from the final pooling layer. The softmax function and the information from the top fully connected layer are used to make predictions in the output layer. Since the accuracy of this CNN model prototype, post-training is near the spray demand; it was chosen as the foundation for optimization. The third part shows how this prototype performs. Orthogonal experimentation is used to optimize CNN model topologies, in which two components and three levels are chosen to test. The amount and the size of convolutional kernels are the two variables. Convolutional kernel sizes are fixed at 10×10, 5×5, and 3×3 in this example. In addition, 8, 16, and 32 convolution kernels are used.

For each of the nine CNN model architectures, the prefix "conv" is followed by the dimension and number of convolution kernels, with "-" as an interval. Layers are used for the pooling and fully connected layers, while "Conv" represents the convolution layer. Under the conditions of this paper, the complexity of a CNN varies using the size of the convolution kernel, as the amount of convolution layers improves as the convolution kernel decreases. True, even if the input information and the entirely connected layer are kept constant. Confusion matrices are employed to aid in model structure optimization after nine different convolution neural network architectures have been obtained. Classification efficiency was evaluated using the CNN model and the confusion matrix as measures. Predicted classes are shown in the columns of a confusion matrix, while the actual classes are shown in the rows. In practice, four values can be determined for a given class: the number of examples correctly assigned to the class (TP = true positives), the number of examples correctly assigned to the class (TN = true negatives), and the number of examples that were either not recognized as class examples (FP = false positives) or wrongly allocated to the class (FN = false negatives). The following metrics were then determined using the data from the confusion matrix:

- Accuracy: A measure of a classifier's accuracy is expressed as a percentage of samples with the expected classification.

$$Accuracy = \frac{TP + FN}{TP + TN + FP + FN} \tag{1}$$

- Precision: The fraction of really positive examples relative to the total number of projected positive examples; this metric measures the classifier's ability to avoid falsely labeling harmful data as positive.

$$Precision = \frac{TP}{TP + FP} \tag{2}$$

- Recall: The proportion properly estimated in all positive information displays the classifier's capacity to identify all positive examples accurately.

$$recall = \frac{TP}{TP + FN} \tag{3}$$

- F1-Score: A metric that considers the costs and benefits of training a classifier and returns a single value that reflects the "goodness" of a classifier when dealing with uncommon classes is called the harmonic mean of accuracy and recall.

$$F1Score = \frac{2 \times Precision \times Recall}{Precision + Recall} \tag{4}$$

A good indicator of the recognition impact, accuracy only reveals the unique circumstances for some groups in test data. Furthermore, throughout the CNN training process, more features will be learned from the data sets with more classes (boll opening crops stage and boll stage) instead of those with fewer classes (bud stage and seedling stage). The F1-score might result in inconsistent overall accuracy at the expense of subpar performance in a few courses. To effectively address this issue, it is crucial to compute indices like precision, recall, and F1-score, which provide clear insight into the performance recognition of every category.

2.5 Tools for Development

2.5.1 Hardware

This study employs the desktop as a platform for development during the learning phase. The workstation's robust processing allows the CNN method to be trained fast. In the prototype testing phase, a Raspberry Pi serves as the central processing unit. Future applications may benefit immensely from Raspberry Pi's ability to serve as both the upper and bottom computer.

2.5.2 Software

Python was used throughout the development of the program used to categorize the cotton growth periods. The image classification model was built on TensorFlow, the current deep learning system with the most comprehensive API. Importantly, it has the most significant community for the users, which will be supportive of future training and the actual implementation of pesticide equipment for cotton, and it covers Linux, Windows, Mac, and many mobile platforms.

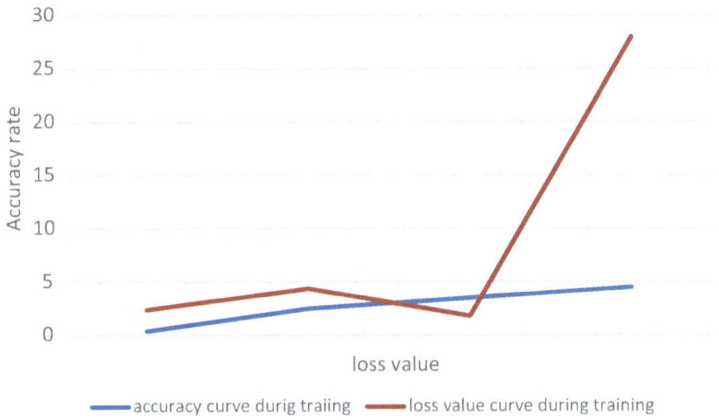

Fig. 5 Evolution of precision and loss of a CNN model prototype

3 Results with Discussion

3.1 *The Training as Well as Evaluation of Models*

Figure 5 depicts the evolution of precision and loss when a CNN prototype was trained to predict the growth period of cotton using processed information sets—training results in a natural convergence of accuracy and loss values. They are applying this trained model to test data results with an 85% accuracy rate. This finding demonstrates that the CNN prototypes can be used to detect the cotton growing season, while more improvements may be made to enhance its effectiveness. As a result, optimized the model's underlying structure.

Figure 6 displays the results of the model optimization procedure, which included the usage of all nine possible CNN model architectures for training. Conv-5 × 5–8 is the most excellent option, with recognition speeds that are only surpassed by structure Conv-10 10–16 and Conv-10 × 10–8, and an accuracy that can reach 94.27%. The poor speed of the Conv-5 × 5–8 model is explained by our analysis showing that the Conv-10 × 10–8 and Conv-10 × 10–16 structures have fewer layers in their convolution neural networks.

3.2 *Optimization of Model Structure Reliability Analysis*

While it is true that the Conv-5X 5–8 model structure may get satisfactory results on the test set, its dependability could be better when training on novel data. To evaluate the efficacy of this model structure, use the K-fold test and modify the data set accordingly. K-fold testing works based on randomly splitting a dataset into k equal halves, then using the first portion as the experiment set and the last part as the

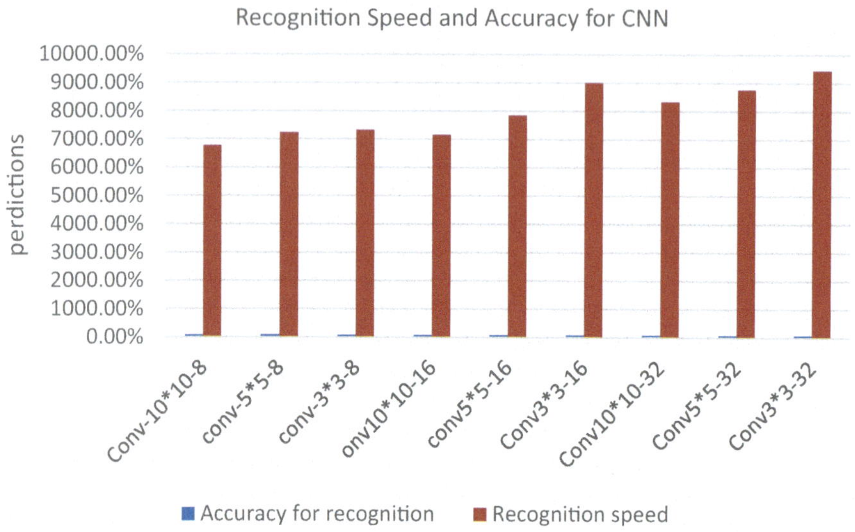

Fig. 6 Results of the model optimization procedure

set for training. There have been K total iterations of the selecting procedure. Then, K models are trained for each dataset and evaluated using the matching test data. The whole dataset is broken down into five sections in this publication. Figures 7 and 8 show the average accuracy, F1-score, and recall of each growth phase achieved after training and testing operations have been performed on separate collections of test and training sets, respectively. Consistent with the findings of the last research, all groups' identification average precision, accuracy, F1-score, and average recall, are over 90%. The model demonstrates the model structure's dependability by demonstrating that it can achieve improved performance across various training and data sets.

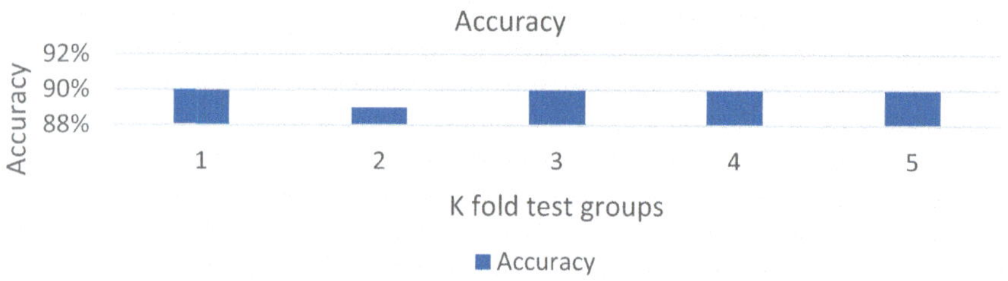

Fig. 7 K-fold test accuracy

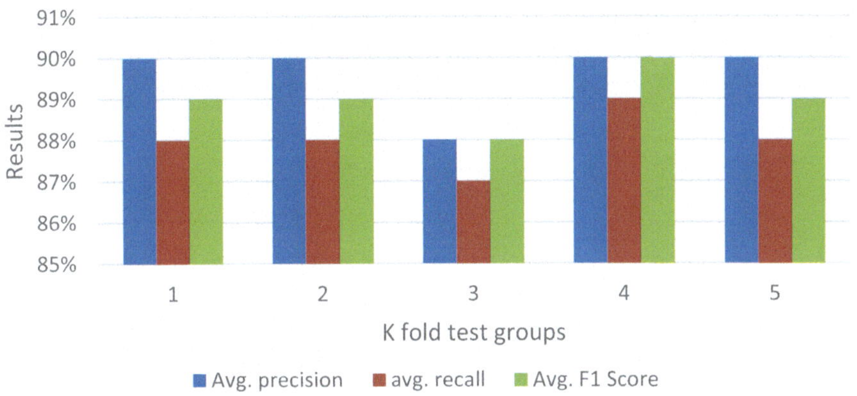

Fig. 8 K-fold test results

3.3 *Model Visualization*

Using images that have been processed via convolution layers, CNN visualization helps to provide light on the network's training procedure. The visualization tool Tensorboard is part of Tensor Flow and may be used to learn more about neural network training. In this work, tensor board code is included in the CNN training script in Python in order to save the filtered pictures for further analysis of CNN's feature learning process.

The optimization model's visualization operation uses the cotton boll stage image. There are a total of six layers in the model: four convolutional and three pooling. In that order, the layers for the pooling are positioned below the next, third, and fourth convolution layers. Pictures filtered by the first and second convolutional layers are presented in Fig. 9a and b, respectively. The image data is retained chiefly, but the convolutional kernels execute comparable edge processes on the picture. Figure 9c displays the filtered images produced by the 3rd convolution layer, whereas Fig. 9d displays the filtered images produced by the fourth convolution layer. Images have a much lower resolution because of the pooling layers before these convolution layers. Figures 9c and d show that as the number of convolution kernels increases, the output becomes more abstract and retains less and less information. The more layers there are, the extra blank content there is; in other words, these convolutional kernels cannot detect the characteristics that are wanted in the input picture. The visualization reveals the color and edge information of importance to low-level convolution kernels. A convolution kernel's interest in abstract and complicated information grows in proportion to the depth of the convolution layer. However, the difficulty of manually selecting such deep features demonstrates that convolution neural networks are preferable for picture identification.

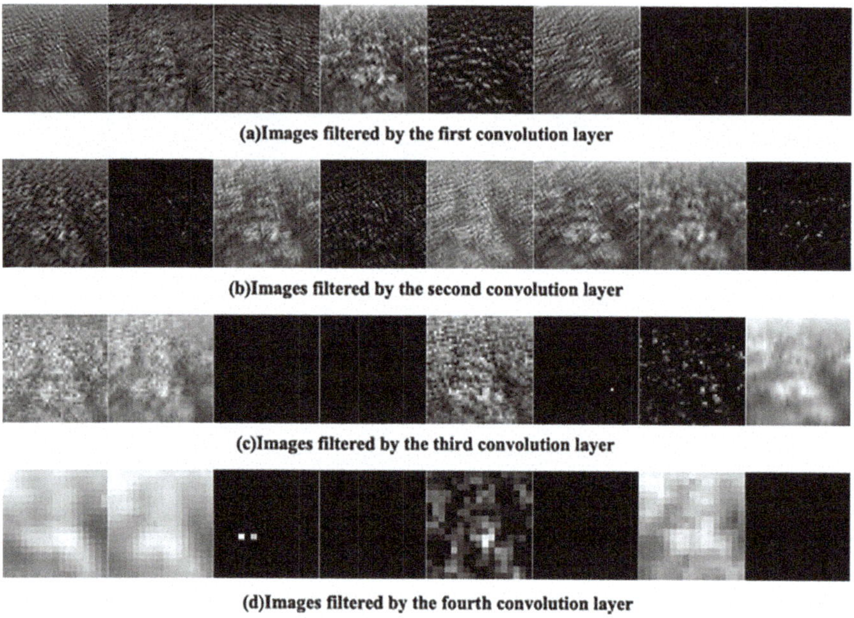

Fig. 9 CNN feature selection visualization

3.4 *Comparison of VGG16 and Alex Net Performance*

Popular deep learning CNN models VGG16 and Alex Net are compared in this study. Figure 10 compares the accuracy and identification speed of VGG16 and Alex Net using model structure conv-5 ~ 5–8, with input picture sizes of 224 ~ 224.

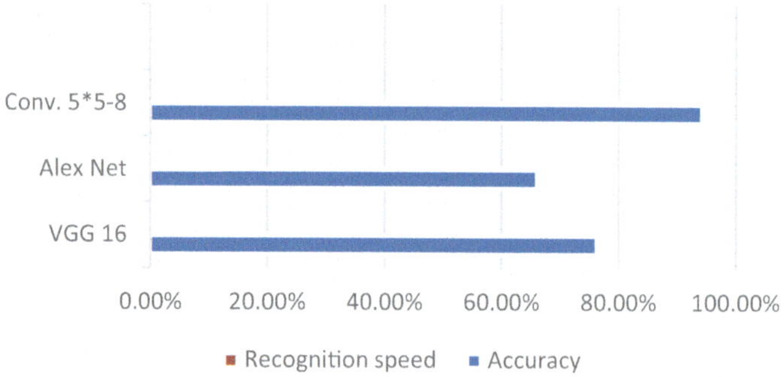

Fig. 10 Comparison of accuracy and recognition speed by CNN, VGG16, and Alex Net

3.5 Accurate Spraying Validation Using Growth Period Recognition

This section conducted tests to confirm the relevance of recognizing cotton growth cycles in directing accurate spraying by spray deposition.

3.5.1 Deposition Spray Tests

The experimental objects consist of cotton plants in their seedling, bud, and boll stages. At each development stage, the spray variables—wind speed and flow rate—are used to determine pesticides' effective utilization rates and deposition equality. The system was configured with a flow rate of 0.22 L/min. Velocities of 14 m per second are being experienced. Figure 11a and b shows the fitting curve and the total number of experimental outcomes. Even when using the same spraying parameters (flow rate and wind speed), the rate of pesticide use and uniformity of deposition across development stages vary. The ideal spray parameters for various development cycles vary regarding cotton, whether optimum spray usage or deposition rates. For instance, while growing cotton seedlings, the ideal wind speed for maximizing pesticide consumption is 14 m/s. For optimal deposition uniformity, the flow rates should be 0.22 L/min.

3.5.2 Optimization and Verification of Spraying Parameters

Using the spray assessment index S, this research completely evaluates pesticide consumption ratio (u) and deposition uniformity (Coefficient of Variation, CV). The S formula is below.

$$S = (w1 \times u + w2 \times (100\% - cv)) \times 100\% \tag{5}$$

The effectiveness of a spray is directly proportional to the rate of pesticide use and the size of the coefficient of variation. The variation coefficient and the pesticide use rate can't be directly weighted summated in this situation. To get uniformity, they take the coefficient of variation and remove 100% from it. Then, find the result using formula (5), which involves weighted summation. Deposition consistency and efficient pesticide consumption rate are the two most important factors to consider during spray operation. Uneven deposition means that no amount of pesticide, regardless of usage, can provide adequate control.

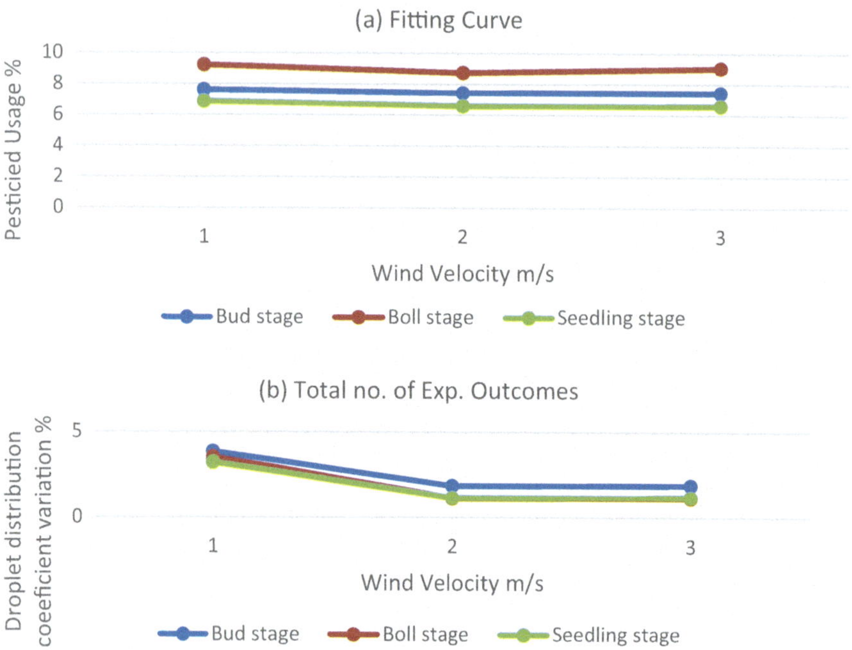

Fig. 11 **a** and **b** Fitting curve and the total number of experimental outcomes

4 Conclusion

Improvements to both the detection technique and the detection objective are detailed in this work. The proposed approach for guiding spraying based on the cotton growing period was implemented using a convolutional neural network-based, low-cost image recognition system. The system achieved an accuracy of 94.27%, precision of 95.31%, recall of 95.76%, F1-score of 96.39%, and recognition speed of 72.46 ms per image, accordingly with a technique derived from the k-fold test and CNN's feature learning process visualization, and assess the recognition model's dependability. Furthermore, the CNN model architecture suggested in this research outperforms VGG16 and Alex Net, demonstrating its superiority for the recognition job at hand and its ability to fulfill the demands of real-time, high precision, and cheap cost. At last, they conducted a control experiment using spray deposition. The experimental findings demonstrated that differentiating between the growth period and the growth stage improved the spray impact, suggesting that identifying the growth phase of cotton might help with variable spraying.

References

1. Shanping W, et al (2021) Recognition of cotton growth period for precise spraying based on convolution neural network. Inf Process Agric 8(2):219–231
2. Xiangpeng F, et al (2023) Deep learning based weed detection and target spraying robot system at seedling stage of cotton field. Comput Electron Agric 214:108317
3. Md Manowarul I, et al (2023) A deep learning model for cotton disease prediction using fine-tuning with smart web application in agriculture. Intell Syst Appl 20:200278
4. Rafael Faria C, Esdras Santiago W, Teruel B (2021) Identification of cotton leaf lesions using deep learning techniques. Sensors 21(9):3169
5. Muhammad T, et al (2021) Identification of tobacco crop based on machine learning for a precision agricultural sprayer. IEEE access 9:23814–23825
6. Alicia A, et al (2022) Precision chemical weed management strategies: A review and a design of a new CNN-based modular spot sprayer. Agronomy 12(7):1620
7. Tanmay T, Patle BK, Kashyap SK (2023) Intelligent insecticide and fertilizer recommendation system based on TPF-CNN for smart farming. Smart Agric Technol 3:100114
8. Dong C, et al (2022) Performance evaluation of deep transfer learning on multi-class identification of common weed species in cotton production systems. Comput Electron Agric 198:107091
9. Naseeb S, et al (2023) Optimization of irrigation timing for sprinkler irrigation system using convolutional neural network-based mobile application for sustainable agriculture. Smart Agric Technol 5:100305
10. Doddamani ST, Karadgi S, Giriyapur AC (2022) Multi-label classification of cotton plant with agriculture mobile robot. In: Data Intelligence and Cognitive Informatics: Proceedings of ICDICI 2021. Singapore: Springer Nature Singapore, 759–772
11. Naseeb S, et al (2022) Semantic segmentation of in-field cotton bolls from the sky using deep convolutional neural networks. Smart Agric Technol 2:100045
12. Pappu Kumar Y, et al (2023) Detecting volunteer cotton plants in a corn field with deep learning on UAV remote-sensing imagery. Comput Electron Agric 204:107551
13. Sharma R, Sharma R, Prabha YB, Rema Devi S, Saxena P, Rajasanthosh Kumar T (2023) Iot monitoring lathe machine performance, Materials Today: Proceedings 80(3):3570–3574. ISSN 2214–7853, https://doi.org/10.1016/j.matpr.2021.07.300. https://www.sciencedirect.com/science/article/pii/S2214785321051592
14. Nazar H (2020) Development of a smart variable rate sprayer using deep convolutional neural networks for site-specific application of agrochemicals
15. Sapkota BB, et al (2022) Use of synthetic images for training a deep learning model for weed detection and biomass estimation in cotton. Sci Rep 12(1):19580
16. Jing Z, et al (2021) Detection of cotton waterlogging stress based on hyperspectral images and convolutional neural network. Int J Agric Biol Eng 14(2):167–174
17. Modi RU, et al (2023) An automated weed identification framework for sugarcane crop: a deep learning approach. Crop Prot 173:106360
18. Weicheng X, et al (2020) Establishing a model to predict the single boll weight of cotton in northern Xinjiang by using high resolution UAV remote sensing data. Comput Electron Agric 179:105762
19. Purnendu Shekhar P, et al (2021) A smart parking system based on IoT technologies. Linguist Antverp 59–64
20. Espejo-Garcia B, et al (2020) Towards weeds identification assistance through transfer learning. Comput Electron Agric 171:105306
21. Congcong G, et al (2022) Predicting F v/F m and evaluating cotton drought tolerance using hyperspectral and 1D-CNN. Front Plant Sci 13:1007150
22. Devi Priya R., et al (2021) Faster R-CNN with augmentation for efficient cotton leaf disease detection. International Conference on Hybrid Intelligent Systems. Cham: Springer International Publishing, 2021

23. Serosh Karim N, et al (2021) Computationally light deep learning framework to recognize cotton leaf diseases. J Intell Fuzzy Syst 40(6):12383–12398
24. Senthil Kumar Swami D, Divya Shamili M (2022) Smart farming using machine learning and deep learning techniques. Decis Anal J 3:100041
25. El-Kenawy, El-Sayed M, et al (2022) Metaheuristic optimization for improving weed detection in wheat images captured by drones. Mathematics 10(23):4421
26. Atluri Vani V, et al (2023) Optimization of irrigation and herbicides using artificial intelligence in agriculture. Int J Intell Syst Appl Eng 11(3):503–518
27. Xiangpeng F, et al Deep learning based weed detection and target spraying system at seedling stage of cotton field. Available at SSRN 4482157
28. Kusuma S, Jothi KR (2022) ECG signals-based automated diagnosis of congestive heart failure using Deep CNN and LSTM architecture. Biocybern Biomed Eng 42(1):247–257. ISSN 0208–5216. https://doi.org/10.1016/j.bbe.2022.02.003

A Robust Optimization Approach for Agricultural Waste Supply Chain Network Design: A Real Case Study

Hossein Kianypor, Ali Husseinzadeh Kashan, Ehsan Nikbakhsh, and Maedeh Hosseinpour

Abstract Nowadays, the preservation of natural resources and their use in optimal ways has attracted both academic study and industrial implementation. The design of supply chain networks plays a pivotal role in effective supply chain management. Optimal decisions in this area ensure the flexibility and responsiveness of the entire system. Also, the role of supply chains in waste management is crucial as it enables the efficient movement and handling of materials, ensuring proper disposal or recycling. An effective supply chain helps the collection, processing, and distribution of waste, optimizing resource use and minimizing environmental impact. This study developed a mixed-integer linear programming (MILP) model to manage agricultural waste in the palm date industry, serving as a case study for producing medium-density fiberboards (MDF). It employs a robust optimization approach to enhance the network's adaptability in accommodating uncertainties related to customer demands. Computational analysis is also provided to present the significance of the model. Also, managerial insights are extracted from the final results, and sensitivity analyses are presented to handle agricultural wastes adequately.

Keywords Supply chain network design · Robust optimization · Waste management

---

H. Kianypor
Faculty of Operations Management, IÉSEG School of Management, Paris, France

A. Husseinzadeh Kashan (✉) · E. Nikbakhsh
Faculty of Industrial and Systems Engineering, Tarbiat Modares University, Tehran, Iran
e-mail: A.Kashan@modares.ac.ir

M. Hosseinpour
Faculty of Industrial Engineering, K. N. Toosi University of Technology, Tehran, Iran

A. J. Kulkarni and M. Hajiaghaei-Keshteli (eds.), *Advances in Intelligent Systems for Sustainable Agriculture*, Lecture Notes in Networks and Systems 1201,
https://doi.org/10.1007/978-981-97-9839-1_20

1 Introduction

Supply Chain Network Design (SCND) involves the strategic planning and optimization of a supply chain's structure to meet business objectives efficiently. In recent years, SCND has gained significant importance across industries globally, emerging as a critical tool for companies consistently seeking enhanced efficiency, reduced costs, and improved customer service. Notably, SCND plays a pivotal role in agriculture by optimizing how goods move from farms to consumers, ensuring efficiency and reducing operational costs. This becomes particularly crucial in waste management and recycling within agricultural cultivation. By integrating waste reduction strategies into network design and strategically placing recycling facilities and waste collection points, SCND enables effective agricultural waste management. This not only minimizes environmental impact but also enhances sustainability, resource efficiency, and the overall resilience of the agricultural ecosystem.

However, there's a clear absence in the available literature when it comes to establishing an efficient supply chain designed explicitly for managing these agricultural wastes. Previous studies have predominantly focused on the minimal contributions of agricultural waste in diverse farming industries. This emphasizes the necessity of introducing efficient network models to optimize these processes. Within industrial societies, the prioritization of waste management and environmental conservation has emerged as a critical focus. Developing a well-structured and efficient system dedicated to collecting, moving, and storing agricultural waste holds significant importance in minimizing waste generation across a range of industries. Consequently, initiatives directed toward the design of such networks serve as a critical factor in mitigating environmental impact and cultivating sustainable practices.

Uncertainty refers to the absence of predictability or the presence of unknown elements within a given situation. It includes a lack of complete knowledge or information about the future, leading to challenges in making accurate predictions or decisions. In various domains, uncertainty arises from sources like incomplete data, randomness, variability, or ambiguity. Within supply chain networks, uncertainty manifests as unpredictability and variability in aspects such as demand, costs, and resource availability. It poses a pervasive challenge, impacting decision-making processes and the overall performance of supply chain systems. This unpredictability often results in increased expenses, compromised product quality, and difficulties in managing real-world variables.

To navigate this uncertain condition, researchers have developed multiple strategies to address uncertainty in supply chains. Stochastic programming, among these methods, utilizes historical data to generate potential scenarios that include diverse outcomes resulting from uncertain parameters. These scenarios are evaluated based on their probabilities of occurrence, enabling decision-makers to plan and make informed choices considering a range of potential situations. Robust optimization, as another approach, concentrates on devising plans or strategies that maintain their effectiveness despite uncertainties. Rather than concentrating on particular scenarios, robust optimization prioritizes crafting solutions that can endure a broad spectrum of potential fluctuations and alterations. It revolves around designing supply chain networks capable of withstanding unforeseen events or shifts in conditions without experiencing significant negative impacts. Furthermore, Two-Stage Stochastic Programming (TSSP) is particularly notable in managing uncertainty. It involves making initial decisions, often substantial in terms of cost or time, before knowing the exact scenarios that might unfold. For instance, decisions related to facility location fit into this category. These decisions are made directly, considering the potential outcomes of different scenarios, allowing for more proactive planning in the face of unpredictability. Overall, uncertainty within supply chain networks necessitates strategies that mitigate the potential impacts of unknown variables. By employing methods like stochastic programming and robust optimization, supply chain managers can proactively address uncertainty, making informed decisions that are resilient to various potential scenarios.

The cultivation of date palms stands as a significant industry in various regions, notably the Middle East and North Africa, carrying substantial economic and cultural importance. However, the process generates considerable waste, including palm fronds, pits, and other byproducts, historically underutilized and posing environmental challenges while missing opportunities for value creation.

Designing an optimal supply chain network to tackle real-world challenges draws insights from numerous valuable research studies in the literature. This involves considering various factors and attributes during the crafting process of these networks. Supply Chain Management (SCM), divided into two primary components, strategic and operational, centers around customers, suppliers, and logistics. At the strategic level, the SCND stands out as a critical decision-making process. It involves crucial strategic determinations like allocating facility locations and delivering products to customers. SCND significantly contributes to sourcing raw materials efficiently, converting them into finished products, and optimizing their delivery to end customers. Moreover, it defines the ideal count of facilities and material movement within the supply chain.

SCM has gained significant focus in recent years [1], leading to multiple operational studies. Depletion of environmental resources and its resulting impacts have prompted policymakers and food planners to actively seek alternative environmental sources [2]. The significance of creating, executing, and overseeing agricultural waste and food supply chains has grown rapidly in recent times. In a parallel study, Yıldız and Ayvaz [3] designed a Mixed Integer Linear Programming (MILP) model to generate energy from waste biomass, particularly considering agricultural waste as a possible resource.

Allen and Schuster [4] designed a model focused on reducing agricultural production waste, particularly aiming to enhance harvesting and storage capabilities. Ahumada and Villalobos [5] investigated the intricate relationship between product quality, pricing dynamics, and pertinent factors like product valuation, labor expenditures, and transportation methodologies. It involved crafting a multi-period Mixed Integer Programming (MIP) model that emphasized select farms situated in distinct locales, aiming to optimize farmers' income through an enhanced product quality focus.

Rahemi et al. [6] suggested a framework to organize a bioethanol supply chain network, taking into account land characteristics and managing the competition between food and biomass feedstock Similarly, Iakovou et al. [7] deliberated on creating sustainable supply chain networks for agricultural food items, including debates on eco-friendly product aspects, packaging, production, and facility design choices. Maria Garcia Perez and Garcia Martinez [8] developed a structure to analyze the design and setup of agricultural food chains. Moreover, Kambli and McGarvey [9] created a specialized supply chain network for regional farming utilizing robust optimization methods to reduce the distance traveled by both farmers and vehicles when transporting agricultural goods from farms to local and central hubs.

Furthermore, Mogale et al. [10] developed a mixed-integer non-linear programming model customized for the wheat supply chain in India, with the objective of reducing transportation, storage, and operational expenses.

Numerous studies in this area have centered on employing multi-objective optimization approaches. For instance, Sarker and Ray introduced the Epsilon-constraint technique as part of their model [11]. Moreover, most academic papers in this field have predominantly highlighted the overall cost and demand aspects.

Baghizadeh et al. [12] created a multi-product, multi-objective MINLP mathematical program to cover the economic, social, and environmental sustainability goals within agriculture, aiming to address all aspects of sustainability in this field. Rafigh et al. [13] Developed a structure for key participants in a sustainable supply chain network, handling uncertainties through the utilization of an economic order quantity model in their investigation.

Numerous studies have integrated agricultural science theories with supply chain management methodologies. For example, Rong et al. [14] introduced an optimization model specifically designed to oversee and manage the quality maintenance of perishable products across various stages within the supply chain, focusing on preserving quality from production to delivery.

Rantala [15] devised a mathematical model focusing on seedling transfers from farms to cold stores, optimizing costs while ensuring quality across the supply chain stages. The objective was to minimize transfer expenses and ensure comprehensive coverage services while accounting for capacity limitations and product perishability. Meanwhile, Van Berlo [16] researched various activities in the vegetable processing supply chain. They used mathematical models to plan the planting, harvesting, processing, and delivery stages for better efficiency. Furthermore, Cheraghalipour et al. [17] suggested a two-level optimization model for the rice supply chain, utilizing Genetic Algorithm (GA) and Particle Swarm Optimization (PSO) as meta-heuristic techniques to solve the model. Ultimately, Motevalli-Taher et al. [18] Created a mathematical model with multiple objectives aiming to decrease network expenses and water usage while enhancing employment prospects within the environmentally sustainable wheat supply chain.

Waste management holds a pivotal role within this domain, with mathematical modeling profoundly influencing various industries. Banasik et al. [19] utilized Mixed Integer Linear Programming (MILP) to address economic and environmental aspects in the mushroom industries. Salehi-Amiri et al. [20] devised a closed-loop supply chain framework tailored for the walnut industry's production network. This model handled both forward and reverse flows, meeting various market demands using precise, meta-heuristic, and hybrid solution techniques. Krishnan et al. [21] introduced a resilient multi-objective optimization model with the aim of designing a sustainable food supply chain network, integrating the optimization of food waste. Hamdi-Asl et al. [22] suggested a mathematical framework for date palm industries involved in delivering palm fruits to factories and customers, emphasizing the most efficient farm-to-factory allocation while reducing overall expenses. Gharye Mirzaei et al. [23] created a MILP model to improve economic goals under uncertainties. Additionally, they introduced the Whale Optimization Algorithm (WOA), Particle Swarm Optimization (PSO), and a combined WOA-PSO hybrid algorithm to tackle the challenges posed by NP-hard problems. Handayati et al. [24] offered an extensive review centered on the agricultural food supply chain, emphasizing coordination challenges and proposing effective alternative methods for managing agricultural waste. Rentizelas et al. [25] proposed a MILP model to optimize the entire supply chain, spanning from waste creation to the ultimate consumer destination. Kamyabi et al. [26] developed a dynamic system and mathematical programming method for the lead-acid battery supply chain amid demand uncertainty. Their research indicated that implementing enhanced policies for recycling could potentially result in underutilized factory capacity, despite intensified recycling endeavors.

Various methods have been suggested to improve agricultural waste supply chains. However, a gap remains in utilizing Date Palm Tree Waste (DPTW) effectively for MDF, handicrafts, and fertilizer production. This study introduces a distinct agricultural supply chain model, differentiating itself from prior research that assumed fixed factory conditions. Instead, it adapts to demand uncertainties by allowing for factory expansions and emphasizes using date palm waste for producing MDF, fertilizer, and handicrafts. Initially, farms dispatch palm tree waste to multiple production sites. These facilities process the waste into MDF, distributing it to warehouses and customers. Domestic customers can also return used MDF to these facilities for reprocessing. This 20-year model uses robust optimization techniques to handle demand uncertainties and conduct sensitivity analyses on costs and parameters.

This paper introduces an innovative mathematical model focused on efficiently managing palm tree waste amid uncertain conditions, particularly centered on uncertain customer demand. Using a robust optimization approach, the study bolsters the network's ability to navigate these uncertainties effectively. Extensive computational analysis underscores the approach's significance and potential implications, providing managerial insights for guidance, alongside sensitivity analyses critical for proper agricultural waste management within this context.

2 Methods and Materials

2.1 Problem Definition

This study involved establishing a system to manage the movement of palm tree waste in the southern regions of Iran. Figure 2 outlines the way of waste movement, originating from farms to factories and handicrafts markets. It also illustrates the transportation route of MDF products from factories to various provinces and neighboring countries.

The waste obtained from date palm trees, known as DPTW, holds value and is employed in creating MDF, fertilizers, and handicrafts. Following this, these waste materials are integrated into MDF production at factories, then dispatched to warehouses and eventually distributed to customers. Furthermore, domestic consumers contribute by supplying worn-out MDF, enabling its recycling as raw material in the factories. This establishes a sustainable cycle for producing and recycling MDF.

The study collected information from 120 primary date palm farms, accounting for 95 percent of Iran's total cultivated date palm area. The model aims to supply MDF to 31 provinces within the country and four neighboring nations for customer distribution. Accurate mapping of all locations was achieved using latitude and longitude coordinates, as shown in Fig. 1. For future cost projections in the mathematical model, inflation rates were factored in to estimate expenses across subsequent periods.

2.2 Model Formulation and Assumptions

The main goal of this study is to improve the closed-loop MDF supply chain network. This network involves date palm farms, handicrafts markets, potential factory locations, warehouses, domestic customers, and international customers.

The first step to design the network, represented in Fig. 2, involves collecting DPTW from farmlands. This system operates across several periods, managing the movement of DPTW to designated factories while allocating a portion to handicrafts markets. Additionally, a percentage of these wastes is utilized for producing fertilizer on farms. Following this, the manufactured MDF is distributed from factories to warehouses and then to respective customers. Moreover, a section of long-term used MDF is gathered and transported back to factories. The model operates under the assumption that transportation occurs via roads and relies on the following presumptions:

- Each farmer annually transports their palm waste to the factories
- The proposed model operates over a twenty-year duration, with a constraint that prevents the opening of any factories during the last t_1 years of this period
- The standard lifespan of MDF is t_2 year; once this period elapses, used MDF can be introduced into the factories as raw material

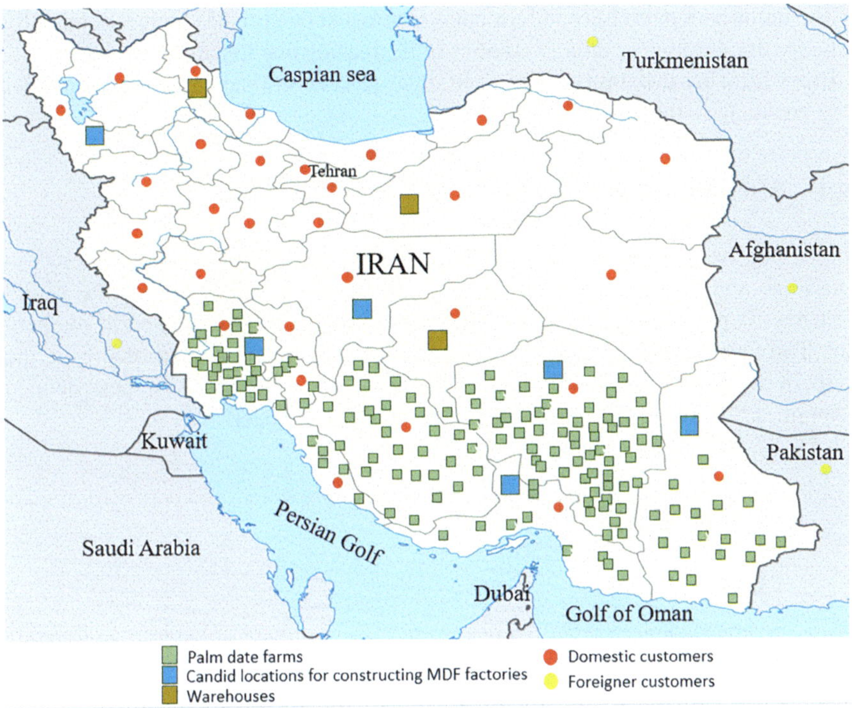

Fig. 1 Distribution of farms, candid locations, warehouses, and related customers

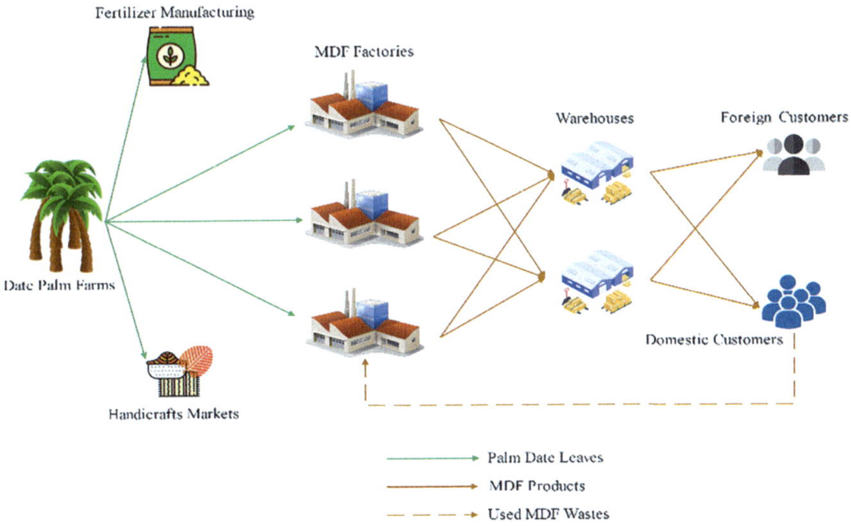

Fig. 2 Network flow from farms to factories and related customer

- The number of palm trees in farmlands is affected by climate change, potentially increasing through planting activities or decreasing due to natural disasters
- Every farm has the ability to burn particular segments of palm tree leaves in order to create agricultural fertilizer

2.2.1 Deterministic Model

Addressing the facility location challenge requires carefully situating factories in a network and aligning their placement with farms to reduce costs. The model evaluates six potential locations for MDF factories specifically situated in southern Iran. Through the MILP model, we explore these six potential factory sites in the southern region of Iran. Our objective is to develop a model based on these defined assumptions. Here are the specific details of the MILP model:

SETS	
i	Index of date palm farmlands
j	Index of candidate location for factories
k	Index of warehouses
l	Index of domestic customers
u	Index of foreign customers
h	Index of handicraft markets
s	Index of scenarios
t	Index of years
Parameters	
f_{it}	Management costs of date palm farmland i year t
f_{jt}	Fixed and management cost of opening factory j in year t
f_{kt}	Management costs of warehouse k in year t
c_{ijt}	Processing and transportation cost from farmland i to factory j in year t
c_{iht}	Processing and transportation cost from farmland i to handicraft market h in year t
c_{jkt}	Processing and transportation cost from factory j to warehouse k in year t
c_{klt}	Processing and transportation cost from warehouse k to domestic customer l in year t
c_{kut}	Processing and transportation cost from warehouse k to foreign customer u in year t
c_{ljt}	Processing and transportation cost from domestic customer l to factory j to in year t
Ω_{jt}	Production cost per unit of product factory j in year t
d_{lt}	The demand for MDF by domestic customers l in year t
d_{ut}	The demand for MDF by foreign customers u in year t
p_{it}	Date palm wastes production of farmland i in year t
p_j	Production capacity of MDF factory j
p_k	Capacity of warehouse k
M	Minimum Production capacity of MDF factory j

(continued)

(continued)

SETS	
ϑ	Minimum number of MDF factories that must be built
v_i	The waste rate of DPTW at farmland i
ff	Co_2 emission rate per by firing DPTW for making fertilizer
em_{it}	Co_2 emission legal rate in farmland i in period t
e_{it}	Expansion rate of farmland i in period t
a_{it}	Reduction rate of farmland i in period t
π	Cost of lost shortages
ξ	Percentage of Transported quantity of DPTW for handicraft markets
Ψ	Percentage of Transported quantity of DPTW for making fertilizer
rr	Convert rate for date palm wastes and used MDF wastes to make MDF
φ	The rate of used MDF boards returning to factories
Decision Variables	
x_{ijt}	Transported quantity of DPTW farmland i to factory j in period t
xh_{iht}	Transported quantity of DPTW from farmland i to handicraft markets h in period t
xf_{it}	Transported quantity of DPTW from i farmland i for making fertilizer in period t
xr_{ljt}	Transported quantity of used MDF wastes from domestic customer l to factory j in period t
g_{jkt}	Transported quantity of MDF products from factory j in to warehouse k in period t
sl_{klt}	Transported quantity of MDF products from warehouse k to domestic customer l in period t
su_{kut}	Transported quantity of MDF products from warehouse k to foreign customer u in period t
pl_{lt}	Shortage quantity of domestic customer l in period t
pu_{ut}	Shortage quantity of foreign customer u in period t
y_{jt}	1 if MDF factory is opened at location j in period t; 0, otherwise

$$\begin{aligned}\text{Min Z} = & \sum_i \sum_t f_{it} + \sum_k \sum_t f_{kt} + \sum_j \sum_t f_{jt}.y_{jt} + \sum_i \sum_j \sum_t c_{ijt}.x_{ijt} \\ & + \sum_i \sum_h \sum_t c_{iht}.xh_{iht} + \sum_j \sum_k \sum_t \left(\Omega_{jt} + c_{jkt}\right).g_{jkt} \\ & + \sum_k \sum_l \sum_t c_{klt}.sl_{klt} + \sum_k \sum_u \sum_t c_{kut}.su_{kut} \\ & + \sum_l \sum_j \sum_t c_{ljt}.xr_{ljt} + \sum_l \sum_t \pi.pl_{lt} + \sum_u \sum_t \pi.pu_{ut} \end{aligned} \tag{1}$$

$$p_{it}.(1 - v_i) = \sum_j x_{ijt} + \sum_h xh_{iht} + xf_{it} \qquad \forall i \in I, \forall t \in T \tag{2}$$

$$\xi.p_{it}.(1-\alpha_i)=\sum\nolimits_h xh_{iht} \qquad \forall i\in I,\forall t\in T \tag{3}$$

$$\Psi.p_{it}.(1-\alpha_i)=xf_{it} \qquad \forall i\in I,\forall t\in T \tag{4}$$

$$xf_{it}.ff\le em_{it} \qquad \forall i\in I,\forall t\in T \tag{5}$$

$$\sum_k g_{jkt}\le p_j.\sum\nolimits_{t'=1}^{t'=t} y_{jt'} \qquad \forall j\in J,\forall t\in T \tag{6}$$

$$\sum_k g_{jkt}\ge M.\sum\nolimits_{t'=1}^{t'=t} y_{jt'} \qquad \forall j\in J,\forall t\in T \tag{7}$$

$$(\sum_i x_{ijt}+\sum_l xr_{ljt}).rr=\sum_k g_{jkt} \qquad \forall j\in J,\forall t\in T \tag{8}$$

$$\sum_j xr_{ljt}=\varphi.d_{lt-t_2} \qquad \forall l\in L,\forall t\in T \tag{9}$$

$$\sum\nolimits_j g_{jkt}\le p_k \qquad \forall k\in K,\forall t\in T \tag{10}$$

$$\sum\nolimits_j g_{jkt}=\sum\nolimits_l sl_{klt}+\sum\nolimits_u su_{kut} \qquad \forall k\in K,\forall t\in T \tag{11}$$

$$\sum\nolimits_k sl_{klt}=d_{lt}-pl_{lt} \qquad \forall l\in L,\forall t\in T \tag{12}$$

$$\sum\nolimits_k su_{kut}=d_{ut}-pu_{ut} \qquad \forall u\in U,\forall t\in T \tag{13}$$

$$\sum_t y_{jt}\le 1 \qquad \forall j\in J \tag{14}$$

$$\sum_j\sum_t y_{jt}\ge\vartheta \tag{15}$$

$$\sum_t y_{jt}=\vartheta \qquad \forall j\in J,\forall t=1 \tag{16}$$

$$\sum\nolimits_{t'=T-t_1}^{t'=T} y_{jt'}=0 \qquad \forall j\in J \tag{17}$$

$$y_{jt}\in\{0,1\} \qquad \forall j\in J,\forall t\in T \tag{18}$$

$$x_{ijt}, xh_{iht}, xf_{it}, xr_{ljt}, g_{jkt}, sl_{klt}, su_{kut}, pl_{lt}, pu_{ut} \geq 0 \quad \forall i \in I, \forall h \in H, \forall j \in J, \forall k \in K, \forall u \in U, \forall l \in L, \forall t \in T \tag{19}$$

Objective function (1) outlines the main goal of the MDF supply chain network. Equation (2) ensures that each farm's production capacity matches the amount of DPTW transported from those farms to factories, handicraft markets, and fertilizer demand.

The size of date palm farms fluctuates over time, either expanding or contracting due to various factors. Farms may decrease in size due to natural disasters like floods, fires, or extreme heat, but they can also increase when new saplings are planted in suitable soil with sufficient rainfall. Equation (20) computes the production capacity of each farm in every period, considering the expansion rate (e_{it}) for farmland I in period t and (a_{it}) represents the reduction rate of farmland I in period t.

$$p_{it} = p_{it-1} + e_{it-1} - a_{it-1} \quad \forall i \in I, \forall t \in T \tag{20}$$

Equations (3) and (4) calculate how DPTW is distributed to handicraft markets and used for fertilizer production, respectively. Meanwhile, Eq. (5) specifies that the combined CO2 emissions generated by burning palm tree wastes for fertilizer production must stay within the legally set limits for emissions. Additionally, Eq. (6) states that the quantity of MDF produced and shipped from each factory shouldn't go beyond the factory's capacity in the current or previous periods. If the factory operates in location j during the current or previous periods, Eq. (7) guarantees that the quantity of MDF produced and shipped from each factory should meet or surpass the minimum production capacity. Moreover, Eq. (8) necessitates a balance between incoming palm wastes, utilized MDF wastes, and the MDF manufactured in factory j. Furthermore, Eq. (9) determines the volume of utilized MDF waste that gets returned to the factories. Equation (10) explicitly states that the quantity of MDF products transported from factories to warehouses must not exceed the warehouse capacity. In addition, Eq. (11) maintains a balance by ensuring that the quantity of incoming MDF products matches the amount of MDF sent out to meet both domestic and foreign demands. Equations (12) and (13) assert that the amount of MDF dispatched from each warehouse should meet or surpass the requirements of both domestic and foreign customers. Equation (14) is focused on ensuring that each possible location is used only once. Equation (15) emphasizes the necessity of a minimum quantity of ϑ factories in operation. Equation (16) requires that a specific number of ϑ factories must be operational at the beginning period. Furthermore, Eq. (17) stops the building of factories within the last t_1 years. Lastly, Eq. (18) represents binary variables, while Eq. (19) guarantees that the decision variables remain positive.

2.2.2 Robust Optimization

When designing logistics networks, uncertainty can be broadly divided into two primary categories. This classification aids in comprehending and managing various unpredictable elements that can impact logistics planning and decision-making processes, as outlined below:

(1) Uncertainty in typical factors, like variations in demand or uncertainties in the supply chain.
(2) Uncertainty arises from potential disturbances within the network. While these disruptions might be unlikely, their effect on network performance can be significant. These disruptions can originate from natural occurrences like floods and earthquakes, or human-induced events such as conflicts and acts of terrorism.

There's another way suggested to classify uncertainty in logistics. This method looks at uncertainty in two separate parts:

(1) Uncertainty in input data can result in adaptability in constraints and specified objectives.
(2) Cognitive uncertainty arises when there's a limited understanding of a particular field or insufficient access to necessary data for defining future problem parameters. To tackle these challenges, probabilistic and fuzzy programming methods are frequently utilized.

This case study deals with uncertainty arising from the variability in input data, particularly concerning the demands for MDF products. To tackle this issue, a scenario-based robust optimization approach was employed, utilizing historical data on uncertain parameters. As highlighted earlier, a specialized scenario-based robust model was developed specifically for this purpose. To convert the deterministic model into a scenario-based robust model, the approach outlined by Malloy and colleagues, subsequently revised by Min and Yu in 2000, is integrated into our initial deterministic model [27].

In a scenario-based robust optimization problem, it is assumed that Z_s is represents the objective function for scenario s, where x_s is the variable dependent on scenario s and y is the independent variable to scenario s. The components (c_s, d_s, K_s, b_s, A_s (are related to scenario s, while (R, q) is independent but associated with scenario s. Additionally, we consider the probability of each scenario occurrence denoted by Pi_s is considered. Following this method, the optimization problem aims to maximize economic profit through the scenario-based model, presented as follows:

$$MaxE(z) - var(z) - feasibility.p = \sum_s Pi_s.z_s - \Lambda \sum_s Pi_s.\left(z_s - \sum_{s'} Pi_{s'}.z_{s'}\right) - \omega \sum_s Pi_s.\xi_s^2 \quad (21)$$

$$z_s = c_s^T.x_s + d_s^T.y \quad (22)$$

$$A_s.x_s + K_s.y \leq b_s + \xi_s \tag{23}$$

$$Ry = q \tag{24}$$

$$y \in Y, x_s \geq 0 \tag{25}$$

In this context, (ξ_s) denotes the variable indicating either shortage or surplus in each scenario, while (ω) represents the stability coefficient within the Malloy model, and (Λ) signifies the importance coefficient for variance across various scenarios. Within this framework, the optimal solution for robustness aims for optimum average system performance (Eq. 21), where variations from the defined goals are regulated across different scenarios. The key difference from the classical scenario-based model lies in the model's adaptability and control, as expressed in Eq. 23, aiming to enhance average performance and minimize discrepancies in the outcomes.

The first level identifies the best quantity and sites for MDF factories, while the second level allocates the most efficient load flow throughout the network. Therefore, the model's objective function considers various operational costs and initial investments within the specified scenarios. The scenario set, denoted as S, includes instances created to address uncertainties in demand parameters. This set comprises four interconnected scenarios, shaped by statistical variations among its members.

Due to uncertainties in customer demands, this study's first objective focuses on minimizing the overall system costs. Moreover, the second objective function incorporates the relative regret index proposed by Kouvelis and Yu in 2013 [28]. This index, particularly applied at the strategic level of supply chain network design using mixed-integer programming, enhances the stability of optimal solutions [29]. As depicted in Eq. 26, the absolute regret index quantifies the disparity between the deterministic objective function value in each scenario Z^*_{s1} and the objective function derived from the scenario-based probabilistic model Z_{s1}.

$$R_s = Z_{s1} - Z^*_{s1} \tag{26}$$

In this research, the non-deterministic model presents the second objective function as defined in Eq. 27. This model aims to minimize the greatest absolute regret from each scenario compared to the optimal value obtained from the deterministic model using data from that scenario. Essentially, this function strives to mitigate the risks arising from uncertainty in the decision-making process.

$$R'_s = \frac{R_s}{Z^*_{s1}} \tag{27}$$

The following section will present the robust model developed based on the deterministic model:

$$\begin{aligned}\text{Min } Z_1 = &\sum_i \sum_t f_{it} + \sum_k \sum_t f_{kt} + \sum_j \sum_t f_{jt}.y_{jt} + \sum_s \sum_i \sum_j \sum_t c_{ijt}.x_{ijts} \\ &+ \sum_s \sum_i \sum_j \sum_t c_{iht}.xh_{ihts} + \sum_s \sum_j \sum_k \sum_t \left(\Omega_{jt} + c_{jkt}\right).g_{jkts} \\ &+ \sum_s \sum_k \sum_l \sum_t c_{klt}.sl_{klts} + \sum_s \sum_k \sum_u \sum_t c_{klt}.su_{kuts} \\ &+ \sum_s \sum_l \sum_j \sum_t c_{ljt}.xr_{ljts} + \sum_s \sum_l \sum_t \pi\ .pl_{lts} \\ &+ \sum_s \sum_u \sum_t \pi\ .pu_{uts}\end{aligned} \tag{28}$$

$$\text{Min } Z_2 = \text{min max } R_s^{'} \tag{29}$$

$$p_{it}.(1 - v_i) = \sum\nolimits_j x_{ijts} + \sum\nolimits_h xh_{ihts} + xf_{its} \quad \forall i \in I, \forall t \in T, \forall s \in S \tag{30}$$

$$\xi.p_{it}(1 - \alpha_i) = \sum\nolimits_h xh_{ihts} \quad \forall i \in I, \forall t \in T, \forall s \in S \tag{31}$$

$$\Psi p_{it}(1 - \alpha_i) = xf_{its} \quad \forall i \in I, \forall t \in T, \forall s \in S \tag{32}$$

$$xf_{its}.ff \leq em_{it} \quad \forall i \in I, \forall t \in T, \forall s \in S \tag{33}$$

$$\sum_k g_{jkts} \leq p_j.\sum\nolimits_{t'=1}^{t'=t} y_{jt'} \quad \forall j \in J, \forall t \in T, \forall s \in S \tag{34}$$

$$\sum_k g_{jkts} \geq M.\sum\nolimits_{t'=1}^{t'=t} y_{jt'} \quad \forall j \in J, \forall t \in T, \forall s \in S \tag{35}$$

$$(\sum_i x_{ijts} + \sum_l xr_{ljts}).rr = \sum_k g_{jkts} \quad \forall j \in J, \forall t \in T, \forall s \in S \tag{36}$$

$$\sum_j xr_{ljts} = \varphi.d_{lt-t_2s}.Pi_s \quad \forall l \in L, \forall t \in T, \forall s \in S \tag{37}$$

$$\sum_j g_{jkts} \leq p_k \quad \forall k \in K, \forall t \in T, \forall s \in S \tag{38}$$

$$\sum\nolimits_j g_{jkts} = \sum\nolimits_l sl_{klts} + \sum\nolimits_u su_{kuts} \quad \forall k \in K, \forall t \in T, \forall s \in S \tag{39}$$

$$\sum\nolimits_k sl_{k,l,t,s} = d_{lts}.Pi_s - pl_{lts} \quad \forall l \in L, \forall t \in T, \forall s \in S \tag{40}$$

$$\sum_{k} su_{kuts} = d_{uts}.Pi_s - pu_{uts} \qquad \forall u \in U, \forall t \in T, \forall s \in S \tag{41}$$

$$\sum_{t} y_{jt} \leq 1 \qquad \forall j \in J \tag{42}$$

$$\sum_{j}\sum_{t} y_{jt} \geq \vartheta \tag{43}$$

$$\sum_{t} y_{jt} = \vartheta \qquad \forall j \in J, \forall t = 1 \tag{44}$$

$$\sum_{t'=T-t_1}^{t'=T} y_{jt'} = 0 \qquad \forall j \in J \tag{45}$$

$$y_{jt} \epsilon \{0,1\} \qquad \forall j \in J, \forall t \in T \tag{46}$$

$$x_{ijts}, xh_{ihts}, xf_{its}, xr_{ljts}, g_{jkts}, sl_{klts}, su_{kuts}, pl_{lts}, pu_{uts} \geq 0 \qquad \begin{array}{c}\forall i \in I, \forall h \in H, \forall j \\ \in J, \forall k \in K, \\ \forall u \in U, \forall l \in L, \forall t \\ \in T, \forall s \in S\end{array} \tag{47}$$

3 Results

This section explains how the proposed scenario-based Robust optimization method is practically applied in the MDF production industry. The model figures out the best locations for factories based on demand fluctuations and transportation costs. We used real case data to work out this model on a computer with an Intel® Core™ i7 8550 CPU, 4.00 GHz, 12.00 GB of RAM, and GAMS 24.1.3 software.

3.1 Modeling Assumptions

We developed the model described in Sect. 3, using specific assumptions, to pinpoint suitable sites for setting up MDF production facilities.

To evaluate our assumptions, we reviewed the maximum MDF production from individual factories, which was determined to be $3 * 10^6 m^3$ per factory. The warehouses, on the other hand, have a maximum capacity of $2 * 10^7 m^3$ for storing MDF. Additionally, considering the extensive presence of date palm farms in Iran, it's reasonable to expect that farmers can sufficiently transport date palm wastes from farms to factories, handicraft markets, and for fertilizer production each period.

3.2 Computational Results

The results within the scenario-based model include 120 farms, 6 potential sites, 3 warehouses, 31 domestic customers, 4 international customers, and the handicrafts market. The uncertainties in demand were acknowledged, and scenario-based planning was integrated into this study. Four scenarios, where nominal customer demand values fluctuate by 25%, were investigated as outlined in Table 1. Furthermore, Table 2 illustrates the different cost elements associated with operating the scenario-based Robust optimization model, detailing the contribution percentage of each cost to the total expenditure.

Through the application of scenario-based Robust optimization, Table 3 and Fig. 3 illustrate the distribution of factories over the 20-year span in the model. In the first year, plans are set to construct three MDF production factories. Subsequently, West Azerbaijan and Isfahan provinces were identified in the following years as potential locations, selected based on considerations of demand patterns and transportation expenses.

Table 1 The result of the deterministic model and robust model

Scenario	Possibility	Total costs	
		Deterministic model	Scenario-based model
1	0.25	310,243	319,049
2	0.25	273,498	
3	0.35	328,143	
4	0.15	293,584	
Average total cost for deterministic model for all scenarios			301,367

Table 2 Various segments contribute to the cost structure of the model

Cost	Value (million toman)	Percentage of each cost
Management costs of date palm farmlands	28,715	9
Fixed and management costs of opening factories	22,332	7
Management costs of warehouses	12,762	4
Transportation costs	255,240	80
Shortage costs	0	0
Total costs	319,049	

Table 3 Locations and year of construction of the MDF factories

Year	1	7	10
Location	Sistan-Khuzestan-Hormozan	West Azerbaijan	Isfahan

Fig. 3 The locations of the factories during the 20-year period

3.3 *Sensitivity Analyses and Discussion*

Sensitivity analyses were performed on select model parameters to evaluate the potential impact of variations on the proposed network. The following section represents the various analyses conducted and their respective influences on the ultimate outcomes.

- ***Sensitivity Analysis of the Total Cost***

The model was developed for a 20-year duration. Additionally, considering inflation rates incorporated into the actual data of each year, the escalating customer demands in subsequent years lead to a rise in the overall costs.

Comparing Table 4 and Fig. 4 show a significant increase in transportation expenses, mainly due to building new factories. Creating more transport routes resulted in more trips between places, raising the overall costs. Also, in the beginning, building many factories led to higher construction expenses.

- ***Sensitivity Analysis of the Allocation of Date Palm Wastes***

The estimated amount of waste produced in palm farms, derived from the case study, might not stay constant. Considering this fluctuation, different scenarios have been examined to understand how these variations could affect the total cost related to palm tree waste.

Table 5 displays how the overall system cost and potential locations are affected by varying percentages of DPTW used for incineration and animal feed, along with allocation to handicraft markets. Additionally, Fig. 5 illustrates that as the waste allocated for handicraft markets increases, there's a corresponding rise in total costs due to increased transportation expenses.

Table 4 Various segments contribute to the cost structure of the model in the first year

Cost	Value (million toman)	Percentage of each cost
Management costs of date farmlands	479	10
Fixed and management costs of opening factories	1,438	30
Management costs of warehouses	192	4
Transportation costs	2,683	56
Shortage costs	0	0
Total costs	4,792	

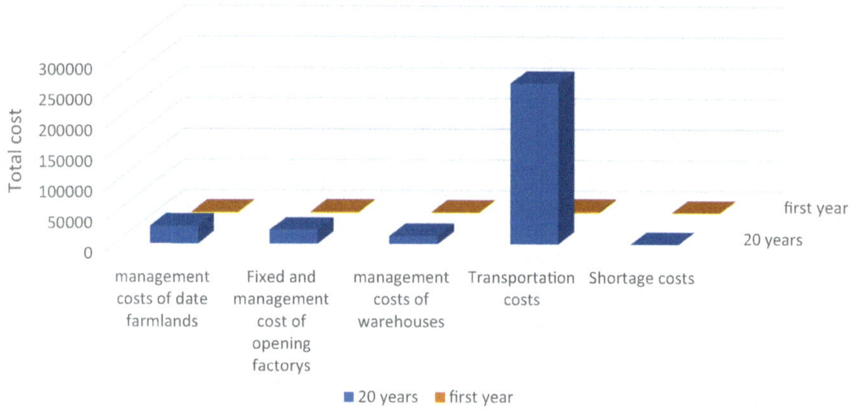

Fig. 4 The different costs of the proposed model in the first year and 20 years

Table 5 Sensitivity analysis of the parameters of palm waste allocation to the handicrafts market and making fertilizer

Parameters		Total cost	Percentage
1	$\xi = 0.6, \Psi = 0.1$	328,143	17
2	$\xi = 0.5, \Psi = 0.2$	356,561	18
3	$\xi = 0.45, \Psi = 0.25$	381,046	20
4	$\xi = 0.4, \Psi = 0.3$	416,327	22
5	$\xi = 0.35, \Psi = 0.35$	438,745	23

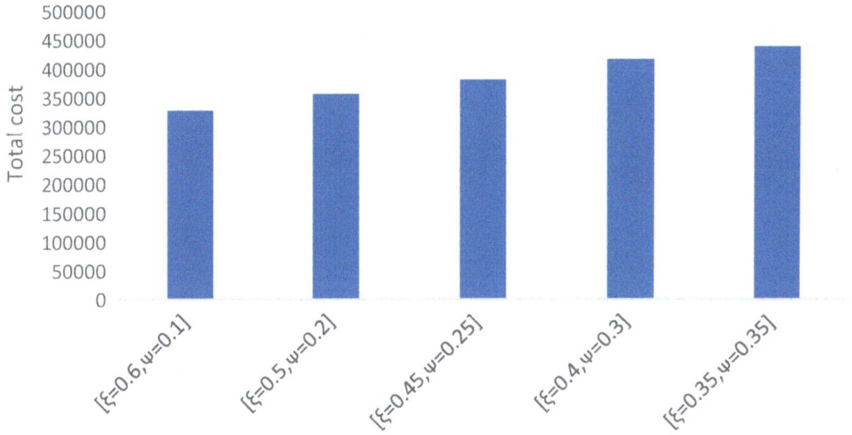

Fig. 5 Changes in total costs due to changes in parameters of palm waste allocation to the handicrafts market and making fertilizer

3.4 *Manegerial Insights*

The primary objective of this study is to create an effective strategy for managing DPTW within the supply chain. This strategy aims to reduce costs associated with producing MDF, handicrafts, and fertilizer, ultimately minimizing operational expenses across date palm farms, MDF factories, and warehouses. Thus, the aim is to optimize operations and lower expenses throughout the supply chain of the date palm industry.

This research holds the potential to offer valuable perspectives to managers in both the wood industry and date palm farming organizations. These insights enhance the management of agricultural waste networks. As a result, the detailed findings and recommendations derived from the case study's outcomes and sensitivity analyses are presented below:

1. Effectively managing the produced waste relies on proposing a highly accurate forecast for the volume of waste generated on farms. Hence, the collection of authentic, real-time data plays a crucial role in achieving precise predictions of

the waste generated. This precision aids managers in making optimal decisions regarding waste management strategies.

2. In recent years, a significant amount of palm tree and wooden board waste is transported using trucks and road transport. This choice is shaped by prevailing attitudes among stakeholders in road transport, including activists, drivers, and companies. There is a perception that expanding the use of alternative transport modes could potentially decrease profits. However, combining various transportation methods within this sector could efficiently reduce associated costs.
3. The principal composition of MDF consists of wood and various wood-based materials. Integrating palm tree waste into wood industries significantly reduces the overall expenses associated with wood products due to the cost-efficient nature of these discarded materials.

4 Conclusion

This paper presents a robust optimization method designed to tackle the challenge of managing DPTW within SCND under uncertain conditions. To address demand uncertainty, various scenarios were created, demonstrating a 25% change in customers' nominal demands. Moreover, the study introduces an innovative approach aimed at effectively managing palm tree waste. This approach emphasizes the production of MDF boards, enhancing both environmental efficiency and productivity for MDF manufacturing entities. Notably, this unique strategy for date palm farms remains unimplemented in the wood industries, showcasing its potential as a promising alternative for promoting environmental sustainability.

The final results highlighted the effectiveness of the model in skillfully managing the distribution of products among specified farms while strategically setting factories and warehouses to fulfill customer needs. A remarkable point was the considerable expense related to transportation, comprising roughly 80% of the overall expenditure. This substantial proportion was largely due to the inclusion of inflation rates spanning different time periods within the model. Moreover, operational expenses related to farm activities and factory construction represented approximately 9% and 7% of the total costs, respectively.

This analysis reflects real-life scenarios, where the creation of an MDF factory naturally leads to higher transportation costs due to increased demand. Furthermore, sensitivity analyses demonstrate a clear relationship: greater amounts of waste in the handicrafts market directly led to higher overall costs.

Furthermore, exploring alternative solutions for SCND offers an intriguing path forward. Future studies could bring considerable benefits by addressing uncertainties in certain parameters related to waste computation from date palm farms. Exploring the impact of multi-modal transportation on the model and its implications could be a promising direction for further investigation. Moreover, expanding the scope to involve broader scenarios, like engaging more foreign customers, considering potential model modifications, and evaluating the necessity of employing a meta-heuristic approach to tackle the potential NP-hard problem, could yield substantial advancements in future research.

References

1. Sabbaghnia A, Razmi J, Babazadeh R Moshiri B (2018) Reducing the Bullwhip effect in a supply chain network by application of optimal control theory. RAIRO-Oper Res 52, 1377–1396
2. Yadav VS, Singh A, Gunasekaran A, Raut RD, Narkhede BE (2022) A systematic literature review of the agro-food supply chain: Challenges, network design, andperformance measurement perspectives. Sustain Prod Consum 29:685–704
3. Yildiz HG, Ayvaz B (2018) Waste biomass based energy supply chain network design. J Int Trade, Logist Law 4:126–137
4. Allen SJ, Schuster EW (2004) Controlling the risk for an agricultural harvest. Manuf Serv Oper Manag 6:225–236
5. Ahumada O, Villalobos JR (2011) Operational model for planning the harvest and distribution of perishable agricultural products. Int J Prod Econ 133:677–687
6. Rahemi H, Torabi SA, Avami A, Jolai F (2020) Bioethanol supply chain network design considering land characteristics. Renew Sustain Energy Rev 119:109517
7. Iakovou E, Bochtis D, Vlachos D (2021) Supply chain management for sustainable food networks, Wiley Online Library
8. Maria Garcia Perez A, Garcia Martinez M (2007) A theoretical framework to study its configuration. Acta Agric Scand Sect C 4:31–39
9. Kambli A, McGarvey RG (2021) Network design for local agriculture using robust optimization. Inf Process Agric 9:469–483
10. Mogale D, Kumar SK, Márquez FPG, Tiwari MK (2017) Bulk wheat transportation and storage problem of public distribution system. Comput Ind Eng 104:80–97
11. Sarker R, Ray T (2009) An improved evolutionary algorithm for solving multi-objective crop planning models. Comput Electron Agric 68:191–199
12. Baghizadeh K, Cheikhrouhou N, Govindan K, Ziyarati M (2022) Sustainable agriculture supply chain network design considering water-energy-food nexus using queuing system: A hybrid robust possibilistic programming. Nat Resour Model 35:e12337
13. Rafigh P, Akbari AA, Bidhandi HM, Kashan A (2022) A sustainable supply chain network considering lot sizing with quantity discounts under disruption risks: centralized and decentralized models. J Comb Optim 44:1387–1432
14. Rong A, Akkerman R, Grunow M (2011) An optimization approach for managing fresh food quality throughout the supply chain. Int J Prod Econ 131:421–429
15. Rantala J (2004) Optimizing the supply chain strategy of a multi-unit Finnish nursery company
16. Van Berlo JM (1993) A decision support tool for the vegetable processing industry; an integrative approach of market, industry and agriculture. Agric Syst 43:91–109
17. Cheraghalipour A, Paydar MM, Hajiaghaei-Keshteli M (2019) Designing and solving a bi-level model for rice supply chain using the evolutionary algorithms. Comput Electron Agric 162:651–668
18. Motevalli-Taher F, Paydar MM, Emami S (2020) Wheat sustainable supply chain network design with forecasted demand by simulation. Comput Electron Agric 178:105763
19. Banasik A, Kanellopoulos A, Claassen G, Bloemhof-Ruwaard JM, van der Vorst J (2017) Closing loops in agricultural supply chains using multi-objective optimization: A case study of an industrial mushroom supply chain. Int J Prod Econ 183:409–420
20. Salehi-Amiri A, Zahedi A, Akbapour N, Hajiaghaei-Keshteli M (2021) Designing a sustainable closed-loop supply chain network for walnut industry. Renew Sustain Energy Rev 141:110821

21. Krishnan R, Arshinder K, Agarwal R (2021) Robust optimization of sustainable food supply chain network considering food waste valorization and supply uncertainty. Renew Sustain Energy Rev 171:108499
22. Hamdi-Asl A, Amoozad-Khalili H, Tavakkoli-Moghaddam R, Hajiaghaei-Keshteli M (2021) Toward sustainability in designing agricultural supply chain network: A case study on date palm. Sci Iran
23. Gharye Mirzaei M, Gholami S, Rahmani D (2023) A mathematical model for the optimization of agricultural supply chain under uncertain environmental and financial conditions: the case study of fresh date fruit. Environ, Dev Sustain 1–34
24. Handayati Y, Simatupang TM, Perdana T (2015) Agri-food supply chain coordination: the state-of-the-art and recent developments. Logist Res 8:1–15
25. Rentizelas A, Shpakova A, Mašek O (2018) Designing an optimised supply network for sustainable conversion of waste agricultural plastics into higher value products. J Clean Prod 189:683–700
26. Kamyabi E, Moazzez H, Kashan AH (2022) A hybrid system dynamics and two-stage mixed integer stochastic programming approach for closed-loop battery supply chain optimization. Appl Math Model 106:770–798
27. Mulvey JM, Vanderbei RJ, Zenios SA (1995) Robust optimization of large-scale systems. Oper Res 43:264–281
28. Kouvelis P, Yu G (2013) Robust discrete optimization and its applications. Springer Science & Business Media
29. Assavapokee T, Realff MJ, Ammons JC, Hong I-H (2008) Scenario relaxation algorithm for finite scenario-based min–max regret and min–max relative regret robust optimization. Comput Oper Res 35:2093–2102

Finding the Non-destructive Monitoring System of Papaya by Machine Learning Algorithm

Sruthi Mohan, N. Prakash, Aparna Chinnaraj, J. Balamurugan, Manish Kumar, and R. Senthamil Selvan

Abstract The economic significance of the tropical fruit papaya (Carica papaya) stems from its excellent nutritional and therapeutic value. A crucial activity in the fruit sector is packing papaya fruit according to its maturity stage. Manually grading papaya fruit using human eyes is laborious and harmful. This research aims to provide a new method for non-destructively determining the maturity condition of papaya fruits. The research proposed two methods for classifying papaya maturity state, both based on transfer learning (TL) and machine learning (ML). Additionally, several transfer learning and machine learning methods are compared. Two hundred photographs of papaya fruits, 50 from each of the three phases of ripeness, were used for the experiments. The machine learning technique comprises three classifiers, each with its unique kernel function and three distinct sets of features. Machine learning methods make use of features and classifiers such as support vector machine (SVM), k-nearest neighbor (k-NN), histogram of oriented gradients (HOG), local binary pattern (LBP), and gray-level co-occurrence matrix (GLCM). Using ResNet50, ResNet101, ResNet18, VGG16, VGG19, Google Net, and Alex Net are

S. Mohan
Department of Computer Science, G.T.N. Arts College (Autonomous), Dindigul, India

N. Prakash
Department of EEE, Kumaraguru College of Technology, Coimbatore, India
e-mail: prakash.n.eee@kct.ac.in

A. Chinnaraj
Department of Information Technology, Rathinam Technical Campus, Coimbatore, India

J. Balamurugan
Department of Master of Business Administration, St. Joseph's College of Engineering, (An Autonomous Institution), Chennai, Tamil Nadu, India

M. Kumar
Department of Physical Education, Lovely Professional University, Phagwara, India

R. S. Selvan (✉)
Department of Electronics and Communication Engineering, Annamacharya Institute of Technology and Sciences, Tirupati, India
e-mail: rsselvan.ece.aitt@annamacharyagroup.org

A. J. Kulkarni and M. Hajiaghaei-Keshteli (eds.), *Advances in Intelligent Systems for Sustainable Agriculture*, Lecture Notes in Networks and Systems 1201,
https://doi.org/10.1007/978-981-97-9839-1_21

the seven previously trained models that make up the transfer learning method. With a training time of just 0.0996 s and a perfect accuracy rate, the weighted k-NN with the HOG feature is the best machine learning-based classification model. With 100% accuracy and 1 min and 52 s of training time (including early stop training), VGG19 outperforms all other transfer learning technique-based classification models. Using a transfer learning methodology, the suggested method (VGG19) for papaya fruit maturity classification obtained 100% accuracy, an improvement of 5% over the current method.

Keywords Fruit papaya · Maturity stage · Machine learning · K-NN · SVM

1 Introduction

Vitamins A, B and C, E abound in papaya, making it a very nutritious fruit [1] (see Fig. 1 for reference). Fatigue, revenge, and prejudice are outside factors that might introduce mistakes in visual fruit classification and quality rating [2]. Due to differences in visual perception, there are errors in the fruit industry's classification system, even when using trained operators. Therefore, an automatic method is necessary for the fruits and offers extra trustworthy data; an auto-maturity state of fruits is vital for two reasons: first, to know how long to store them before eating them, and second, to ensure that they are of good eating quality [3]. Assisting human operators in determining these qualities is both harmful and time-consuming. Because of this, methods in this field of application must be quick, innovative, and non-destructive [4]. This study sorts papaya fruits into three groups according to their ripening nature. The maturity status prediction of various fruits has been the subject of several study reports. One may use a fuzzy model to classify the pineapple fruit according to its degree of ripeness [5]. These characteristics were retrieved from the hue channel and the CIE Lab. Once again, a 94.12% accuracy rate was attained by improving the model for fuzzy with particle swarm optimization.

Two automatic techniques, the area feature algorithm and the mean color intensity procedure, are utilized to analyze the three stages of maturity categorization of the banana fruit [6]. Geometric and the GLCM feature are used for classification. One hundred pictures of papayas, 30 of which are immature, 40 of which are ripe, and 30 of which are over mature are used to test the systems. According to the results, average color and area are the two most essential factors in banana fruit classification. The three phases of papaya ripeness were accurately classified with accuracy of 99.2% using the average color intensity system. In addition, it was noted that the area feature algorithm had a limitation: it couldn't distinguish between banana fruits that were fully matured and those that were overripe. The mango fruits were sorted into four groups based on their ripeness degree using a technique of fruit sorting [7]. The footage was recorded from the conveyor belt where the papayas are stored using a charge-coupled device camera (CCD). The obtained video signal is then used to extract 25 characteristics from each frame. When time for classification, the team

Fig. 1 Papaya nutritious food (vitamins)

turned to a recursive feature removal strategy employing support vector machines (SVMs). Using 15,000 photos, the system achieves an accuracy rate of 97%. The fruits of oil palm in the bunch were distributed into three types according to their level of maturity using a developed portable four band sensing system [8]. There are four different wavelength bands in the system: 571,671,751, and 871 nm. Using 100 oil palm fruit bunches considered the discriminant analysis using the Mahalanobis distance classifier reached accuracy of 86% utilizing spectral characteristics. Estimating quality attributes and categorizing strawberry maturity states were addressed using an alternative methodology [9] that relies on multispectral photography. Two approaches were used in this case: principal component analysis and the support vector machine with a backpropagation neural network (PCA-BPNN). According to reports, SVM achieved a perfect score of 100% when tested on a dataset consisting of 250 photos for ripeness stage categorization. A non-invasive sensing system is suggested to predict when durian fruit will ripen [10]. The approach attained an accuracy of 93.8% by using the notion of wireless message with the Ricardian k-factor for ripeness prediction. A model for determining when plums are ripe is suggested [11] using an image processing approach. Researchers looked at their size, texture, color, and other outward characteristics to determine when plums were ripe. Color characteristics (RGB indices) were also shown to be more critical than texture and size variables in the investigation. The experimental findings showed a correlation of 99.67% between the chemical analysis and picture analyses.

The fruit sector has seen a plethora of research in the last 10 years covering topics including detection of defect [12], grading [13], fruit quality evaluation [14], and prediction of mass and volume [15]. Machine learning [16], image processing [17], and computer vision [18, 19] are commonly utilized for fruit classification [20]. In addition, the fruit industry extensively uses deep learning, machine learning, and computer vision [21]. Fruit on-tree identification, categorization, and grading have been the substance of much study in current years [22]. Based on our study, just one publication discusses using random forest (RF) as an image processing approach to forecast when papaya fruits would mature [23]. The approach attained an accuracy of 95.8% using a hundred samples and ten handmade characteristics to categorize them into three sets according to their developmental stages.

This research aims to provide a model that uses machine learning and transfer learning to categorize papaya fruits into three states: immature, partly ripe, and mature. Three classes of features and classifiers make up the machine learning strategy. The characteristics include a co-occurrence matrix for gray levels, a histogram of gradients, and local binary patterns. Naive Bayes, support vector machine (SVM), and k-nearest neighbors (k-NN) are the classifiers with unique functions of kernel. The rest of the article is organized as follows: Sect. 2 covers the technique, Sect. 3 discusses the findings and discussion, and Sect. 4 concludes with further scope.

2 Methodology

Papaya fruit maturity level categorization is accomplished using transfer learning and machine learning. Table 1 and Fig. 2 show that the visual characteristics of skin color characterize the maturity level. The relevant sections include the specifics of the sample collection and the proposed classification models for both methods [24, 25].

2.1 *Collect Samples*

After picking them fresh from the tree, the papaya fruits are carefully put on white paper to be photographed. Between 10 in the morning and 3 in the afternoon, the photos were taken using a 13-megapixel smartphone camera in natural daylight [17, 26, 27]. An indoor method of picture capturing is used, using a mobile phone carrying stand to keep an approximate distance of 36 cm. Every effort was made to ensure that the space had enough natural light without any direct sunlight that may cast shadows or reflections on the backdrop. As a bonus, a white backdrop eliminates visual clutter and stresses clarity. All the pictures have been shrunk to a

Table 1 Visual characteristics of skin color characterize the maturity level

Maturity level	Summary
Immature	Without yellow stripe, green skin
Partially mature	Clear yellow line on green skin
Mature	Light green patches may be seen on yellow skin

Fig. 2 Papaya maturity levels

227 by 227 by 3-pixel size. Using the visual features shown in Table 1, the papaya fruit photos are categorized into three groups: immature, somewhat developed, and mature. Moreover, corresponding transverse and cross sections of longitudes are used to verify all samples. A total number of 300 samples, 100 from each group, are analyzed. 80% of the samples are utilized for training purpose, 10% for validating, and 10% for test purpose as shown in Fig. 3.

2.2 *Applying Machine Learning*

Feature extraction and secondary classification are the two phases that make up the machine learning method. Features such as LBP, HOG, and GLCM are retrieved from the pictures used as samples. There are 62 characteristics in the LBP, 38 in the HOG, and 14 in the GLCM. Both the HOG and the LBP try to take advantage of gradients around pixels. Their utilization of gradient information is the main distinction. In contrast to HOG, which only uses one direction per pixel, the LBP uses all nine. The 14 features of GLCM are as follows: energy, homogeneity, variance, smoothness,

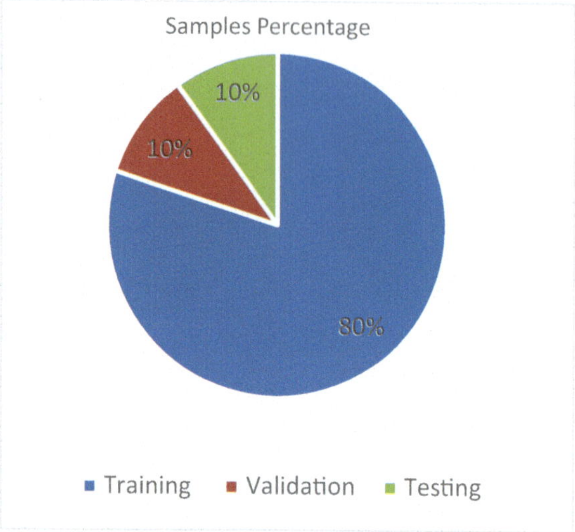

Fig. 3 Sample percentage for training, testing, and validation

kurtosis, skewness, RMS, entropy, and contrast. The characteristics are used by the classifier both during training and testing. The classifiers are SVM, k-NN, and Naive Bayes, each using various functions of kernel as shown in Fig. 4. Figure 5 shows papaya fruit ripeness categorization using machine learning.

2.3 *The Transfer Learning Method*

The goal of transfer learning, a method in machine learning, is to apply the skills learned from one issue to another by re-stating the problem from the beginning. The present research used transfer learning to improve the previously learned convolutional neural network (CNN) models. The transfer learning-based papaya fruit maturity state classification model is shown in Fig. 6.

Seven previously trained networks are evaluated here to classify the papaya fruit development phases. The workload is little with ResNet50, ResNet101, and VGG19, ResNet18, VGG16, Google Net, and Alex Net already pre-trained. With over a million photos culled from the Image Net collection, the previously trained net is a type of neural network for convolution. "Skip connections" is a feature that the previously trained Residual Neural Network (ResNet) network brought to the table. It is common practice to use the activation function of ReLU to compute a transformation for the matrix input. Bypassing the first alteration and adding the results in the last function of ReLU is how Residual Networks operate. Depending on the number of layers, ResNet comes in three flavors: ResNet101, ResNet50, and ResNet18.

Three different networks can categorize photos into 1000-item categories: ResNet101, ResNet50, and ResNet18. Each of these networks has 101, 50, and 18

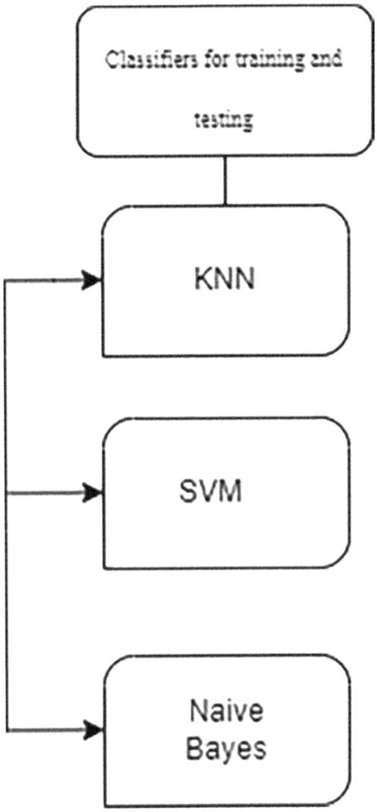

Fig. 4 Training and testing classifiers

layers. After training on 227 × 227 × 3 pictures, Alex Net moves on to a convolution layer that uses a 96 × 12 × 12 filter with a stride of 3. There is a 55 × 55 reduction in the output volume's dimensions. The pooling layer follows, using the stride two and maximum pool by 3 × 3 filters. It continues to the FC layer, which has 9217 parameters, and the two FC levels that follow it, each having 4097 nodes. Finally, it employs the Soft Max function with a thousand distinct output classes. Sixty million parameters are available. The simplicity of the VGG network is its defining feature; it employs a simple stacking of three-by-three convolutional layers. Maximizing pooling handles the reduction of volume size. A Softmax classifier follows two fully connected layers, each having maximum nodes. The variants of the VGG network denoted as VGG16 and VGG19, respectively, indicate the total number of weight layers in the network. The Image Net Large-Scale Visual Recognition Challenge (ILSVRC) uses Google Net, a pre-trained model trained on a portion of the Image Net database. Google Net's architecture is distinct from Alex Net's and VGG's architecture. It employs a mix of modules for inception, including convolutions at various sizes, pooling, and concatenation operations. Additionally, it employs the selector for features like 1 × 1 feature convolutions.

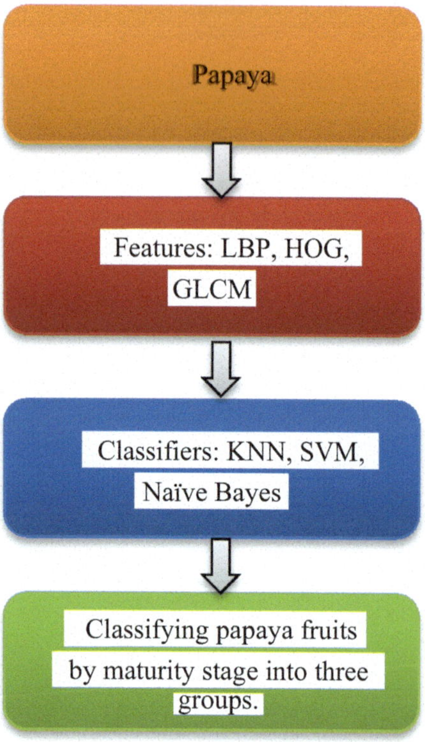

Fig. 5 Papaya fruit ripeness categorization using machine learning

3 Results and Discussion

Transfer learning and machine learning were used in this research to categorize the fruit papaya according to its level of maturity. MATLAB 2019a is used to conduct the experiments. An Acer Predators Helios 300 Core i5 8th Gen laptop with 8 GB of RAM, 1 TB of hard drive space, Windows 10 Home, 128 GB of solid-state drive space, 4 GB of graphics, and an NVIDIA GeForce GTX 1050Ti was used to run all programs. The accuracy, area under the curve, and time needed for trainings are the metrics by which each classifier is evaluated within the machine learning method. One of the most crucial measures for evaluating the effectiveness of any classification model is the area under the curve (AUC). Evaluation of the performance of the transfer learning-based classification model takes into account, once again, the number of iterations needed to obtain maximum validation accuracy, accuracy, and the time for training is necessary. In the sections that follow, people will compare the classifiers' performance.

Fig. 6 Papaya ripeness categorization using transfer learning

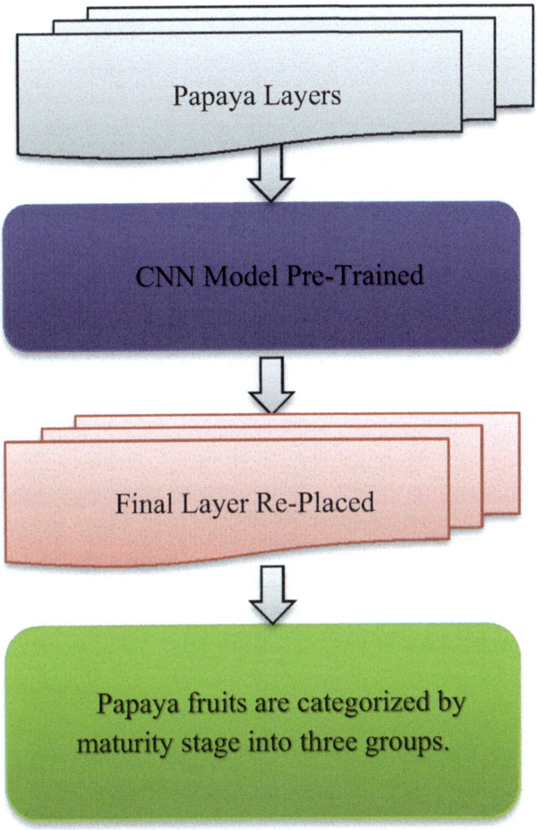

3.1 Outcomes from Machine Learning Method

The classifier is trained and tested using features derived from the sample pictures, such as the local binary pattern (LBP), histogram of oriented gradients (HOG), and gray-level co-occurrence matrix (GLCM). A collection of nine classifiers, including medium k-NN, fine k-NN, weighted k-NN, coarse k-NN, quadratic SVM, linear SVM, cubic SVM, fine Gaussian SVM, and kernel Naïve Bayes, are considered in total as shown in Fig. 7. According to the experiments, weighted k-NN outperformed its peers when trained independently using HOG, LBP, and GLCM features. The accuracy rate is 100% when using the weighted k-NN with the HOG, LBP, and GLCM features separately. Furthermore, k-NN using LBP features requires 0.10999 s training time, HOG with 0.099549 s, and GLCM with 0.085653 s. Due to its superior accuracy and training duration, the feature known as HOG with weighted k-NN model for classification outperformed other models. Figure 8a–c shows the performance scores of several machine learning-based categorization models. The findings are

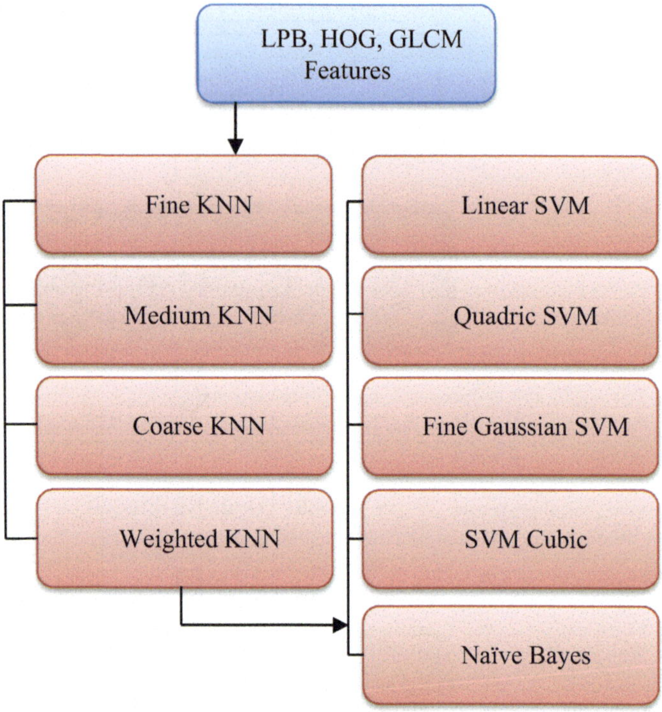

Fig. 7 Collection of nine classifiers for training and testing

derived by averaging the values of 20 separate runs, with training, validation, and testing variables chosen at random.

3.2 *Results from Transfer Learning*

Using previously trained networks for CNN, they carried out transfer of learning-based fine-tuning in this part. The training settings for transfer learning were set to 32, 5, 10, and 0.001 for validation frequency, minibatch size, maximum epoch, and starting rate for learning, respectively. Another technique used for learning is the stochastic gradient descent with momentum (SGDM). Using three non-incremental accuracy validation levels, early stop training is used here. Figure 9 displays the performance scores of the previously trained convolutional neural network methods. Performance is evaluated by calculating the maximum validation accuracy, training duration, and the number of iterations needed to obtain that accuracy.

Based on the results of the experiments, VGG19 outperforms the other transfer learning classification models with a perfect score. Training VGG19 only takes 1 min and 52 s, while other classification models like ResNet50, ResNet18, VGG16, and

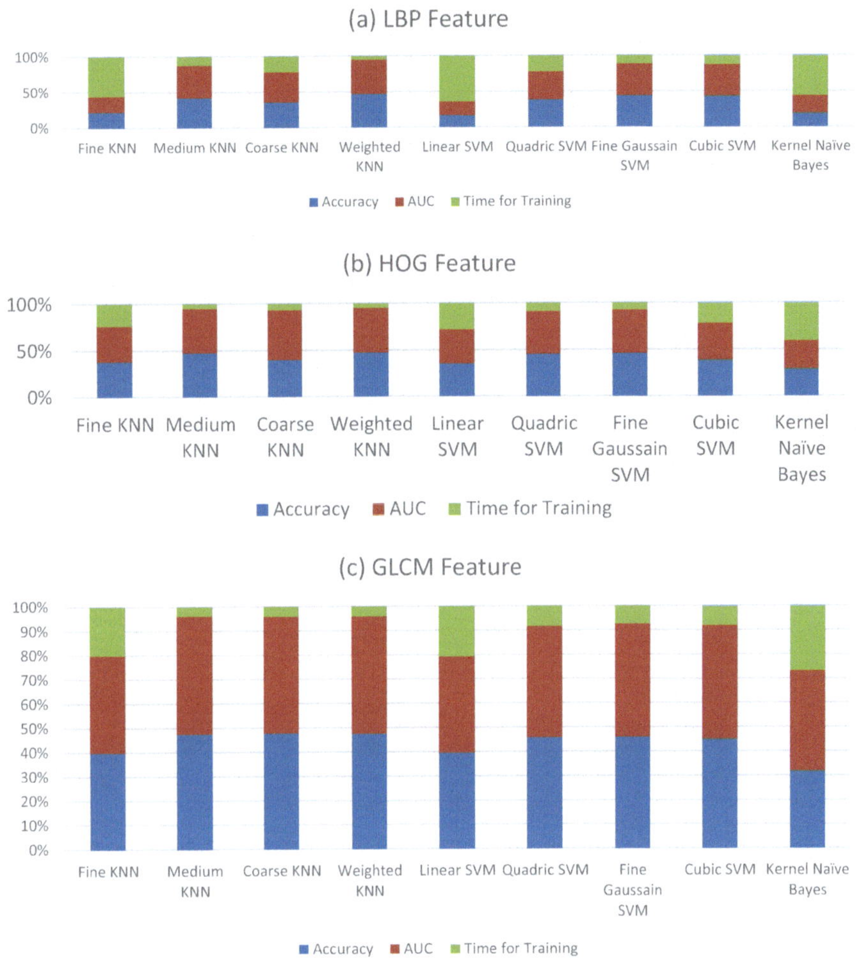

Fig. 8 a–c Performance scores of several machine learning-based categorization models

Google Net reached 100% accuracy. Furthermore, VGG19 achieves maximum validation accuracy sooner than other classification models before the tenth iteration. Figure 10 shows the classification model's confusion matrix when using a transfer learning strategy and based on VGG19.

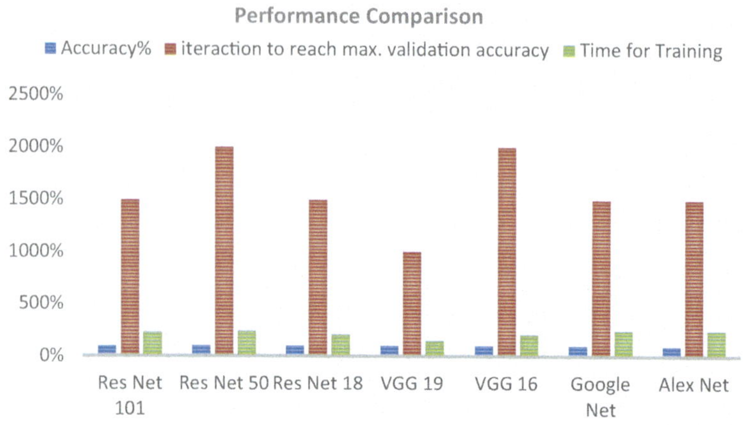

Fig. 9 Performance scores of the previously trained convolutional neural network models

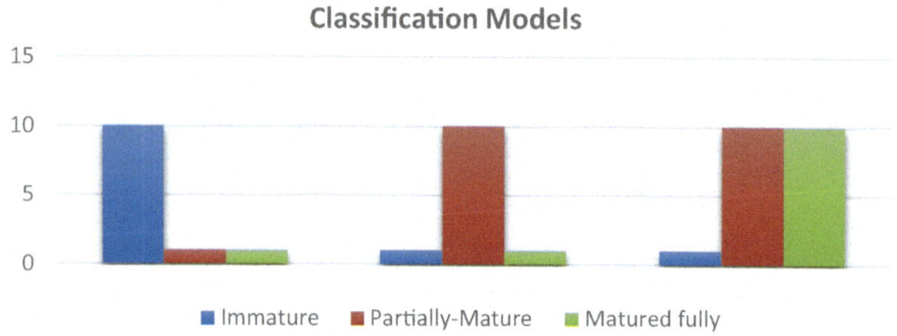

Fig. 10 Classification model's confusion matrix when using a transfer learning strategy and based on VGG19

4 Conclusion and Further Scope

This study suggests a methodology combining machine learning with transfer learning to determine when papaya fruits are ripe. The machine learning method evaluates nine classifiers, including LBP, HOG, and GLCM features, using k-NN, SVM, and Naïve Bayes with various kernel functions. The weighted k-NN with HOG achieves remarkable results, with a training time of just 0.099549 s, an area under the curve of 1 (AUC), and a 100% accuracy rate. With a training duration of under 1 min and 52 s, VGG19 achieves perfect accuracy when using a transfer learning strategy. VGG19 is superior to other machine learning and transfer learning approaches. There is no need to extract and choose features since VGG19 is built on transfer learning. Even though the transfer learning technique requires intricate architecture, extensive training time, and massive datasets, it is a one-and-done deal. Regardless, machine learning still has certain limits, such as its limited data handling capabilities and the need for segmentation. However, the results using machine learning and transfer learning are accurate to a 100% level, which is better than the prior method's 94.8% accuracy. This effort might have gone much farther if a prototype had been developed and integrated into the fruit sector to speed up the sorting process.

References

1. Jean Frederic Isingizwe N, Linus Opara U (2020) Machine learning applications to non-destructive defect detection in horticultural products. Biosyst Eng 189:60–83
2. Kasbe MS, et al Freshness study of non-destructive papaya by using an electronic nose. vol 5, IJCAES
3. Amir M, et al (2022) Non-destructive measurement and real-time monitoring of apple hardness during ultrasonic contact drying via portable NIR spectroscopy and machine learning. Infrared Phys Technol 122:104077
4. Dhritiman S, et al (2023) Rapid and non-destructive detection of hard to cook chickpeas using NIR hyperspectral imaging and machine learning. Food Bioprod Process 141:91–106
5. Li Bo, Lecourt J, Bishop G (2018) Advances in non-destructive early assessment of fruit ripeness towards defining optimal time of harvest and yield prediction—A review. Plants 7(1):3
6. El-Mesery HS, Mao H, Abomohra A-F (2019) Applications of non-destructive technologies for agricultural and food products quality inspection. Sensors 19(4):846
7. Aradhana T, et al (2020) Investigating the carotenogenesis process in Papaya fruits during maturity and ripening by non-destructive spectroscopic probes. Anal Lett 53(18):2903–2920
8. Mei M, Li J (2023) An overview on optical non-destructive detection of bruises in fruit: Technology, method, application, challenge and trend. Comput Electron Agric 213:108195
9. Preety B, et al (2023) Non-destructive banana ripeness detection using shallow and deep learning: a systematic review. Sensors 23(2):738
10. Sowmya N, Ponnusamy V (2023) Development of a smart sensor array for adulteration detection in black pepper seeds using machine learning. Int J Adv Comput Sci Appl 14(3)
11. Bhargava A, Bansal A, Goyal V (2022) Machine learning–based detection and sorting of multiple vegetables and fruits. Food Anal Methods 15(1):228–242
12. Armacheska RM, Chiang J (2021) Non-invasive grading system for banana tiers using RGB imaging and deep learning. In: 2021 7th International Conference on Computing and Artificial Intelligence
13. Farahmand B, Luisa Amodio M (2020) The potential use of non destructive optical-based techniques for early detection of chilling injury and freshness in horticultural commodities
14. Krishna Kumar P (2020) Studies on destructive and non-destructive quality evaluations of mango fruit. Notion Press
15. Vinta SR, Koshariya AK, Aditya SK, Gottimukkala A (2024) Investigation of early symptoms of tomato leaf disorder by using analysing image and deep learning models. EAI Endorsed Trans IoT 10
16. Gantela P, Ilankumaran S, Arunachalam M, Selvaprasanth P, Senthamil Selvan R (2022) Analysis of alzheimer disease with K means algorithm and PSO segmentation. In: 2022 IEEE 2nd Mysore Sub Section International Conference (MysuruCon), Mysuru, India, 1–6, https://doi.org/10.1109/MysuruCon55714.2022.9972409
17. Nikitha M, et al (2021) Emerging non-destructive methods for quality and safety monitoring of spices. Trends Food Sci Technol 108:133–147
18. Jiangbo L, Zhang Z (Eds) (2021) Nondestructive evaluation of agro-products by intelligent sensing techniques. Bentham Science Publishers
19. Wilson Daniel Caicedo C, et al (2022) Image analysis applied to control postharvest maturity of papayas (Carica papaya L.). J Food Process Preserv 46(11):e16999
20. Surendra Reddy V, et al (2023) Identification of significant clinical attributes for developing heart disease prediction system. Int J Intell Syst Appl Engi-Neering 11(8s):552–559

21. Purnendu Shekhar P, et al (2021) A smart parking system based on IoT technologies. Linguist Antverp 59–64
22. Sanchez, Philip Donald C, et al (2020) Applications of imaging and spectroscopy techniques for non-destructive quality evaluation of potatoes and sweet potatoes: A review. Trends Food Sci Technol 96:208–221
23. Tina N, et al (2023) Non-destructive monitoring of quality attributes kinetics during the drying process: A case study of celeriac slices and the model generalisation in selected commodities. Food Chem 424:136379
24. Syed Sohaib Ali S, et al (2021) Mango maturity classification instead of maturity index estimation: A new approach towards handheld NIR spectroscopy. Infrared Phys Technol 115:103639
25. Singh C, Rao MSS, Mahaboobjohn YM, Kotaiah B, Kumar TR (2022) Applied machine tool data condition to predictive smart maintenance by using artificial intelligence. In: VE Balas, GR Sinha, B Agarwal, TK Sharma., P Dadheech, M Mahrishi (Eds) Emerging Technologies in Computer Engineering: Cognitive Computing and Intelligent IoT. ICETCE 2022. Communications in Computer and Information Science, vol 1591. Springer, Cham. https://doi.org/10.1007/978-3-031-07012-9_49
26. Eduard L, et al (1999) Non-destructive banana ripeness determination using a neural network-based electronic nose. Meas Sci Technol 10(6):538
27. Sun R, Zhou J-y, Duo Yu (2021) Nondestructive prediction model of internal hardness attribute of fig fruit using NIR spectroscopy and RF. Multimed Tools Appl 80:21579–21594
28. Umuhoza A, et al (2023) Advancement of non-destructive spectral measurements for the quality of major tropical fruits and vegetables: a review. Front Plant Sci 14
29. Tejaswi B, Ghosh P, Singh Rana S (2023) Mechanical, optical, and acoustic methods: Noninvasive quality assessment of fruits and vegetables. Qual Control Fruit Veg Process: Methods Strat-Egies 283

Crop Yield Prediction and Price Forecasting Using Machine Learning

Nihar Ranjan Swain, Vaishali Choudhary, Shivam Silswal, Apoorva S. Shastri, and Sotirios Spanogiannopoulos

Abstract The rising population necessitates meeting the demand for maximum crop yield to ensure a stable life. However, traditional farming methods may not be suitable for every crop, especially in challenging conditions such as low water availability and soil infertility. This makes it imperative to explore alternative approaches to address these agricultural challenges. Traditional farming gives less yield as it mostly depends on natural resources like soil and water, however technology can help to increase quality of soil by providing precise fertilizer leading to healthy crop condition. Artificial Intelligence (AI) plays crucial role for making smart decision approach to increase the crop yield. AI can enhance soil quality through precise fertilizer application, promoting optimal crop conditions. Additionally, deciding which crops to cultivate involves considering various factors. The crop yield prediction model utilizes Bayesian Ridge Regression, K-Nearest Neighbours (KNN), Support Vector Machine (SVM), Decision Tree, Random Forest, Gradient Boosting Machine (GBM), Extreme Gradient Boosting (XGBoost), Light Gradient Boosting Machine (Light GBM), and Artificial Neural Network (ANN). XGBoost and Light GBM outperformed with an R-Square of 99.2631% and 99.9963% respectively by reducing overfit. In crop price forecasting models, gated recurrent units (GRU) outperform over Long Short-Term Memory (LSTM) and Autoregressive Integrated Moving Average (ARIMA).

Keywords Crop yield · Crop price forecasting · Machine learning · Deep learning

N. R. Swain (✉) · V. Choudhary · S. Silswal · A. S. Shastri
Institute of Artificial Intelligence, Dr. Vishwanath Karad MIT World Peace University, 124 Paud Road, Kothrud, Pune MH 411038, India
e-mail: swainnihar1@gmail.com

A. S. Shastri
e-mail: apoorva.shastri@mitwpu.edu.in

S. Spanogiannopoulos
Khalifa University, Center for Autonomous Robotic Systems (KUCARS), Abu Dhabi, UAE
e-mail: sotirios.spanogiannopoulos@ku.ac.ae

A. J. Kulkarni and M. Hajiaghaei-Keshteli (eds.), *Advances in Intelligent Systems for Sustainable Agriculture*, Lecture Notes in Networks and Systems 1201,
https://doi.org/10.1007/978-981-97-9839-1_22

1 Introduction

Crop yield prediction is a challenging task as it requires high accuracy and efficiency in various aspects spanning from efficient use of resources, and seasonal crop management, to an informed decision-making process. Efficient decision-making for Crop handling in accordance to seasons involves the utilization of various datasets including those related to weather, soil conditions, seed varieties, fertilizer usage, and other dynamic influencing factors. Likewise, biological elements like crop genetics and plant health fluctuations also impact the prediction of crop yields. Natural calamities like droughts and floods are other factors that substantially influence the production of crops [1]. Machine learning (ML) concentrates on acquiring knowledge from datasets where outcomes are derived from previous experiences. ML models can be categorized as descriptive and predictive. Descriptive models analyse datasets to provide insights, while predictive models make predictions based on various features. Two frequently used methods for predicting crop yields are statistical model and process-based modelling. Statistical model involves assessing direct associations amongst different predictive variables such as seasons, soil factors, irrigation type, and crop yield within a specific dataset. While process-based model for crop yield simulates the physiological aspects of crop growth considering environmental conditions and management practices [2]. There are many potential benefits of ML in agriculture. It helps in decision-making for optimizing plant growth which has a direct effect on crop yield. ML can help farmers to implement strategies to increase yield. Also, it can assist in using resources efficiently such as water, land, and fertilizers leading to more sustainable agriculture practices that contribute to reducing environmental impact. Predicting crop yields can help farmers mitigate risks related to adverse weather conditions, market fluctuations, and crop failure [3].

In any prediction model, a dataset is a crucial element of the intelligent system. The system is applied to the dataset representing the state of Rajasthan, India. Since climate parameters vary by districts, the secondary data taken into consideration is based upon district parameters. After further analysis and cleansing, regression models were developed to predict a continuous variable, namely, crop yield. Consideration is given to factors such as irrigation type, soil characteristics, season, and crop type. Parameters undergo preprocessing to align with training data, and then a regression model is applied. Various machine learning models, known for their proficiency in regression, are used to achieve improved results. Their effectiveness is assessed using metrics such as MAE (Mean Absolute Error), Mean Square Error and R-Squared. Additionally, a separate time series model is developed to forecast crop prices, enhancing the comprehensiveness of the predictive system.

2 Literature Review

Iniyan et al. [3] proposed the application of machine learning techniques to forecast crop yields considering environmental and soil parameters. The objective was to offer valuable insights to farmers, aiding them in making informed decisions regarding crop selection and anticipated yield. Various regression models, including multiple linear regression, decision tree regressor, Lasso (Least absolute shrinkage and selection operator) regression, Ridge regression, partial least square regression, gradient boosting regression, and LSTM, were employed. Among these, LSTM proved to be the most efficient model, achieving an accuracy of 86.3% [3]. Jeong et al. [2] use random forest regression to predict crop yields for wheat, maize, potato, and silage. The authors trained the model using thirty years of data on maize grain yield and maize silage and potato tuber from the northeastern seaboard region. The research assessed the effectiveness of random forest regression by comparing it to multiple linear regression (MLR) as a benchmark model [2]. Fatima et al. (2015) proposed a framework based on KNN algorithm. The main goal is to enhance the worth and productivity of the farming area. The framework consists of input, data acquisition, data preprocessing and output. Inputs considered are soil nutrients, weather conditions, and previous crop production. Data acquisition include collecting information about weather conditions and soil properties involves extracting data from weather records and a database of soil information, respectively. In the data processing step, the system used multiple linear regression to forecast the crop. In the output the system predicts the most profitable crop and offers the user various crop suggestions based on the crop duration [4]. Ashapure et al. [5] proposed a ANN based machine learning framework. The primary goal was to forecast crop yield for automatic selection of cotton genotypes. This prediction is based on three types of crop features. Data was derived from UAS (Unmanned aircraft system). First type of data consists multi-temporal features including canopy cover, height and volume, overall features of vegetation of crop. The second type of data comprises non-temporal features, such as the size, count, and volume of cotton balls. Third type of data is irrigation status. ANN model performed better than other models, like Support Vector Regression (SVR) and Random Forest Regression (RFR). The proposed model can accurately predict yield as early as 70 days after planting [5]. Veenadhari et al. proposed a tool named 'Crop Advisor Tool,' relying on the decision tree algorithm to pinpointed the most crucial climatic factor impacting crop yields [6]. Pant et al. [7] used Support Vector Machine, Gradient Boosting Regressor, Random Forest Regressor, and Decision Tree Regressor. Crop yields for the four most grown crops in India are gathered from the Food and Agriculture Organization (FAO) of the United Nations data repository. The prediction model is constructed using features such as temperature, pesticide usage, average rainfall, and annual production. A model for predicting crop yield is developed for four crops: rice, wheat, maize, and potatoes. The results indicate an excellent prediction score for potatoes [7].

3 Role of Machine Learning Algorithms

Several factors play a pivotal role in influencing the crop yield. The paper undertakes Regression based Machine Learning Algorithms for the prediction and forecasting of crop yield. This section gives a brief overview of the implemented Machine Learning algorithms.

3.1 Bayesian Ridge Regression

This approach employs a regression model that integrates Bayesian principles and incorporates ridge regularization. In this method, model parameters are regarded as probability distributions, leading to the formation of a posterior distribution encompassing all parameters. Like ridge regression, a regularization term is introduced in the objective function to address overfitting, especially when multicollinearity is present. Prior distributions are assigned to the regression coefficients, which are then updated to create posterior distributions. Additionally, the model features a strength parameter that governs the extent of regularization and influences the selection of prior distribution [8].

3.2 Decision Tree

Decision tree algorithm is a Tree inspired machine learning algorithm. The algorithm is shaped like an inverted tree which mainly possesses three components: Nodes, Roots and leaves. Each node comprises of a parametric expression based on the pattern analysis of the intricacies in the training data. Based upon the satisfaction or dissatisfaction of these numerical equations the input is transferred further onto the respective successive node or leaf, this transfer is done through the channel name roots. During the initialisation stage, a single decision node is initialised with a numerical expression which is the successively subdivided into two child nodes, based on the outcome of the parent decision node the input variable is transferred to either node using roots, this successive node or the so-called child node then again bifurcate further into two this process is continued until the input value finally reaches the end leaf (The Final Outcome). There are multiple steps one can opt for to optimise this process like pruning where the tree is pruned to reduce the internal complexity and simultaneously improve predictive productivity of the model by reducing overfitting. In context of a regression model, this model assesses the standard deviation of the target variable. Along with that, the standard deviation for each additional variable in relation to the target variable is calculated. The decision of splitting is based upon choosing the variable with the highest standard deviation, provided it exceeds the specified threshold (e.g., a coefficient of variation of 10%). The final prediction value for each leaf node is determined, the average target value within that node is used as the final value. This process iterates for all potential leaf nodes to construct the prediction model [9].

3.3 Random Forest Algorithm

Random Forest Algorithm is another supervised learning technique. It is fundamentally one of the Bagging type of Ensemble Learning Algorithm, which is the process of combining multiple classifiers to solve a certain complex problem with improved performance and accuracy. Under this technique a number of Decision Trees are compiled together creating a forest-like structure each with their own parametrical expressions and complexities, each based on various differential subset of the training dataset in an attempt to further improve the predictive accuracy of the model. Hence instead of relying on a single Decision Tree Model, it takes predictions from several models and outcome with the majority takes precedence and is considered as the output. During the initialization stage, n number of Decision Tree Models are introduced. Each model is trained with its respective subset of the training dataset. Then, they produce their outputs, which are collected and analysed. The outcome with the majority of votes is considered as the final output of the predictive model [10].

3.4 K-Nearest Neighbours

K-Nearest Neighbours (KNN) is a non-parametric, supervised learning-based algorithm which uses proximity or in other words the distance between the new observation and all other preexisting observations of the training data to identify the possible outcome by taking the average of the target variable values of closest neighbouring observations and considering it as the final outcomes. During the initialization stage, the model plots all the training data on a plot based on the values of the common variable in the input value and the training data. It then analyses the distance between the new observation and other observations, selecting k number of neighbours based on the shortest distance between them. The predicted value of the new observation is then calculated as the average of the target value of these K nearest neighbours. The cardinal advantage of KNN is its inter predictability and simplicity, although it also imposes a vitality of the model over the value of K as choosing the value of k way too small or way too large might cause catastrophic results due to overfitting or underfitting. For a regression prediction, the average of the K nearest neighbours target values is used [11]. Hence, instead of relying on a single Decision Tree Model, the algorithm takes predictions from several models, and the outcome with the majority takes precedence and is considered as the final output.

3.5 Support Vector Machine

SVM is a Support Vector Machine based supervised learning model in which the model develops a hyperplane which best represents and bifurcates the underlying intricate pattern in the data similar to normal linear regression. But the SVR model uses several other key concepts like the kernel trick to plot the input data on a higher dimensional space in an attempt to locate the nonlinear relationship among the observations in the data. During the initialisation stage the model maps the training data on the higher dimensional plane and analyses the plot in an attempt to locate a hyperplane which can accommodate the data, around that hyper plane the model margins out an epsilon tube (A tube like structure around the hyperplane which acts like a margin of tolerance.). The model aims to fit the majority of the data in the epsilon tube to minimise the error between the actual values and predicted value The observations that fall inside the tube are disregarded. The ones outside the epsilon tube are considered as errors, and their distance from the margin is calculated for further utilization to minimize the error and optimize the prediction capacity of the model. The observations that fall below the hyperplane are often denoted as C*, and those above are represented as C. These vectors are considered as the support vectors because they support the structure of the tube and determine its positioning. Hence, it is called support vector regression [12, 13].

3.6 Gradient Boosting Machines

GBM is an ensemble based supervised learning algorithm which attempts to accurately predict the outcomes based on combining the estimates of sets of weaker and simpler models. The model is based upon the ideology of gradient descent where in each iteration the model learns from its past mistakes and losses and trains a new model to overcome those deviations. In the initialization process, the developer chooses an ML model aimed to boost, for example, Decision Tree. The GBM model starts building decision trees in an iterative manner to gather prediction residues or the deviations in the targeted outcome and the real outcome at hand. Then the model analyses the residue and trains another Decision Tree model to fill the deviant gap and attaches the later one to the original one to make it more efficient. This process is then carried out until the maximum deviation error is minimised. The model updates and improves the quality of outcome as the model moves to each new tree in the right direction via small updates and additions [14].

3.7 XGBoost

Extreme Gradient Boosting is another ensemble-based supervised learning algorithm, which is an exceptionally advanced Gradient Boosting Machine that performs the task of boosting simple ML models to the next level. The XGBoost model performs similar to the other ensemble models but in an optimised manner, it is a machine learning model which uses several key features to optimise the outcome validity and reliability. XGBoost model inculcates features like L1 and L2 regulation which helps the model to deal with the overfitting and underfitting problems which are often faced while dealing with large and highly dimensional data sets, The model also possess built in capabilities to handle missing values which eases the extensive pre-data processing, along with that the model is also capable to perform cross validation without the need for external libraries and also comes in handy to stop the training when there is no significant improvement and simultaneously handout the optimum number of iterations based on the output of the model. Unlike its predecessor GBM, XGBOOST has the pruning ability where it can cut off or remove the unnecessary or less significant branches to further optimise the ultimate model. The model also supports multiple parallelization allowing the use of several processors and distributed computer frameworks to function more efficiently [15].

3.8 Light Gradient Boosting Machine

LGBM is an ensemble based supervised learning model which follows the GBM framework, where new models of weak learners are developed to rectify the deviations or errors made by the previous models. The new features which set the model apart from others are its two unique features which are Gradient based One Side Sampling (GROSS) and Exclusive Features Bundling (EFB). The functional utility of GROSS is to exclude the less significant part of the data and primarily focus on the portion with more gradients to estimate overall information gain, this technique grants the model ability to accumulate significant outcomes even with less data. While EFB increases the model's efficiency and optimises the process by combining the mutually exclusive features together to reduce the number of features, and reducing the overall impact on the outcome without hampering or compromising the accuracy of the output value [16].

3.9 Artificial Neural Network

Inspired from Biological Neural Network, this ML model inculcates the understanding of the neural network connection and its functionality into the domain of Artificial Intelligence. The model is basically divided into 3 combinational layers, namely Input Layer, Hidden Layer, and the Output Layer, each filled with a set number of nodes which performs as the building block of the model. The information is transferred from one layer to successive layers using the feed forward network. ANN is one of the most commonly used Linear regression model where each node in the Hidden layer comprises of weight assigned to it which determine how much influence each input has on the outcome, This Hidden Layer is also referred to as the black box because of it the mathematical complexities imposed by the model due to the intensity of calculus equations on each node, which are often incomprehensible by the human mind. This Black Box issue imposes a huge amount of distrust due to its intense complexity which makes the model a bit less trustworthy, hence models like these require rigorous testing and training to ensure acceptable success rate. During the initialisation stage the model takes several inputs (the influential factors which play a vital role in shaping the trends of the regression model) these values are instilled into the input layer nodes which are then transferred to the Hidden layer through feed forward networking channel where each input values interacts and multiplies with a subsequent randomly generated weight. After this the values are finally received by the nodes of the hidden layer where a predefined bias based on the number nodes is present. The value is then summed with the respective bias value, then the input is passed through an activation function to transform the summed-up value to an output value which is then either passed to the next hidden layer or is used as the final output [12].

3.10 Long Short-Term Memory

Based on the modified structure of the Recurrent Neural Network, LSTM is an advancement and modification to the hefty recurrence in the RNN. Where RNN constantly processes the new information and the old data together over and over again LSTM provides a solution to this long-term dependency problem by introducing a new component referred to as internal state to the RNN node which includes three so called gates, namely Forget Gate, Input Gate and Output Gate. These gates perform like a checkpoint cell where all the new information and least statements from previous steps are checked and analysed. During this step the information is interpreted and its utility and viability is analysed, The forget gate sorts the information and states what part of the data present in the internal state can be forgotten, while input gate securitize the data and gives out what new information should be added or updated in this new into is this working added information, after that the output gate finalises the new updated relevant information and gives out the output which will be used in the next iteration. This constant pruning and updating of new and old information make the model far more robust. This model also makes it easier to store more important and vital information, both long-term and short-term, for future context [17, 18].

3.11 Gated Recurrent Units

GRU is a higher level of advancement to the LSTM model which makes it another type of RNN model which introduced more compact, portable, and easy to understand version of than the LSTM model, by narrowing the three gates of LSTM to only two gates, namely Update Gate and Reset Gate. Inherent to the predecessor gate the gate performs similar functions to the LSTM gates but in a more efficient way. The task of Update Gate is to analyse the input information and update the memory with the important information, on the other hand the job of Reset Gate is to resent or forget the unnecessary information and clear out the memory space for better efficiency and robustness of the model. Hence, in the initialization stage, the GRU model functions similarly to the LSTM and RNN. However, the tables turn when the efficient gates take the lead and vanguard the model ahead [17, 18].

3.12 Autoregressive Integrated Moving Average

ARIMA is a statistical model for time series forecasting. For any time-series data, a three-observation pattern is required to identify the structural and behavioural characteristics of the data over time. Trend is the long-term movement of a time series, which can be upward, downward, or stable. Seasonality is the period fluctuation of data at a fixed interval. Cyclicality is a long-term pattern that does not stick to time. Noise or randomness is an irregular or unpredictable pattern that follows data. To forecast an ARIMA model, data must be stationary, which means that its static properties, such as mean, variance, and autocorrelation, do not change over time. ARIMA combines autoregression (AR), differencing (I), and moving averages (MA). Autoregression is the correlation of a current value to a previous observation. The integrated operation is used to calculate the difference between two consecutive observations. Moving Average establishes a relationship between an observation and a residual error that is applied to lagged observations. ARIMA (p, q, d) denotes the order of (AR), (I), and (MA), respectively [18].

4 Research Methodology and Problem Formulation

In this research investigation, the utilization of machine learning algorithms enhances the anticipation of crop yield and forecasts crop prices, resulting in more substantial profits for farmers compared to conventional farming methods. The goal of this experiment to provide intelligent solutions to farming industry in particular water scarcity and infertile soil. The Kaggle agriculture datasets[1] from 2018 to 2019 in Rajasthan, India, demonstrate the effectiveness of the ML algorithm in predicting crop yield

[1] https://www.kaggle.com/datasets/suraj520/agricultural-data-for-rajasthan-india-2018–2019.

and forecasting prices. The datasets include four data files containing information on crop production, including district name, crop type, season, area (hectares), yield (quintals), and production (tonnes). The paper identified the ten most commonly cultivated crop yield during the kharif and rabi seasons using this data. All available data were considered for constructing the prediction model. The second dataset pertains to water usage and comprises features such as district, crop type, irrigation method employed, water consumption (liters/ha), and water availability (liters/ha). From this data percentage of water consumption for each crop in every district based on the irrigation method used are shown. The third dataset involves soil analysis, featuring information on area, soil type, pH, organic matter percentage, nitrogen content (kg/ha), phosphorus content (kg/ha), and potassium content (kg/ha). For soil analysis purposes, only the organic content for each soil type is considered, which subsequently influences other characteristics. The crop price dataset encompasses district, crop type, market, date, and price (INR/100 weight approx.). In the context of crop price forecasting, only the date and price variables are utilized. For crop price prediction first three datasets of size are merged which construct (433,580 $\times$ 15) size. On the splitting of dataset yield value is represent label value for the ML model.

The Pandas library is utilized to identify various issues such as soil types, crop types, and irrigation methods. During the preprocessing stage, the data distribution of key features is carried out. For the standard testing procedure, the datasets are divided into 80% for training and 20% for testing. Since the data does not contain any missing values, there is no need for missing value imputation. One-Hot Encoding is selected over Label Encoding to transform categorical features into numerical ones, as it assigns equal importance to all features. Although Label Encoding is useful for averting multicollinearity by substituting values with the same size, it is more appropriate for ordinal data [19]. The data is scaled using a standard scaler method to enhance the performance of the model.

Crop yield prediction models, including Bayesian Ridge Regression, KNN, SVM, Decision Tree, Random Forest, GBM, XGBoost, LightGBM, and ANN, have been effectively implemented in this study. In the process of hyperparameter tuning, each model employs Randomized Search instead of Grid Search for more efficient convergence due to its faster computational performance. For the training of the ANN model, a deep architecture is employed, utilizing rectified linear unit activation function and the 'He Normal' kernel initializer method. To mitigate overfitting, a dropout of 20% is applied, and Batch Normalization is incorporated. The model is compiled with a mean square error loss function, and the Adaptive Moment Estimation (Adam) optimizer is utilized for weight optimization. Training is conducted over 100 epochs with a batch size of 64, applying the TensorFlow and Keras modules. Despite the superior performance of the ANN model, the challenge lies in setting optimal parameters for a deep neural network. In this experiment, Keras Tuner is initially used for hyperparameter tuning; however, due to suboptimal results, a manual procedure is subsequently implemented to fine-tune the parameters and construct the ANN model.

Crop prices forecasting involves the utilization of three models: ARIMA, and two deep learning models are GRU and LSTM. The initial step in the forecasting process is the application of the Dickey-Fuller test to ascertain whether the data exhibits stationarity or non-stationarity. The dataset is partitioned into training and testing sets, with 80% of the total data allocated for training and the remaining 20% reserved for testing. The ARIMA model is trained directly on the training data and subsequently used to make predictions on the test data. In contrast, for both the GRU and LSTM models, the data is transformed into a time series format. In this experiment, the data is segmented into 12 timestamps to forecast the subsequent five values. Each model incorporates the rectified linear unit activation function, employs the mean square loss function, and utilizes the Adam optimizer. The models are trained over 30 epochs, with a batch size of 32.

5 Results and Discussion

Certain crops do well in the rainy season, while others thrive in the winter. The Kharif season commences from the rainy period, typically spanning from June to September, while the Rabi season begins during winter, covering the period from October to March. Figures 1, 2, 3, 4, 5, 6, 7, 8, 9, 10 depicts the crop yield of the major ten crops during the Kharif and Rabi seasons in the mentioned district. The selected crops are those whose yield significantly depends on the seasons. Wheat, for instance, tends to yield more during the Rabi season in all districts except Jodhpur and Nagaur, due to influences from climatic factors. The district-wise yield depicted in the graphs can assist farmers in making season-specific decisions for each crop. Additionally, factors such as climate conditions, soil parameters, and water management should also be taken into account for assessing crop yield. The outcomes depicted in the combined plot will assist the farmer in choosing which crop to cultivate, taking into account past yields. Further data is needed for monitoring the health of crops that yield the highest output. Another approach involves identifying soil types across districts and determining the most suitable crops for each. The goal of this plot gives analytical solution for smart farming rather than traditional farming.

Figure shows organic matter present in soil. The presence of organic matter is directly linked to the soil's quality, impacting various other soil components. When organic matter increases, there is a corresponding increase in nitrogen and phosphorus in the soil. Additionally, the pH level of the soil also rises, providing essential nutrition for the health of plant crops [20]. While the increase in organic matter leads to a decrease in potassium in the soil, it's noteworthy that this doesn't significantly affect the yield of grain and pulse crops, which are the primary crops considered. The figure indicates that clay soil has the highest organic matter compared to other soil types, suggesting that clay soils require fewer fertilizers. Rajasthan's predominant soil types include desert soil, dune soil, brown soil, black soil, sierozems, red loams, saline-sodic soil, among others, which are generally less fertile, there is a need to address soil quality, including NPK composition, to improve crop yields through increased fertilizer application in traditional farming. However, in the context of smart farming,

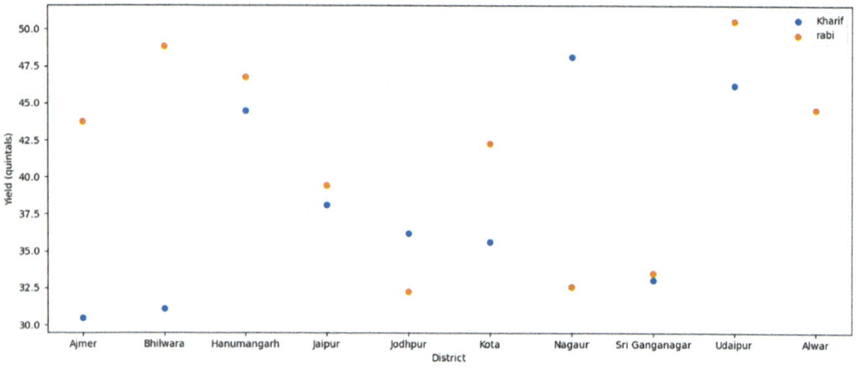

Fig. 1 Wheat seasonal crop yield in each district

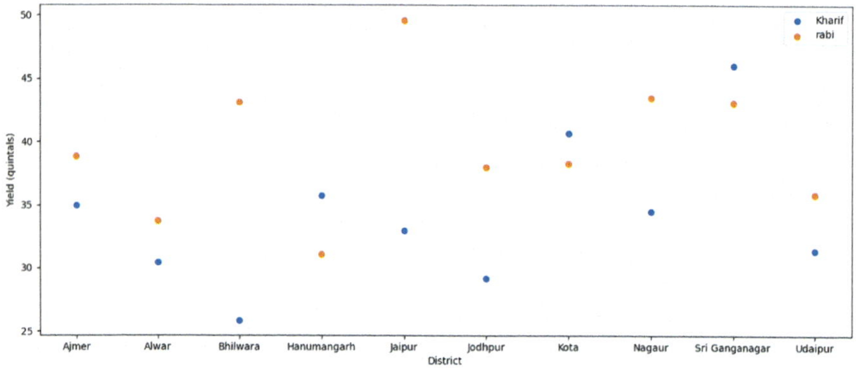

Fig. 2 Pulses seasonal crop yield in each district

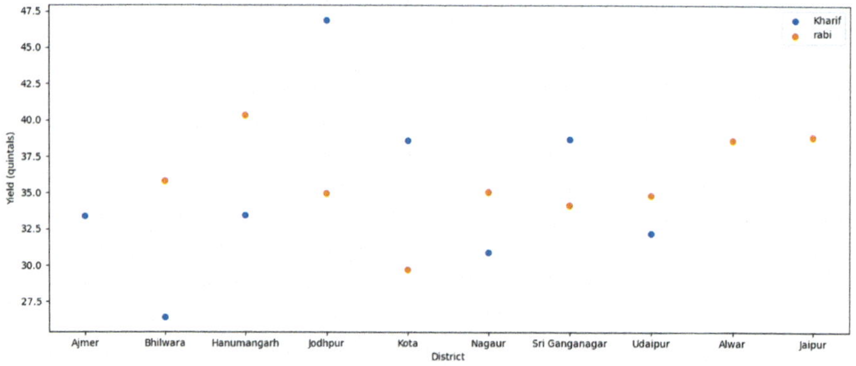

Fig. 3 Onion seasonal crop yield in each district

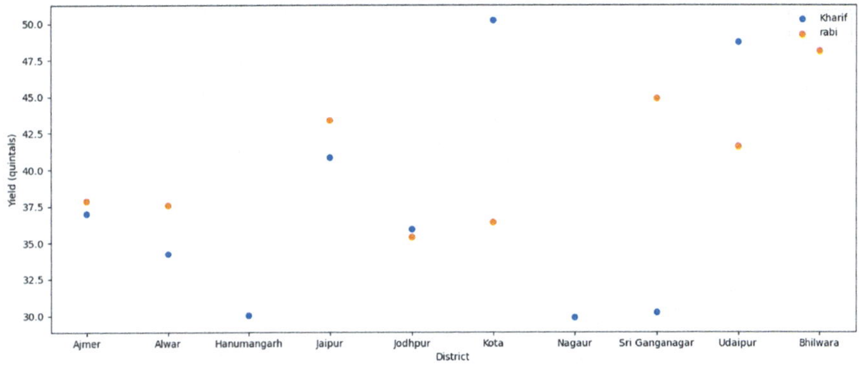

Fig. 4 Oilseeds seasonal crop yield in each district

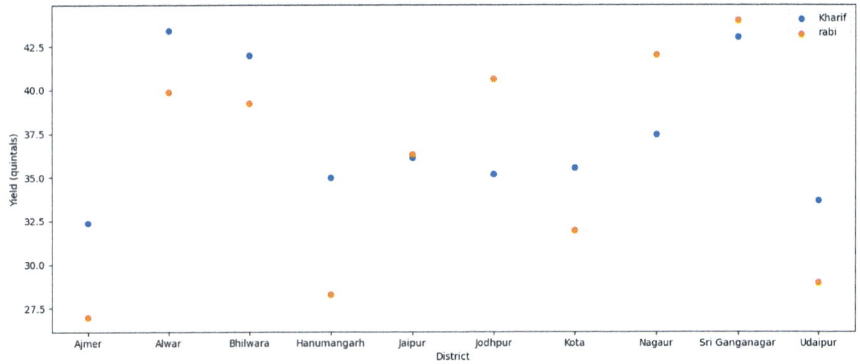

Fig. 5 Mustards seasonal crop yield in each district

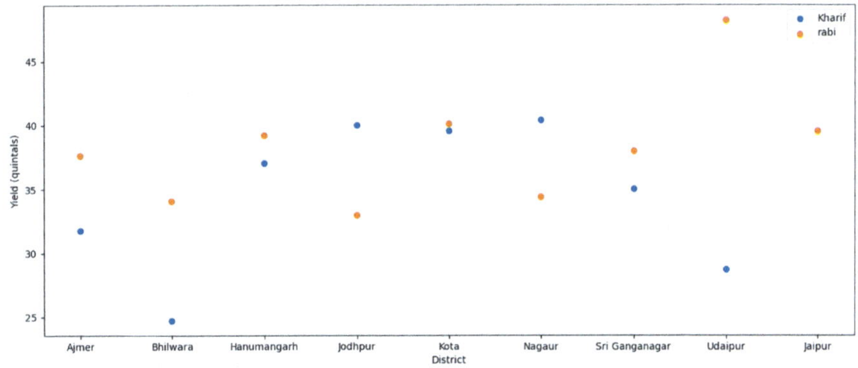

Fig. 6 Maize seasonal crop yield in each district

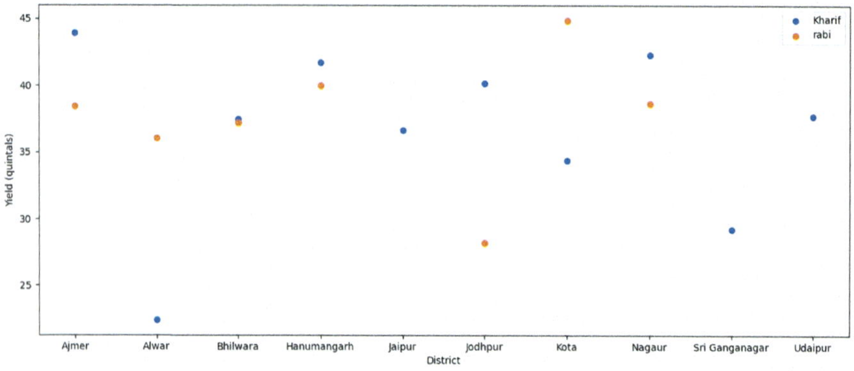

Fig. 7 Garlic seasonal crop yield in each district

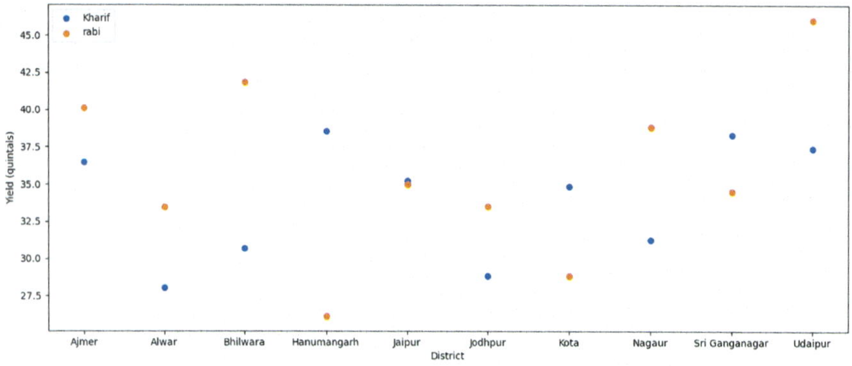

Fig. 8 Cumin crop yield—seasonal crop yield in each district

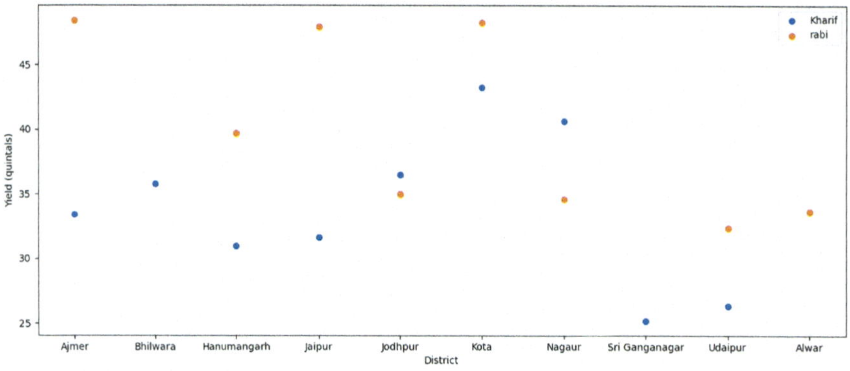

Fig. 9 Cotton seasonal crop yield in each district

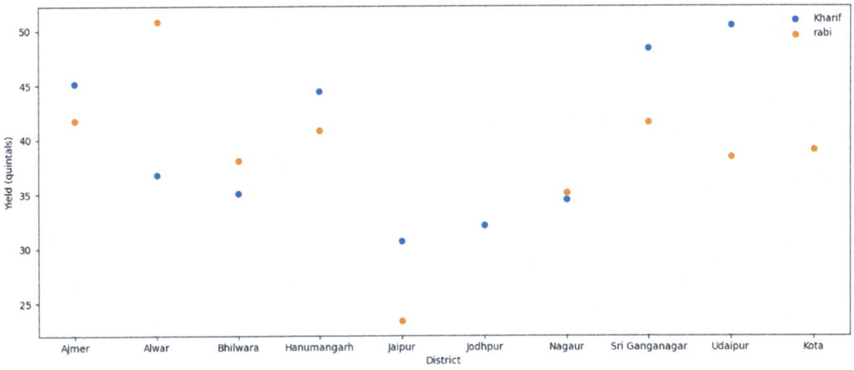

Fig. 10 Bajra seasonal crop yield in each district

it necessitates periodic soil testing, either weekly or at specific intervals, to ensure optimal plant health maintenance (Fig. 11).

Figures 12, 13, 14, 15, 16, 17, 18, 19, 20, 21 depicts the percentage of water consumption for each crop with irrigation method across the district. The plot describes each colour bar for the irrigation method. This will make farmers to take decision to apply the irrigation method for maximum yield. More than 100% of water consumption shows rainfall rate in each district, however, Udaipur district shows maximum rainfall than others. The purpose of this plot is to manage the water allocation for individual crops, rather than providing a continuous water supply. It advocates for a more intelligent approach to water distribution tailored to the specific needs of each crop. With minimal water availability, crops often consume water to their maximum capacity. By analysing the plots, it becomes evident as to which districts have lower water content. Another solution for farmers is to opt for crops that require less water.

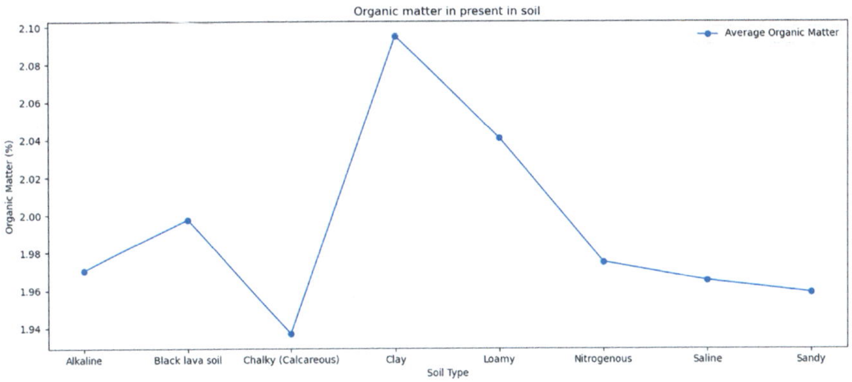

Fig. 11 Quantity of organic matter present in each soil type

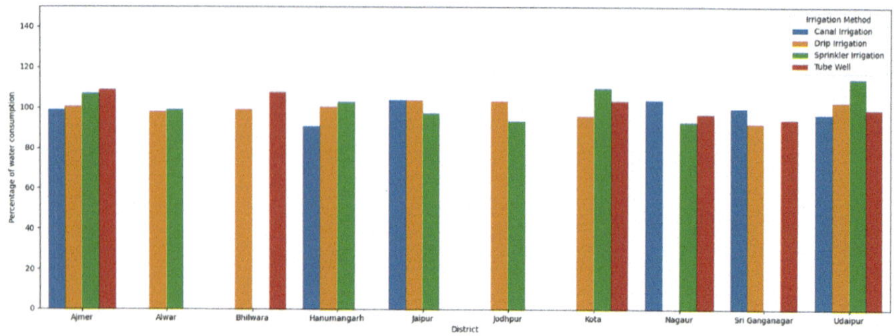

Fig. 12 Percentage of water consumption in each district for wheat production

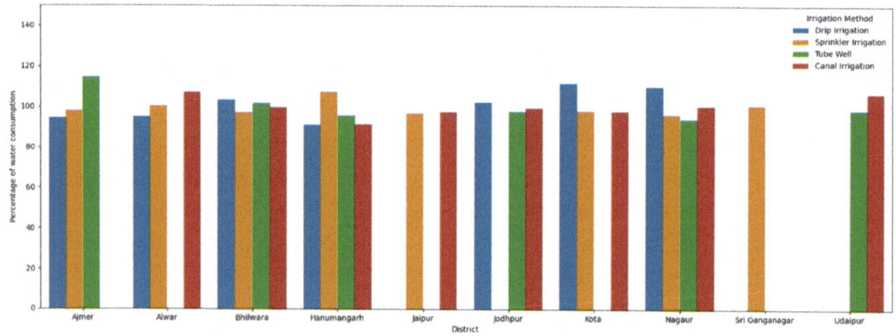

Fig. 13 Percentage of water consumption in each district for pulses production

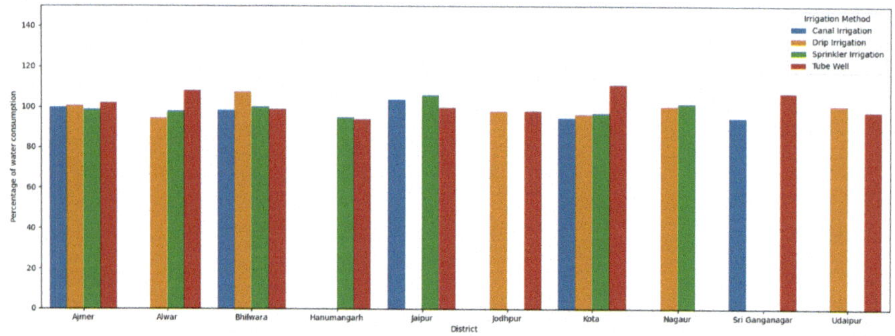

Fig. 14 Percentage of water consumption in each district for onion production

5.1 Crop Yield Prediction Results

Figures 22 and 23 explain the mean absolute error and mean square error for both training and testing on each predictive model. Out of all the model decision Tree, Random forest, Xgboost and LightGBM outperformed in mean square error. A

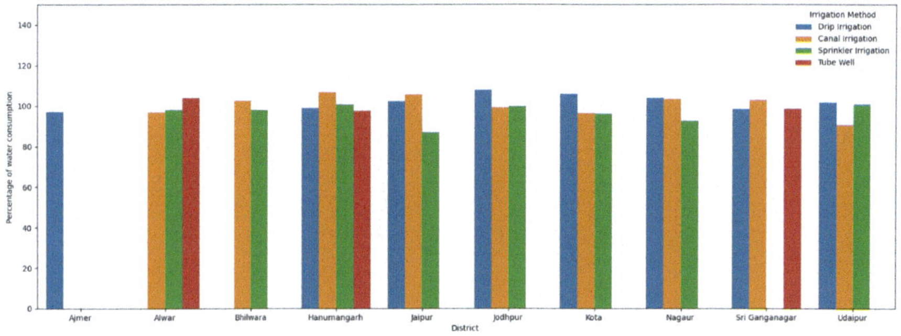

Fig. 15 Percentage of water consumption in each district for oilseeds production

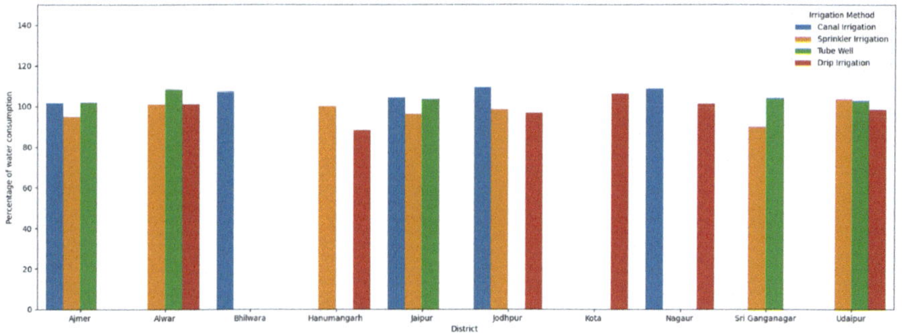

Fig. 16 Percentage of water consumption in each district for mustards production

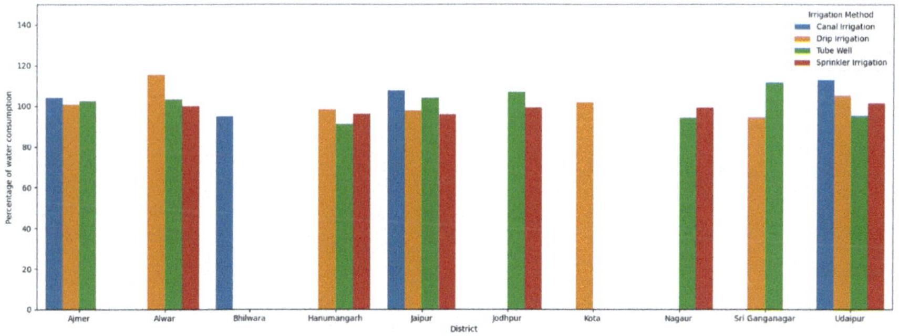

Fig. 17 Percentage of water consumption in each district for maize production

tunned ANN model can be performed better. Although the decision tree and random forest show maximum R-square and adjusted R-square value. As these models are overfit in nature while other methods are implemented to create a perfect prediction model. XGBoost and LightGBM model shows better prediction result as shown in Figs. 24 and 25. The execution time for model training a deep neural network

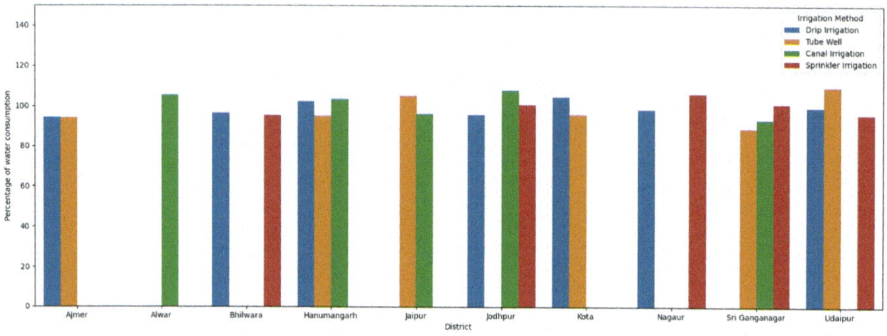

Fig. 18 Percentage of water consumption in each district for garlic production

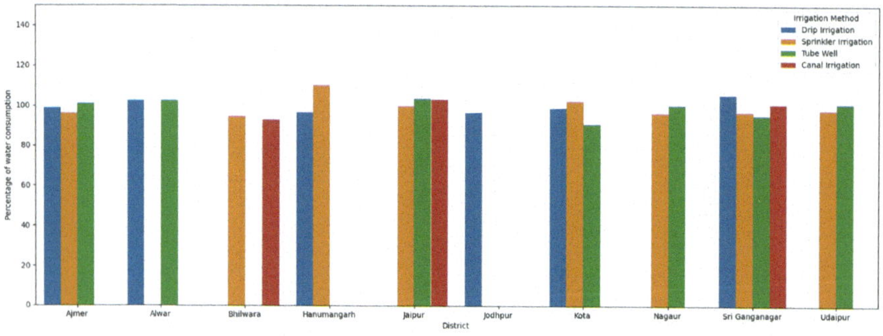

Fig. 19 Percentage of water consumption in each district for cumin production

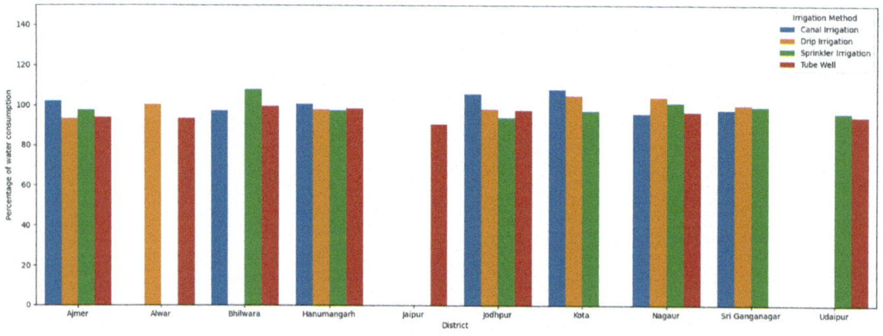

Fig. 20 Percentage of water consumption in each district for cotton production

was very high but for its ability to capture all intricate pattern. For hyperparameter tuning XGBoost and LighGBM models are faster and accurate. Figure 26 explains time execution for each ML model. After all the preprocesses the feature increases to finding the pattern to the particularly fitting model was complex, so in this experiment all possible regression models are tested to find an optimal solution as well as learn

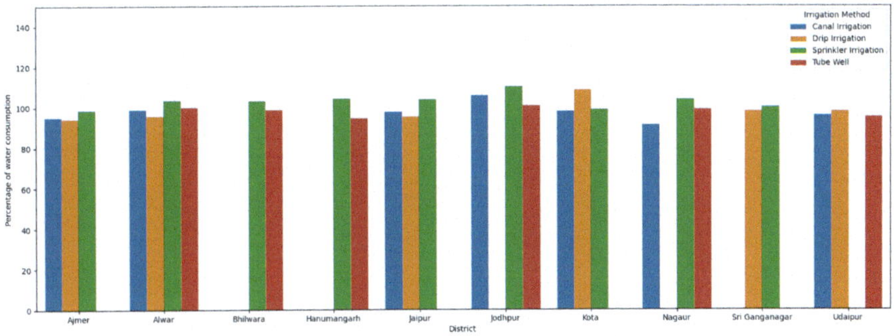

Fig. 21 Percentage of water consumption in each district for bajra production

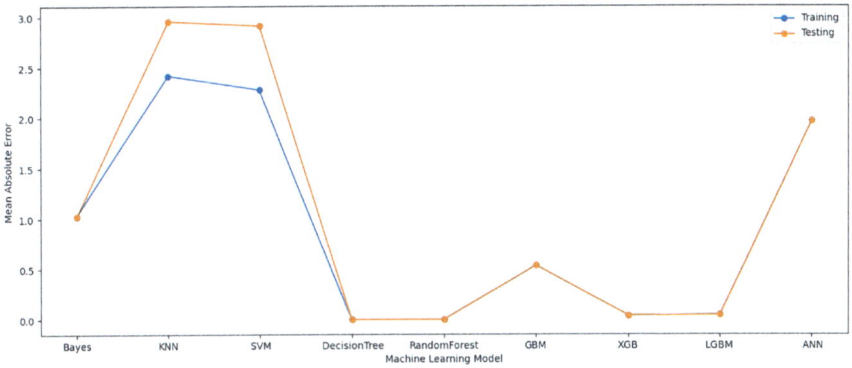

Fig. 22 Mean absolute error on each model

each model's pros and cons during training. Table 1 shows test dataset prediction results.

Training a model presents challenges, particularly with hyperparameter tuning, which can prolong the process of achieving an optimal model. Additionally, determining the appropriate range for hyperparameters poses difficulties, especially when longer parameters result in increased training times and necessitate higher RAM capacity. Improving model training can involve techniques like feature selection and addressing the course of dimensionality, which becomes problematic as the feature size grows, potentially leading to multicollinearity issues. In the case of training a deep neural network (DNN) model, the maximum achieved Rsquare value was 90%. Nonetheless, various approaches can enhance DNN models, such as increasing hidden layers, adjusting dropout sizes, and experimenting with different activation functions.

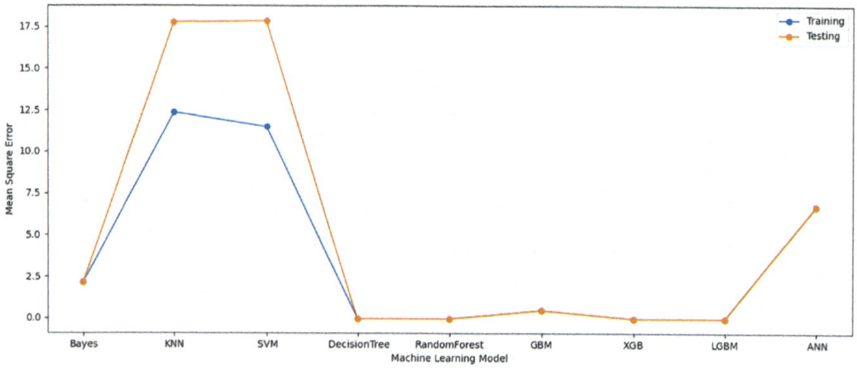

Fig. 23 Mean square error on each model

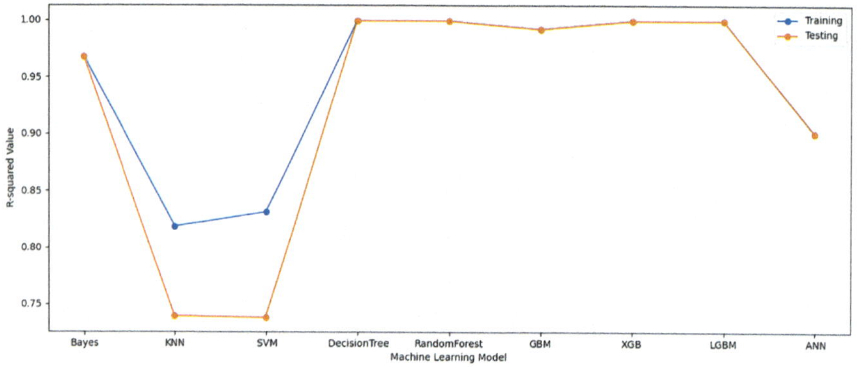

Fig. 24 R-squared value on each model

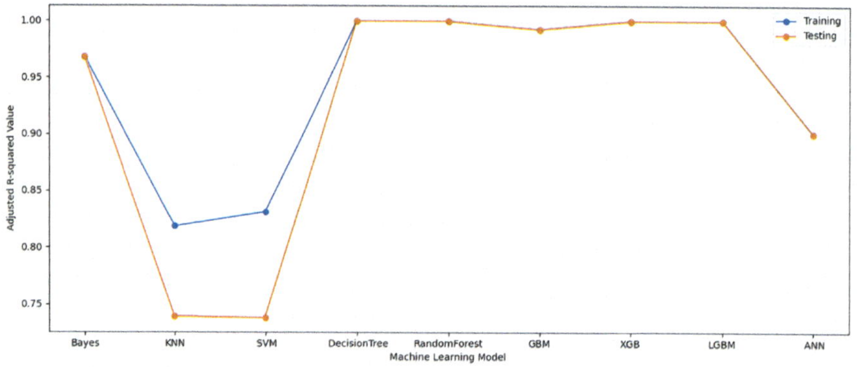

Fig. 25 Adjusted R-squared value on each model

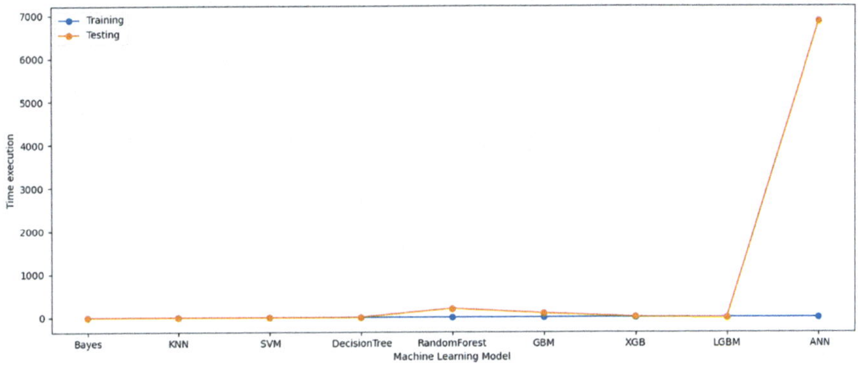

Fig. 26 Time execution on each model

Table 1 Performance of machine learning model

Model	MAE	MSE	RMSE	R-squared value	Adjusted R-squared value	Time execution in seconds
Bayesian ridge regression	*1.030402*	*2.167756*	*1.47233*	*0.968305*	*0.968285*	*1.568284*
K-nearest neighbors	2.961132	17.7853	4.217262	0.739962	0.739794	0.04118
Support vector machine	*2.917512*	*17.86788*	*4.227042*	*0.738755*	*0.738586*	*0.036773*
Decision tree	1.32E-13	3.57E-26	1.89E-13	1	1	4.795943
Random forest	*6.74E-14*	*8.46E-27*	*9.2E-14*	*1*	*1*	*202.0738*
Gradient boosting machine	0.535763	0.50398	0.709916	0.992631	0.992627	95.84308
Extreme gradient boosting	*0.038469*	*0.002556*	*0.05056*	*0.999963*	*0.999963*	*8.226263*
Light gradient boosting machine	0.046343	0.003724	0.061028	0.999946	0.999946	3.66459
Artificial neural network	*1.973378*	*6.751076*	*2.598283*	*0.901293*	*0.901229*	*6864.542*

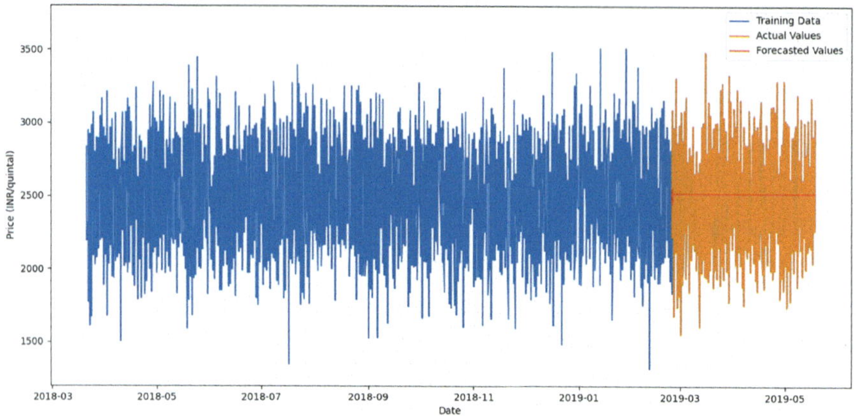

Fig. 27 ARIMA forecasting model

5.2 Crop Price Forecasting Results

For forecasting crop price only date and price column are considered. The ARIMA model outperformed in the prediction of test data. However, the forecasting model shows better in GRU model. Also, the LSTM model performed a forecast model comparatively less accurate. Outcome of this analysis will give results for farmer counting their profit from a particular crop on any season. For making a forecasting model for GRU, LSTM makes modified data which tends to date column to timeline. Figures 27, 28, 29 show the crop price forecast model using ARIMA, GRU and LSTM respectively. While the forecasting outcomes of the GRU and LSTM models demonstrate improvements compared to the ARIMA model, their predictive performance falls short. A significant contributing factor could be the limited input data and the potential for further optimization of the model through fine-tuning.

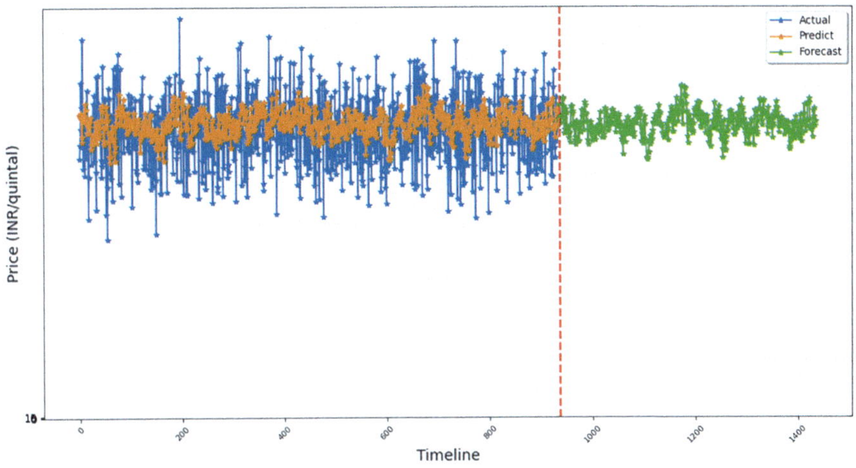

Fig. 28 GRU forecasting model

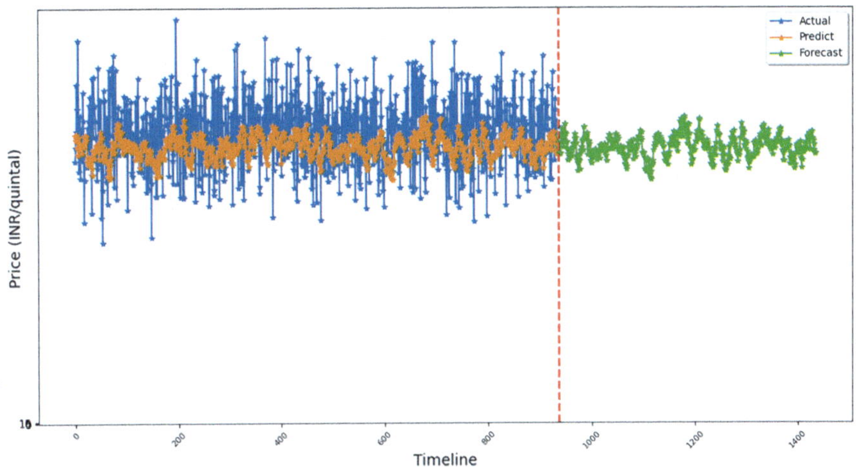

Fig. 29 LSTM Forecasting model

6 Conclusion

To surpass the expected crop quantity demand, this study introduces an intelligent predictive model for both crop yield and crop price forecasts over time. The analysis of graphs enables informed decision-making for farmers throughout the Kharif and Rabi seasons. Additionally, alternative approaches such as evaluating soil quality and employing smart irrigation methods contribute to increased crop yield. In the crop yield prediction model, both LightGBM and XGBoost demonstrated superior performance, achieving an impressive R-Square value of 99%, thereby mitigating the tendency for overfitting observed in decision tree and random forest models. Interestingly, for crop price forecasting, the ARIMA model yields a more accurate representation, whereas the GRU model outperforms in the overall forecasting model.

Based on the findings of this experiment, the other intelligent solution involves leveraging data to predict the pesticide requirements and overall plant health. Continuous monitoring of soil conditions and water levels enables the anticipation of water needs or the identification of the most effective irrigation methods based on the plant's health status. Optimal temperature for cultivating a crop.

References

1. Van Klompenburg T, Kassahun A, Catal C (2020) Crop yield prediction using machine learning: A systematic literature review. Comput Electron Agric 177:105709
2. Jeong JH, Resop JP, Mueller ND, Fleisher DH, Yun K, Butler EE, Timlin DJ, Shim KM, Gerber JS, Reddy VR, Kim SH (2016) Random forests for global and regional crop yield predictions. PLoS ONE 11(6):e0156571
3. Iniyan S, Varma VA, Naidu CT (2023) Crop yield prediction using machine learning techniques. Adv Eng Softw 175:103326
4. Ramesh D, Vardhan BV (2015) Analysis of crop yield prediction using data mining techniques. Int J Res Eng Technol 4(1):47–473
5. Ashapure A, Jung J, Chang A, Oh S, Yeom J, Maeda M, Maeda A, Dube N, Landivar J, Hague S, Smith W (2020) Developing a machine learning based cotton yield estimation framework using multi-temporal UAS data. ISPRS J Photogramm Remote Sens 169:180–194
6. Veenadhari S, Misra B, Singh CD (2014) Machine learning approach for forecasting crop yield based on climatic parameters. In: 2014 International Conference on Computer Communication and Informatics (pp 1–5). IEEE
7. Pant J, Pant RP, Singh MK, Singh DP, Pant H (2021) Analysis of agricultural crop yield prediction using statistical techniques of machine learning. Mater Today: Proc 46:10922–10926
8. da Silva FA, Viana AP, Correa CCG, Santos EA, de Oliveira JAVS, Andrade JDG, Ribeiro RM, Glória LS (2021) Bayesian ridge regression shows the best fit for SSR markers in Psidium guajava among Bayesian models. Sci Rep 11(1):13639
9. Rathore SS, Kumar S (2016) A decision tree regression based approach for the number of software faults prediction. ACM SIGSOFT Softw Eng Notes 41(1):1–6
10. Li Y, Zou C, Berecibar M, Nanini-Maury E, Chan JCW, Van den Bossche P, Van Mierlo J, Omar N (2018) Random forest regression for online capacity estimation of lithium-ion batteries. Appl Energy 232:197–210
11. Goyal R, Chandra P, Singh Y (2014) Suitability of KNN regression in the development of interaction based software fault prediction models. Ieri Procedia 6:15–21
12. Chen WH, Hsu SH, Shen HP (2005) Application of SVM and ANN for intrusion detection. Comput Oper Res 32(10):2617–2634
13. Yao K, Chen Y, Li Y, Zhang X, Zhu B, Gao Z, Lin F, Hu Y (2024) Water quality prediction of small-micro water body based on the intelligent-algorithm-optimized support vector machine regression method and unmanned aerial vehicles multispectral data. Sustainability 16(2):559
14. Guo R, Fu D, Sollazzo G (2022) An ensemble learning model for asphalt pavement performance prediction based on gradient boosting decision tree. Int J Pavement Eng 23(10):3633–3646
15. Dhaliwal SS, Nahid AA, Abbas R (2018) Effective intrusion detection system using XGBoost. Information 9(7):149
16. Ke G, Meng Q, Finley T, Wang T, Chen W, Ma W, Ye Q, Liu TY (2017) Lightgbm: A highly efficient gradient boosting decision tree. Adv Neural Inf Process Syst 30

17. Fu R, Zhang Z, Li L (2016) Using LSTM and GRU neural network methods for traffic flow prediction. In: 2016 31st Youth academic annual conference of Chinese association of automation (YAC) (pp. 324–328). IEEE
18. Pirani M, Thakkar P, Jivrani P, Bohara MH, Garg D (2022) A comparative analysis of ARIMA, GRU, LSTM and BiLSTM on financial time series forecasting. In: 2022 IEEE International Conference on Distributed Computing and Electrical Circuits and Electronics (ICDCECE) (pp 1–6). IEEE
19. Hancock JT, Khoshgoftaar TM (2020) Survey on categorical data for neural networks. J Big Data 7(1):1–41
20. Schjønning P, Jensen JL, Bruun S, Jensen LS, Christensen BT, Munkholm LJ, Oelofse M, Baby S, Knudsen L (2018) The role of soil organic matter for maintaining crop yields: Evidence for a renewed conceptual basis. Adv Agron 150:35–79

Optical Remote Sensing and Machine Learning Algorithm Used to Access the Leaf Area Index and Canopy Water Content

Surendra Reddy Vinta, Gireesh Kambala, Anzar Ahmad, and Pramoda Patro

Abstract Precise evaluation of crop canopy characteristics requires optical remote sensing (RS) equipped with solid procedures. Despite much research on the subject, algorithm performance employing RS still needs improvement. In a short amount of time, five distinct algorithms partial-least-squares regression (PLSR), support vector regression (SVR), and random forest regression (RFR) with feature selection were used to evaluate the LAI and CWC of rice canopy data. Results from the visible to short-wave electromagnetic spectrum of canopy reflectance were input into the algorithms. Researchers used two sets of repeated field trials in the 2014–15 and 2015–16 rice growing seasons to gather data on leaf area index (LAI) (600) and Canopy water content (CWC) (480). The coefficient of determination (R^2) was used to assess the efficiency of each method. Compared to other algorithms, PLSRLW outperformed them with R^2 values of 0.77 for LAI and 0.66 for CWC. In addition, for every model, created a kernel concentration estimator of the root mean formed error values using a bootstrapping strategy. By giving calibration samples, which have the same canopy assembly as the test samples, more weight, the findings indicated that an appropriate method might be used to increase the forecast accuracy of LAI and CWC. Calibration of the model should be done using an entire season's worth of canopy spectral data since sub-setting it leads to high error values in the test dataset.

S. R. Vinta (✉)
School of Computer Science and Engineering, VIT-AP University, Amaravati, Andhra Pradesh, India
e-mail: vsurendra.cse@gmail.com

G. Kambala
Teach For America, 2140 Hillside Derby Run, 30040 Cumming, GA, USA
e-mail: Gireesh.kambala@teachforamerica.org

A. Ahmad
Department of Electronics and Communication, Graphic Era to be deemed University, Dehradun, India

P. Patro
Department of Mathematics, Koneru Lakshmaiah Education Foundation, Hyderabad, India

A. J. Kulkarni and M. Hajiaghaei-Keshteli (eds.), *Advances in Intelligent Systems for Sustainable Agriculture*, Lecture Notes in Networks and Systems 1201,
https://doi.org/10.1007/978-981-97-9839-1_23

Keywords Canopy water content · Support vector · Leaf area index

1 Introduction

Remote sensing (RS) is quickly becoming a go-to method for assessing canopy properties without causing any damage. In RS, the biophysical and metabolic features of plants may be assessed using a single spectral designation that spans the noticeable, near-infrared, and shortwave electromagnetic (360–2400 nm) parts of the spectrum [1, 2]. The Remote sensing method requires extensive crop-specific spectrum libraries and fast, precise, and resilient algorithms [3, 4]. Over the past few decades, researchers have utilized a variety of retrieval algorithms to study and utilize the relationship between reflectivity spectra and in-situ vegetation data [5, 6]. Some examples of these algorithms include radiative transfer models, experiential regressions, PLSR, SVR, and ANN [7, 8]. Because physical-based approaches are less suited for retrieval applications and have an ill-posed model inversion, empirical regression algorithms are the focus of this work [9, 10]. In the supplemental materials, you may see a few research on the effectiveness of empirical algorithms in predicting canopy characteristics of vegetation [11]. The obtained coefficients of determination (R^2) range from somewhat good to very bad, even after applying several algorithms to various vegetative factors. This means that vegetation parameter tests should use effective methods that have proven effective in previous spectroscopic investigations [12, 13]. To make spectrum reflectance-based algorithms more accurate predictors, crop-specific spectral libraries have been created in the last decade [14]. Spectral reflectance from a single leaf or the whole crop canopy may be collected via crop-specific spectral libraries, contingent on the parameter of interest [15, 16]. The sensors employed, the quality of the radiation source, the environment, the plant canopy, the soil nutrients, the availability of water, and other elements all make these spectral library areas unique [17, 18]. Factors such as crop type and sampling site selection impact the reliability of data calibrating spectral models from field campaigns. Several crop properties are predicted to alter as time and space pass [19]. Algorithms relying on linear transformations may not work 100% of the time in crop spectrum libraries since they deal with complicated datasets that display nonlinear attribute correlations. Algorithms that handle linearity well, such as SVR and RFR, and PLSR-combined techniques, like PLSR with feature selection (PLSRFS) and locally weighted (PLSRLW), are available for this issue. Canopy water content (CWC) and leaf area index (LAI) are two primary plant attributes that may be assessed with the use of highly sensed spectral reflectance data [20, 21]. A plant's growth and harvest are sensitive to atmospheric fluctuations in water and energy, which are affected by both factors. Crop-specific spectral libraries have been created in the last period to make spectrum reflectance-based algorithms more accurate predictors. Spectral reflectance may be measured from individual leaves or the whole crop canopy using crop-specific spectral libraries; the choice is based on the parameter of interest [22].

Several algorithms, including PLSR, SVR, and RFR, have been used to estimate the canopy water content and leaf area index. The coefficient of determination for LAI and CWC ranged between 0.62 and 0.94 among studies that examined various vegetation types and environmental factors. It is clear from the literature above that more research into spectroscopic data and testing it with newly created algorithms and other kinds of vegetation is necessary for predicting LAI and CWC [23, 24].

To significantly enhance the accuracy of predictions, optical remote sensing applications have been created that estimate soil parameters using PLSR modified into locally-weighted PLSR (PLSRLW) and feature selection-based PLSR (PLSRFS). Due of their novelty, these algorithms have not yet been tested in agricultural research. So, we need to see how well the aforementioned algorithms can forecast the traits of crop canopies. We have developed five separate methods to rapidly evaluate LAI and CWC for rice coverings by analyzing canopy reflectivity spectra that cover the noticeable to short-wave infrared spectrum. Using R^2 and root mean squared error (RMSE), we compared and contrasted the suitability of each technique. To make sure that the data was consistent, a repeated field research was conducted using three different kinds of rice and five different soil water possible levels.

2 Methodology

2.1 *Experimental Information and Study Location*

Field trials were conducted many times from 2014 to 2016 at the Vishakhapatnam, Andhra Pradesh, Indian agriculture Farm. These tests were carried out throughout the dry season, which is December–April [25, 26]. The sandy loam soils that were used for the experiments are acidic. According to the USDA soil taxonomy system, Typic-Haplustalf is the classification of soil types. There is an average of about 700 mm of pan evaporation each year and 1600 mm of annual rainfall in this humid subtropical climate, defined by short winters and long, scorching summers. Figure 1 shows the experimental plots used to cultivate three different types of rice: two kinds that are drought-sensitive and one drought-tolerant variety. While Vandana is an upland variety with a short duration of 90 days, Satabdi and IR36 are medium-duration types ideal for lowland irrigated environments. In 2014–2015, 28-day-old rice seedlings were transplanted into the puddled plots, whereas in 2015–2016, 18-day-old seedlings were transplanted. Soil water potential values of 30, 50, 70, 120, and 140 kPa were used as water stress treatments in a randomized block design with two replications. During the vegetative development phases, plants were subjected to water stress treatments, including not watering and letting the soil dry naturally. Various field experiments were conducted at the Experimental Farm, which is a part of the Agricultural and Food Engineering Department of the Indian Institute of Technology, from 2014 to 2016. We utilized the right amounts of chemical insecticides to keep the bugs and other pests at bay [27, 28].

(a)

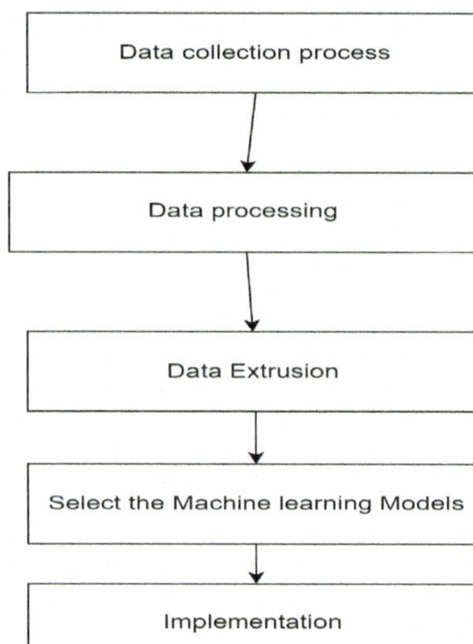

(b)

Fig. 1 **a** The experimental location as seen from the field **b** Work flow

2.2 *Initial Processing of Spectra*

Every spectrum was subjected to the sensor modification. We acquired smoothing utilizing the third-order pressing method with a 9 nm span length, and we reached an excellent signal-to-noise ratio to guarantee that all canopy traits followed a normal distribution, we used the Kolmogorov–Smirnov test with a significance threshold of 5%. Water absorption bands were present in all reflectance spectra between 1350 and 1460 nm, 1804 and 1971 nm, and there were no longer any noisy reflectance values between 400 and 2447 nm, or after 400 nm, or after 2447 nm. Using the remaining 1809 wavebands, analysis was carried out. We used the standard normal variate (SNV) procedure to clean up the spectra [29, 30].

2.3 Descriptions of Algorithms

This research made use of three distinct algorithm types: (a) conventional PLSR, SVR, and RFR; (b) locally weighted-PLSR (PLSRLW); and (c) PLSRFS, which combines PLSR with feature collection. The PLSR method aims to get the maximum possible covariance between the input characteristics and the response variable by projecting them into a space. The SVR method uses an estimated function from a predetermined training set so that the dissimilarity between the predicted and observed responses for each training set is kept to a minimum. The sensor was adjusted for each spectrum. With a 9 nm span length, we achieved an outstanding signal-to-noise ratio and obtained smoothing using the third-order Savitzky–Golay pressing technique. Perform the Kolmogorov–Smirnov test with a significance level of 5% to ensure that all canopy features followed a normal distribution. Between 1350 and 1460 nm, 1804 and 1971 nm, and after 2447 nm, there were no more noisy reflectance values between 400 and 2447 nm, or after 400 nm. All reflectance spectra showed water absorption bands. Analysis was conducted using the remaining 1809 wavebands. For this purpose, we cleaned up the spectra using the SNV technique.

Resembling a network of classification trees, RFR is a classifier that uses the "votes" of each tree to assign a label to a sample. The most popular classification is chosen by the forest. Spectrometric data is an example of a problem with a low sample-to-feature ratio; RFR is resilient to overfitting and often achieves well in such cases. Given the similarity in modeling outcomes between the SVR, PLSR, and RFR methods, efforts were also undertaken to enhance the PLSR model's performance utilizing the PLSRFS and PLSRLW techniques. An ordered predictor selection (OPS) method was used to apply the PLSRFS technique, which allowed for selecting a parsimonious set of forecasters for a given regression model.

In this method, the variables that are chosen are those that are dependent on PLSR (regression coefficients, squared residual, VIP, net analyte signal, and so on) and those that are independent of PLSR (correlation vector, weight mid correlation vector, AMI, StN, and CovProc). We detailed the calculation of the spectral variable indicators and provided further information about them. Furthermore, new sets of spectral variable pointers were produced by combining (a) indicators that were completely reliant on PLSR (six combinations) and (b) indications that were both reliant and independent of PLSR (20 combinations). The absolute values of the indicators for the normalized spectral variables were multiplied element by element to form the composite indicator. Hence, 35 spectral variable indicators were considered in this research, consisting of 9 single indicators and 26 combinations. To choose a reasonable collection of predictor variables to include in the regression model, we used the OPS method to each group. This allowed us to reduce the number of variables while also improving their attractiveness. The forecast accuracy of the final PLSR model was enhanced when this feature selection strategy was implemented. For every spectrum, the PLSRFS method was secondhand. Study the results pertaining to the most complex feature combinations in relation to the root-mean-squared error (RMSE) and the number of optimal spectral variables (NSV) for prediction accuracy models where the standard deviation of the root-mean-squared errors (RMSEs) was less than 5%. For each crop parameter, the optimal model was determined by comparing its NSV to that of the related variable indicator. In PLSRLW, PLSR model calibration takes into consideration, via a weighting mechanism, the similarity between each calibration sample and its corresponding validation sample. Here is an example of how the similarity metric may be used to calculate the reserve between the correction and justification samples:

$$d_{c,u} = \sqrt{(xc - xv) - (xc - xv)}$$

If you want to know how similar the observer and predictor variables are, you may use the identity matrix h. The validation sample, xv (p 1), and the correction sample, xc (p 1), are represented here by dc and v, respectively. Finally, p represents the total of all the predictor variables. Two different approaches were used to get the value of dc, v. "Correlation between response and predictors" is the first approach to finding h. Another approach uses the predictor-response covariance to calculate h:

$$\theta\,\text{CORR} = \text{diag}\left\{\frac{||xty||}{||x||y||}\right\}$$

In this context, diag{ } denotes the transverse matrix, X is a predictor matrix with dimensions (n p) and sample size (n), and y is the answer variable's support vector. Using the predicted dc v, the following weights were computed for each sample:

$$wc,\ v = \exp\left(\frac{dc,\ v}{ds}\right)$$

Here, wc, v represents the mass differential between the cth correction sample and the vth authentication model. The shorthand for the localization limit is /. In relation to the vth validation sample, this distance, ds, represents the standard deviation of each calibration sample. The user specifies the degree of localization by setting the localization parameter, represented as /. When you set / to zero, you essentially use the PLSR approach, which gives each calibration sample the same weight.

The Partial Least Squares with Ridge and Lasso Weights (PLSRLW) technique is anticipated to outperform the Partial Least Squares Regression (PLSR) method due to its ability to effectively capture non-linear relationships in the data.

2.4 Development of Spectral Models

The spectral models connecting the properties of the rice canopy to the reflectance spectra were constructed using the techniques outlined above. the MATLAB copilot tool to search for extreme values in the residuals (5% significance level) between the response variables that were seen and those that were predicted using principal component regression. Following the elimination of outliers, a validation dataset was generated using 30% of the initial data and a calibration dataset was formed using 70% of the original data. Divide the data using a stratified sample strategy. The weight difference between the *c*th correction sample and the *v*th authentication sample is represented by wc, v in this figure. As a shorthand, the localization parameter is [/]. This distance, denoted as ds, describes the standard deviation of every calibration sample with respect to the *v*th validation sample.

2.5 Analyzing the Excessive Fitting of the Model

In situ, canopy spectra measurements taken at various phases of rice development throughout two growing seasons were used to compile the spectral reflectance dataset. This kind of dataset records rice spectra across time. The second point is that the canopy spectra came from only one duplicated experiment. It was beyond the scope of this work to critically validate the haunted model regulated from such a dataset using trials spanning several sites, varieties, and even seasons. While this study's spectrum model was being examined for overfitting, three different simulation methodologies were employed:

- Spectral dataset cross-year authentication: The purpose of this workout was to calibrate and validate the model using datasets from specific years in order to verify its transferability.
- Filtering off ghostly data for specific dates, then sorting the remaining data into calibration and authentication samples using a 70:30 ratio: This experiment aimed to determine if specific models can be adequately evaluated using a condensed dataset, which was achieved by removing data from one particular observation day.
- This year's validation sample will consist of spectral data from one sampling day. In contrast, the remaining samples will serve as calibration: In this experiment, we used a training dataset consisting of a subset of observation dates to see if a calibrated model could accurately predict specific crop characteristics for a new set of observation dates outside the training set.

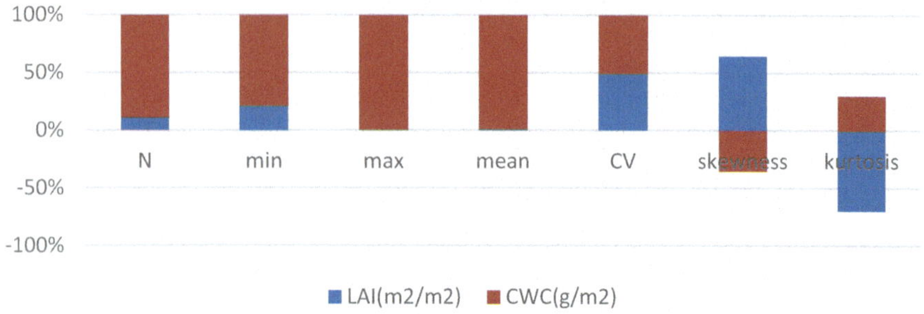

Fig. 2 Canopy water content and leaf area directory descriptive data

3 Results

3.1 *Information About LAI and CWC*

According to the descriptive numbers shown in Fig. 2, the average values of both LAI and CWC in the rice sample were within the predictable range. It was also noted that the outcomes for CWC and LAI were comparable. Throughout the experiment, the experimental technique allowed for a large range of values for LAI (m^2 m^2) and CWC (g m^2). The average loss area in square meters (LAI) was 4.05 m^2 m^2, with a range of 1.21–8.59 m^2 m^2. Additionally, there was considerable variation in the CWC values; they averaged 382.4 g m^2 and ranged from 61.9 to 1130.7 g m^2. Changes in canopy architecture and smaller leaves caused by water stress may contribute to the high variability in CWC and LAI. Such variation can have been caused by other variables such as various kinds, stages of development, and the state of the soil. The data acquired in this research offers suitable variability for developing regression models, as seen by the more significant constant of variation (CV) values for LAI and CWC statistics gathered for both seasons (CV = 44.9 and 45.6%, respectively). Both vegetation characteristics had flatter and less skewed frequency distributions. According to the Kolomogorov-Smirnov test, which was conducted at a significance level of 5%, both vegetation characteristics were found to follow a normal distribution. Therefore, when they were calculated, no modification was applied to these two crop attributes.

3.2 *Spectral Reflectance of Rice Canopies*

Several elements related to the canopy that may be detected may influence its reflectivity. The average, lower, and upper quartile values of phantom reflectivity across the visible (VIS), near-infrared (NIR), and first and second halves of the shortwave infrared (SWIR) bands are displayed in Fig. 3, along with the regular canopy spectra

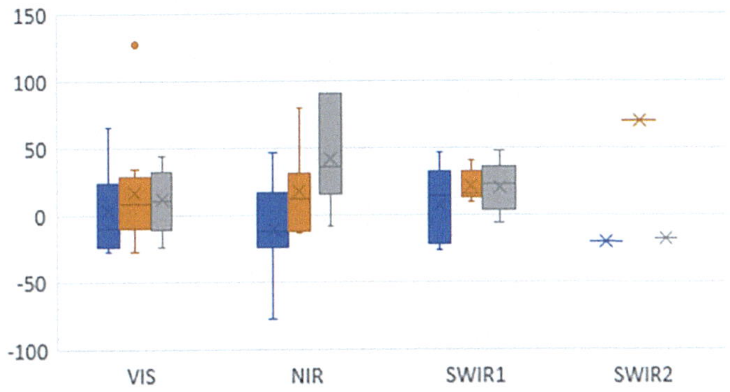

Fig. 3 Spectrum reflectance on average

(n = 600) and their differences. The wavelengths of these bands are 350–700, 700–1300, and 1971–2440 nm, respectively. These findings show that canopy characteristics affect the optical properties of leaves, with low reflectivity in the VIS and SWIR2 areas and in height reflectivity in the NIR and SWIR1 spots. The experimental site's surface soil properties were consistent throughout all 30 plots. Canopy spectra are, therefore, quite variable, and this diversity is likely due to optical properties and design changes.

3.3 *Analyzing Several Modeling Techniques for LAI and CWC Prediction*

Figures 4 and 5 show the validation R^2 and RMSE values of CWC and LAI obtained from the pooled dataset using various techniques. According to the algorithms, LAI may be predicted more accurately than CWC. Several techniques for retrieving LAI from multispectral data have been developed because of the functional relationship between canopy spectral reflectivity and LAI.

With R^2 values of 0.77 and 0.73, respectively, for LAI and CWC, the PLSR combination techniques of PLSRLW and PLSRFS outperformed the PLSR, SVR, and RFR algorithms. The PLSRLW and PLSRFS algorithms demonstrated strong predictive power for both vegetative metrics, LAI and CWC, as shown in Fig. 4's scatter plot comparing actual and anticipated values. The employment of PLSR, SVR, and RFR did not improve prediction accuracy due to a combination of improper waveband selection and erratic weight adjustment.

In diagram 6, prior work has shown that LAI prediction on rice canopies using MLR, PLSR, and LS-SVM models may get RMSE values ranging from 0.85 to 1.04, which are similar to or even better than the RMSE values obtained in our study. Reduced root-mean-squared errors (RMSEs) of 0.22–0.41 for vegetative water content (VWC) were seen in canopy spectra obtained by pot studies, PLSR, and

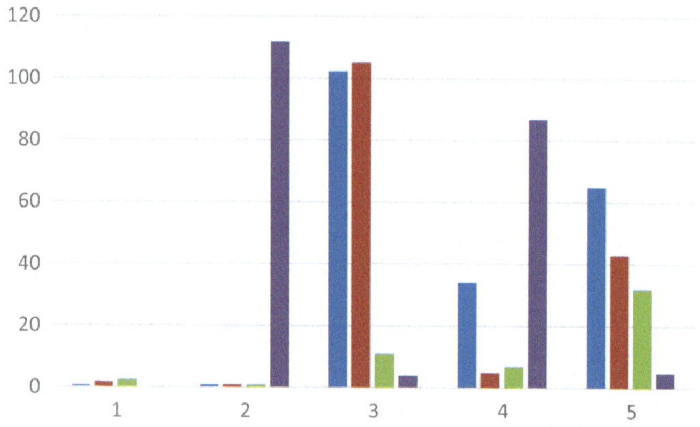

Fig. 4 LAI and CWC scatter plots compared to predictions

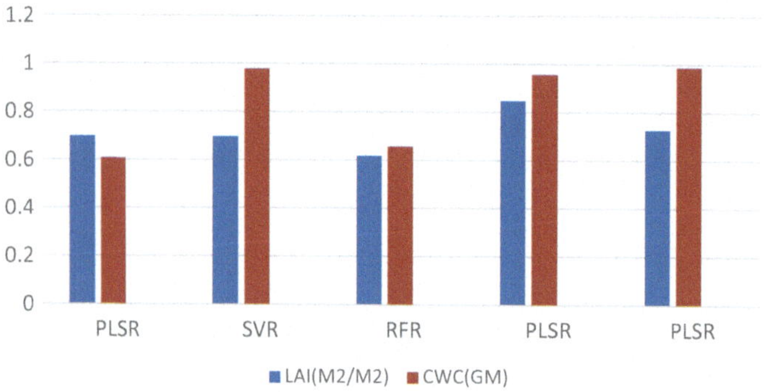

Fig. 5 Regression statistics for LAI/CWC prediction

artificial neural network methods. While PLSR, SVR, and RFR had RMSE values as high as (0.97–1.05), LAI's RMSE values reached from (0.95–0.94) in PLSRLW and PLSRFS. Just as the RMSE values for PLSRLW and PLSRFS were 96.25 to 99.92, the other algorithms had RMSE values as high as (101.2–105.3). Based on these findings, the PLSRLW and PLSRFS algorithms can predict LAI and CWC from canopy spectral data with acceptable accuracy (Fig. 6).

3.4 Model Performance Evaluation and Significance Test

Both vegetation parameters had their RMSE values calculated. The results demonstrate that the suggested PLSRLW algorithm outperformed the underperforming SVR model by about 15.3% after RMSE reduction in LAI. Similarly, CWC shows a 9.4% decrease in RMSE. PLSRFS, on the other hand, boosts model performance by 4.3%

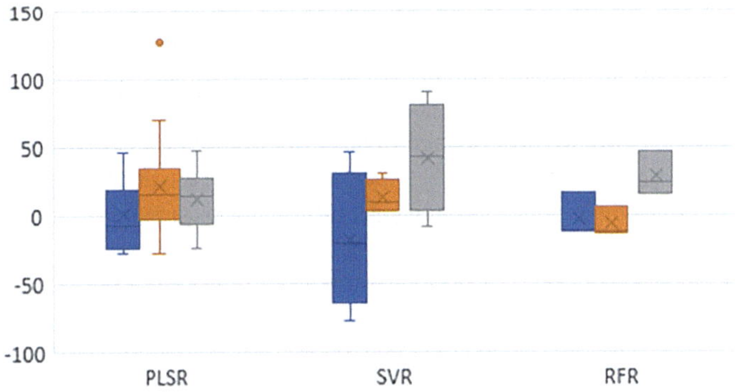

Fig. 6 Boxplots containing median, lower, and higher quartiles from all boot strapped calibration methods for LAI estimation

for LAI and 5.5% for CWC. The density functions that have been kernel-smoothed are shown in the left panels of these figures. Additionally, the boxplots that emerge from applying the calibrated model to these two parameters display the potential error distributions in the right panels. The full-spectrum model is compared to optimal models using a variety of variable indicators for both crop parameters using the PLSRFS approach. Figure 7 displays the variation in RMSE values as a percentage. The full-spectrum model has a baseline RMSE of zero. In terms of the accuracy of predictions made using the OPS method, a falling bar shows progress while a rising bar shows regression (maybe as a result of information loss). Based on our research, CovProc is the most reliable variable indicator for LAI. On the other hand, the best variable indicators for CWC are the OPS approaches done by b-StN and N-StN. For both crop parameters, the best model outperformed the full-spectrum model. The greatest reduction in RMSE value (6.37% for LAI) was observed when adopting the OPS technique. For CWC, there was a 4% reduction in RMSE.

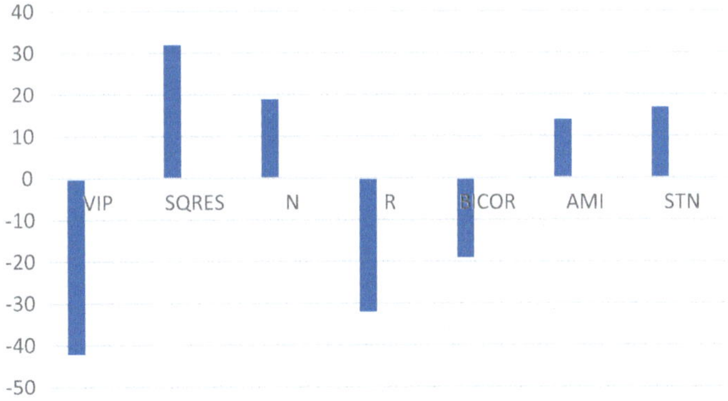

Fig. 7 Full-spectrum RMSE optimal spectral models employing variable indicators

3.5 Effectiveness of Algorithms in Computation

The time it took to execute each method was also used to compare their computational efficiency. While SVR, RFR, and neural networks were competitive among non-parametric algorithms, no clear winner emerged. Because the technique used various distance computations for each iteration to adjust the parameters, PLSRLW often required more processing time to execute the procedure. The computational load of PLSR coupled with feature selection is comparable; after all, it takes more time to choose the optimal feature space from among 35 distinct mutable pointers, including regression coefficients, VIPs, and adjacency factors based on mutual information and their combinations. Local and variable indication-based techniques may have a low presence in remote sensing literature due to the high processing time they demand. These techniques have the potential to be used for real-time estimates of undergrowth canopy metrics, thanks to the advancements in rapid processing technology.

4 Conclusion

The purpose of this work was to employ algorithms to examine optical remote sensing in order to find the best empirical model for evaluating LAI and CWC for dry rice canopies. The technique of reflectance spectroscopy was practical. With R^2 values of 0.77 and 0.66, respectively, the results demonstrated that LAI and CWC were successfully predicted by locally weighted PLSR. It is feasible to improve PLSRLW's performance by decreasing the estimated inaccuracy in visible and near-infrared space as well as its projected space by using suitable weighting algorithms. Selecting a suitable distance measure to construct the weighting scheme is a crucial component of PLSRLW that enhances its effectiveness in crop canopy feature modeling. This work shown that locally weighted PLSR is a valid and fast method for remote LAI and CWC assessment on rice canopies. Prioritizing variables that were geographically near together inspired this approach. A major limitation of our study is that our strategy was only tested on one experimental region. The present approach has to be evaluated in numerous environments and with diverse crops due to the fact that soil and crop characteristics might change. More agricultural features may be studied using PLSRLW and remote sensing data. To add insult to injury, the study's modeling of canopy spectra for LAI and CWC estimates did not consider background soil reflectance. Spectral reflectance values may vary substantially during the early mounting season when the crop canopy is not yet fully covering the soil surface due to variations in soil moisture, even though there may be a slight fluctuation in soil characteristics in these experimental plots.

References

1. Féret J-B et al (2019) Estimating leaf mass per area and equivalent water thickness based on leaf optical properties: potential and limitations of physical modeling and machine learning. Remote Sens Environ 231:110959
2. Pan H et al (2018) Modeling winter wheat leaf area index and canopy water content with three different approaches using Sentinel-2 multispectral instrument data. IEEE J Select Top Appl Earth Observ Remote Sens 12(2):482–492
3. Liu S et al (2021) Estimating leaf area index using unmanned aerial vehicle data: shallow vs. deep machine learning algorithms. Plant Physiol 187(3):1551–1576
4. Chen Z et al (2020) Leaf area index estimation algorithm for GF-5 hyperspectral data based on different feature selection and machine learning methods. Remote Sens 12(13):2110
5. Sun Q et al (2023) Machine learning algorithms for the retrieval of canopy chlorophyll content and leaf area index of crops using the PROSAIL-D model with the adjusted average leaf angle. Remote Sens 15(9):2264
6. Verma B et al (2022) Investigation of optimal vegetation indices for retrieval of leaf chlorophyll and leaf area index using enhanced learning algorithms. Comput Electron Agricul 192:106581
7. Fang H et al (2023) Leaf area Index Inversion of Spartina Alterniflora using UAV hyperspectral data based on multiple optimized machine learning algorithms. Remote Sens 15(18):4465
8. Brown LA, Ogutu BO, Dash J (2019) Estimating forest leaf area index and canopy chlorophyll content with Sentinel-2: an evaluation of two hybrid retrieval algorithms. Remote Sens 11(15):1752
9. Song C (2013) Optical remote sensing of forest leaf area index and biomass. Prog Phys Geogr 37(1):98–113
10. Raj R et al (2021) Leaf water content estimation using top-of-canopy airborne hyperspectral data. Int J Appl Earth Obs Geoinf 102:102393
11. Mao H et al (2019) Comparison of machine learning regression algorithms for cotton leaf area index retrieval using Sentinel-2 spectral bands. Appl Sci 9(7):1459
12. Yu T et al (2023) Potato leaf area index estimation using multi-sensor unmanned aerial vehicle (UAV) imagery and machine learning. Remote Sens 15(16):4108
13. Reisi Gahrouei O et al (2020) Estimation of crop biomass and leaf area index from multi-temporal and multispectral imagery using machine learning approaches. Can J Remote Sens 46(1):84–99
14. Hosseini M et al (2021) A comparison between support vector machine and water cloud model for estimating crop leaf area index. Remote Sens 13(7):1348
15. Mananze S, Pôças I, Cunha M (2018) Retrieval of maize leaf area index using hyperspectral and multispectral data. Remote Sens 10(12):1942
16. Ilniyaz O, Kurban A, Du Q (2022) Leaf area index estimation of pergola-trained vineyards in arid regions based on UAV RGB and multispectral data using machine learning methods. Remote Sens 14(2):415
17. Zhang Y et al (2022) Toward multi-stage phenotyping of soybean with multimodal UAV sensor data: a comparison of machine learning approaches for leaf area index estimation. Remote Sens 15(1):7
18. Chowdhury S, Sesharao Y, Abilmazhinov Y (2021) IoT based solar energy monitoring system
19. Cheng M et al (2022) Estimation of soil moisture content under high maize canopy coverage from UAV multimodal data and machine learning. Agricul Water Manage 264:107530
20. Sai Pavan KV, Deepthi K, Saravanan G, Rajasanthosh Kumar T, Vinay AV (2020) Improvement of delamination spread model to gauge a dynamic disappointment of interlaminar in covered composite materials &to forecast of material debasement . PalArch's J Archaeol Egypt/Egyptol 17(9):6551–6562. https://archives.palarch.nl/index.php/jae/article/view/5214
21. Jothi KR, Mamatha VL (2020) A Systematic Review Of Machine Learning Based Automatic Speech Assessment System To Evaluate Speech Impairment. In: 2020 3rd international conference on intelligent sustainable systems (ICISS), Thoothukudi, India, pp 175–185. https://doi.org/10.1109/ICISS49785.2020.9315920

22. Nandan R et al (2022) Evaluating optical remote sensing methods for estimating leaf area index for corn and soybean. Remote Sens 14(21):5301
23. Srinet R, Nandy S, Patel NR (2019) Estimating leaf area index and light extinction coefficient using Random Forest regression algorithm in a tropical moist deciduous forest, India. Ecol Inf 52:94–102
24. Kganyago M, Mhangara P, Adjorlolo C (2021) Estimating crop biophysical parameters using machine learning algorithms and Sentinel-2 imagery. Remote Sens 13(21):4314
25. Borzuchowski J, Schulz K (2010) Retrieval of leaf area index (LAI) and soil water content (WC) using hyperspectral remote sensing under controlled glass house conditions for spring barley and sugar beet. Remote Sens 2(7):1702–1721
26. Zhang J et al (2021) Leaf area index estimation model for UAV image hyperspectral data based on wavelength variable selection and machine learning methods. Plant Methods 17(1):1–14
27. Yamaguchi T et al (2020) Feasibility of combining deep learning and RGB images obtained by unmanned aerial vehicle for leaf area index estimation in rice. Remote Sens 13(1):84
28. Shiwakoti P, Jothi KR, Kalyanaraman P (2023) Revolutionizing the farm-to-table journey: a comprehensive review of blockchain technology in agriculture supply chain. In: Swarnalatha P, Prabu S (eds) Handbook of research on deep learning techniques for cloud-based industrial IoT. IGI Global, Hershey, PA, pp 338–353. https://doi.org/10.4018/978-1-6684-8098-4.ch020
29. Rosso P et al (2022) Processing of remote sensing information to retrieve leaf area index in barley: a comparison of methods. Precis Agricul 23(4):1449–1472
30. Vinta SR et al (2023) Identification of significant clinical attributes for developing heart disease prediction system. Int J Intell Syst Appl Eng 11.8s:552–559

Incorporating LoRa Based Wireless Sensor Network and Machine Learning Technologies to Improve Precision Agriculture

Aman Shaikh, Nihar M. Ranjan, Satayush Rai, Pranil Ashok Rao, and Ganesh Shinde

Abstract In the contemporary landscape, the agricultural sector has undergone transformational changes, embracing progressive dynamics through the integration of cutting-edge technologies and methodologies. The global adoption of IoT, AI/ML, and similar innovations reflects their capacity to furnish precise, reliable, and consistent outcomes, enabling data-driven decision-making. The amalgamation of IoT and ML models emerges as a strategic choice to enhance data analytics, aiming at a substantial improvement in both the quantity and quality of agricultural yields. This paper seeks to elucidate the synergistic interaction of various agrarian elements, orchestrating the computational prowess facilitated by ML models and data derived from IoT sensors transmitted through LoRa-based Wireless Sensor Network (WSN) infrastructure. The proposed system advocates a comprehensive three-tier architectural framework, wherein ML and IoT systems collaborate seamlessly to process data for field management. The envisioned system extends its utility by translating these predictions into visually comprehensible representations. The integrated approach signifies a departure from conventional methods, offering a novel pathway to ensure the sustainability of yield, encompassing both production and grading standards. This paper thus contributes to the discourse on leveraging advanced technologies for agricultural optimization in response to evolving global demands.

A. Shaikh (✉) · N. M. Ranjan · S. Rai · P. A. Rao · G. Shinde
Department of Information Technology, JSPM's Rajarshi Shahu College of Engineering, Pimpri-Chinchwad, India
e-mail: amanshaikhw@gmail.com

N. M. Ranjan
e-mail: nihar.pune@gmail.com

S. Rai
e-mail: raisatayush7232@gmail.com

P. A. Rao
e-mail: pranil3005@gmail.com

G. Shinde
e-mail: ganeshshinde72620@gmail.com

A. J. Kulkarni and M. Hajiaghaei-Keshteli (eds.), *Advances in Intelligent Systems for Sustainable Agriculture*, Lecture Notes in Networks and Systems 1201,
https://doi.org/10.1007/978-981-97-9839-1_24

Keywords LoRa · WSN · Precision agriculture · Machine learning · Embedded systems · IoT

1 Introduction

The world has seen a tremendous rise in population. It has swelled up to about 8.05 billion in 2023 from 7.25 billion back in 2013, which is about 800 million more people on the earth, consequently, there has also been a steady rise in terms of the food supply demand. Although for the past decade or so the growth rate has dipped from 1.24 to 0.88%, there hasn't been a proportionate rise in food production to be able to satisfy the need under current circumstances of consistent increase in population [1, 2]. As for the year 2050, it is estimated that the population will draw closer to the 10 billion zeniths [3]. With the major focus shifting toward being able to produce an exorbitant amount of food, the quality-to-quantity proportion has become unbalanced causing an occurrence of what is called hidden hunger to come into existence. Then again, these unorthodox patterns of production and consumption have led to unsustainability, consequently leading to a triple planetary crisis [4]. The pandemic of 2020 was responsible for exposing the lagging advancements in our agrarian growth which has added to the worldwide food crisis [5, 6]. The dynamics of the natural world keep shifting in terms of climate recorded and the associated disasters have compelled us to improvise. This gave rise to exploring approaches in precision agriculture which focuses on pinpointing the possible improvements to be made for noteworthy progress in production while maintaining quality. Precision agriculture is grounded in the foundational technologies of Artificial Intelligence (AI), Machine Learning (ML), Internet of Things (IoT), and Cloud Computing, among other key components. IoT sensors gather data and transmit it over the network to be stored which is then fed into the ML model to process. In [7, 8] authors have presented an analysis of ML algorithms for yield prediction. The authors have scrutinized the H-ANN model for the same and exemplified its advantages over the others. In [8], the authors present an optimized approach to water usage for agricultural irrigation using the ML model. The authors [9] have made use of LoRa technology for communication, it provides good trade-offs comparatively in terms of overall costs and its usage. The authors in [10] have advocated a mesh network wherein the LoRa server serves as the master and the LoRa gateway acts as the slave to transmit data to the MQTT which then subsequently transfers data to a distant cloud server. In [11], the authors have put forth a machine learning model to process sensory data in order to intimate the irrigation needs of the farmland. This research paper aims to provide much-needed insights into effectiveness, outlay, and precision that can be obtained using mentioned modern techniques.

The major focus lies on the application of IoT sensors and ML algorithms in precision agriculture for data gathering and as for the latter abstracts regarding yield prediction, and optimized irrigation have been gone over. Also, a synopsis of adjustments made in order to incorporate IoT and ML has been investigated. To conclude, the potential challenges and forthcoming trends within precision agriculture have been elucidated.

Section 2 gives the survey on existing technologies in the field of Precision Agriculture. Section 3 describes the proposed system architecture followed by Sect. 4 with the experimentation details of the proposed system. Section 5 outlines the results and discussions obtained from the experimentation details followed by the conclusion and future scope in Sect. 6 and the references.

2 Literature Survey

Precision agriculture is supported by an expansive set of literature. Within the survey we have studied, ML-based approaches for predicting crop yield and providing recommendations for optimizing farming practices, another approach utilizing IoT/WSN devices to gather environmental data and monitor agriculture, and to conclude an approach that integrates WSN and ML-based solutions holistically has been included.

In this paper, Jayanarayana et al. [8] have investigated models that utilized algorithms such as CNN, KNN, MCNN, H-ANN, Random Forest, LSTM, and others for yield prediction. The features such as soil content, and other environmental conditions such as humidity, temperature, rainfall, and air pressure among others were taken into consideration. Though CNN and MCNN reduced prediction error it resulted in an extraneous relative error being introduced leading to reduced efficiency, the RMSE stood at 1396.4 and the RE was 8.465. H-ANN yielded an RMSE value of 4.72. The other approach was a blend of ANN and MLR, resulting in an RMSE value of 0.051, while DNN reflected a validation RMSE value of 12.79. The obtained accuracy was significantly shaped by the availability of data and the judicious selection of features.

Maya Gopal et al. [9] have made use of a combination of MLR with ANN to enhance the precision of the model, assimilating the MLR parameters and their corresponding bias. Subsequently, it underwent preprocessing procedures involving the treatment of missing values and outliers. Different algorithms were employed to derive a feature set, utilized as input for the MLR model to identify the optimal subset. As a result, variables like area, canal length, count of open wells, tank availability, temperature, etc., were taken into consideration. Compared to H-ANN, algorithms such as SVR, KNN, RF, conventional MLR, and conventional ANN were outperformed. To put it into perspective the H-ANN gave an RMSE value of 0.061 which is 0.037 less than MLR and conventional ANN, with inaccuracy rates of 0.041 and 0.064 respectively. The obtaining of near-global optimal minima was significantly influenced by the decisive selection of initial weights through the application of MLR.

The authors, Sharma et al. [10] have presented ways of ML utilization to enhance irrigation efficiency through soil moisture prediction. Their work involves incorporating essential parameters like temperature, humidity, and internet-sourced data regarding weather forecasts to make informed and timely crop irrigation possible.

Algorithms such as MLR, ENR, GBRT, and RFR have been examined for the same purposes. Two distinct approaches were employed for soil moisture prediction one involved inclusion of soil temperature the other excluded it. The soil temperature was predicted using ENR and MLR models, with ENR outdoing MLR with an MSE of 2.25 in contrast to 3.4. Now, in the first approach soil temperature was included as a parameter; the GBRT model performed the best with MSR, and R-squared values of 4.04, and 0.94 respectively in comparison to 19.26, and 0.78 respectively in the case of RFR. In the second strategy excluding soil temperature as a parameter for the GBRT model outperformed others with an MSR of 5.78 and an R-squared value of 9.93 was obtained. However, it is evident the first approach yielded better results.

In this paper, Suji Prasad et al. [11] have suggested an IOT system based on LoRa communication to be deployed as a smart agriculture system. Authors in [12] have compared various communication technologies such as LoRa, Wi-Fi, Zigbee, & Bluetooth and it was found that LoRa has lower power consumption, is affordable, and has better distance coverage. The sensory nodes deployed in the farm measured soil temperature, moisture, and humidity. The analysis of the collected data from these nodes revealed a negative correlation between temperature and humidity. The system enables farmers to monitor the farm and aids in improving the harvest. The existing system can be enhanced further by making use of unsupervised clustering algorithms to analyze sensor data and, as a result, have automated control over the farm. Mesh networks could be used to improve performance to a certain extent.

Authors Chaopisit and Kaewpraek [13] have put forth a mesh networking system that involves LoRa technology in WSN for agrarian farmlands. The solution provided makes use of master–slave architecture. The nodes act as slaves that send data to the LoRa server (Master). The data is transmitted by the server to the LoRa Gateway which transfers the data to a distinct server on the cloud using MQTT. The nodes collect data on temperature, soil moisture, and humidity across the farm, each placed at 400 m^2 intervals within the farmland. These nodes operate in three states: running, transmission, and deep sleep. During the running state, a node lasted for a duration of 55 h and 24 min, while in transmission mode, it lasted for 39 h and 8 min. In the deep-sleep mode, a node can persevere for 193 h and 54 min. The proposed system exhibited effective communication for distances up to 850 m, with a signal strength of 101 dBm. Future enhancements could focus on optimizing the antenna design to reduce power consumption and extend the communication range. Also, the system could benefit from the addition of a web-based dashboard for real-time monitoring. Additionally, considering solar energy-based rechargeable batteries for the nodes would contribute to sustainable and prolonged operation.

Bhoi et al. [14] introduced and operationalized a three-tier architecture for ML-based irrigation recommendations. The architecture consists of the Crop Field level, Cloud level, and User level. At the Crop Field level, various sensors are diversely positioned to monitor soil moisture, soil temperature, air temperature, and humidity. The data is gathered by the sensors twice daily, and transmitted to an Arduino. The Arduino acts as an intermediary, forwarding the collected data to cloud-level storage. The stored data undergoes analysis by the ML models. A regression model was used to anticipate parameters in advance. Subsequently, the values are directed to a clustering model employing a binary classification algorithm to categorize data into two clusters: one indicating irrigation is required and the other indicating its absence. At the User level, a real-time mobile application has been developed to present comprehensive information about the crop field to the user. This architecture of the ML model offers an effective and optimized solution for recommending irrigation requirements in the field.

The authors Kanuru et al. [15] in this paper have delved into the use of IoT, ML, and a GPS module to uncover the fertilizer requirements. Considerations such as soil type, region, and crop type are factored in to tailor the fertilizer choice. Various ML models such as KNN, the Naïve Bayesian model, and the linear SVC have been investigated. The implementation involves 4 stages: dataset gathering, processing data, establishing a correlation between soil class and chemical utilized, and lastly the analysis of data using ML. After execution of the 3 ML algorithms, KNN fared the best with an accuracy of 0.8145, next was linear SVC with 0.777 accuracy, and Naïve Bayes with 0.759. Adopting a geographic-centric approach to fertilizer selection proves effective in reducing wastage and optimizing costs.

Ranjan et al. [16] have proposed the use of Logistic Regression and Naive Bayes. Logistic Regression is being used for irrigation automation based on soil moisture and Naive Bayes for nutrient recommendation by using the N, P, K values based on crop input from application. However, the authors didn't provide much concluding proof about the results as discussed in the paper.

The author Akshay et al. [17] have tested k-Nearest Neighbor (KNN) algorithm to predict ground parameters like soil moisture, temperature humidity, and water level over existing machine learning techniques like K-means and SVM since they are facing overfitting problems. And proposed that KNN is one of the best suited for accurate prediction of soil moisture with low cost.

Khutsoane et al. [18] proposed the use of LoRa-based communication between IOT devices for long distances. The authors opted for the given approach as the efficiency provided by LoRa-based communication was higher compared to the other communication technologies. However, the proposed solution discussed in the paper is expensive to deploy on a large scale and the applications discussed are dominated by devices for monitoring purposes. Also, the necessity to develop a cross-platform device for monitoring and control based on LoRa has been stated.

The authors, Kumar et al. [19] have implemented a machine learning model for yield production. The end model created makes use of Random Forest as it reduced overfitting and improved forecast accuracy as compared to Decision Trees which lagged in these particular scenarios. The model lacks methods to handle unstructured data which is necessary for an independent system.

From the extensive review of the existing literature survey, we conclude that precision agriculture has potential benefits but its adaptation is hindered by various challenges such as limited access to technologies, skills, knowledge, and high costs, however critical observation of the need for a standardized architecture has been necessitated to provide interoperable and enhanced technological components working in synergy.

3 Proposed System

With the food demand skyrocketing it has become essential for a system to be crafted. In response to the diminishing farmland, farming requirements are increasingly being met by farmlands in remote locations where advanced technologies such as GPS/ GSM/GPRS are not yet prevalent. In recognition of these challenges, there arises a demand for a precision agriculture system adaptable to farms in remote areas, to ensure scalability. The conceived solution introduces a three-tier architecture, illustrated in Fig. 1.

Figure 1 shows the three-tier architecture of the system along with the components present at each tier.

Tier 1 Components:

1.1 Weather API.
1.2 Predictive AI Model.

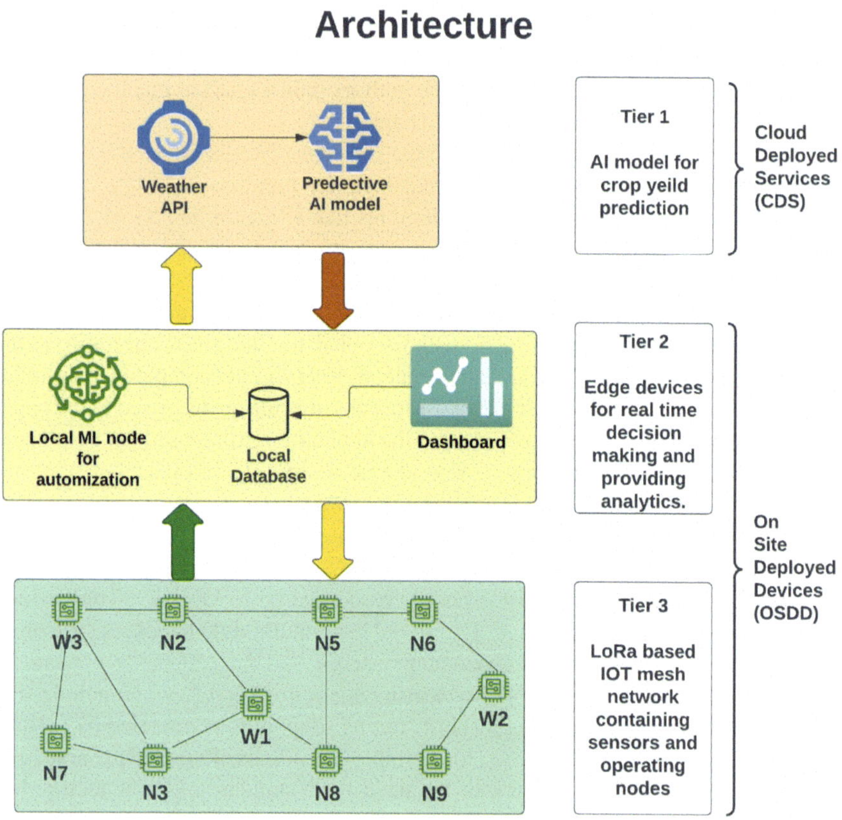

Fig. 1 System architecture

Tier 2 Components:

2.1 Real-Time Local Database.
2.2 Local ML Models.
2.3 Analytical Dashboard.

Tier 3 Components:

3.1 Sense Node.
3.2 Actuator Node.
3.3 Sink Node.

Tier 1:

1.1 Weather API: Timely requests are made to the weather server for live data of the climate for given coordinates. The data gathered includes humidity, temperature, rain chance percentage, etc.
1.2 Predictive AI Model: The ML model acquires data from the global server. The input data includes two sources: the global source, where data is sourced from weather stations through an API, and the on-field data, including soil temperature, humidity, etc. Suggestions are provided to the farmer based on predictions made by the model after it has analyzed the data fed to it with relevant parameters in place.

Tier 2:

2.1 Real-Time Local Database: The system uses MS SQL Server as the local database to store the data collected by the WSN nodes and the data through the Weather API. The database combines the data gathered from the on-field sensors and API.
2.2 Local ML Models: The local model consists of two ML models to produce corresponding output for decision-making. One of the models is the yield prediction model, wherein various features such as soil quality, environmental components, agricultural practices, and farm size are taken into account to predict the yield from the farm. Another model is the soil moisture prediction model to optimize water usage, which makes use of the weather forecast data for judicious irrigational methods to be employed. Lastly, a model to provide data on the soil fertility for making decisions on the fertilization needs of the farm based on parameters such as crop type and associated soil type.
2.3 Analytical Dashboard: The dashboard is hosted via Microsoft PowerBI which can be accessible via all mobile devices, desktops, etc. It shall reflect both live data as well as monthly historical data. Parameters such as total precipitation in a month, cloud cover percentage, wind speed, soil moisture, soil temperature, yield predicted, and total fertilizers/pesticides used to avoid overuse of these to prevent crop damage and soil leaching. It helps to get live on-field data and take action accordingly.

Tier 3:

3.1 Actuator nodes: It is a WSN node responsible for receiving the commands from the sink node, to take further actions. An actuator node can operate devices such as motors, pumps via an electromagnetic switch.

3.2 Sense Nodes: This node has a LoRa transceiver which is responsible for the sensory data communication within the mesh network. The node has a micro-controller connected to the probe to measure soil moisture and temperature, a DHT11 sensor, etc. The node keeps timed tracking of sensed values.

3.3 Sink Nodes: These nodes within the interwoven network transmit data from the sensing nodes to the internet. The other nodes need to find an optimized path to reach the sink node. The network must have at least 4 sink nodes.

4 Experimentation Details

4.1 Dataset Description

The dataset used for soil moisture prediction is a combination of Bragg's farm dataset and corresponding weather data collected from the visual crossing weather API service. Bragg's is a farm site located in Alabama. Bragg's dataset is a public report provided under the Natural Resources Conservation Service by the U.S. DEPARTMENT OF AGRICULTURE. The dataset contains hourly entries of soil moisture and soil temperature collected using Hydra Probe Analog (2.5 V) at a depth of 4 inches and meteorological parameters such as temperature, humidity, precipitation, wind speed, cloud cover, and weather conditions for the year 2022. The data consists of 8761 entries. The entries are split into training and testing datasets of sizes 7445 and 1314 respectively after preprocessing.

The dataset for yield production has been obtained from the Food and Agriculture Organization (FOA) and the World Data Bank. The end dataset is created as a result of combining datasets specific to data regarding pesticides, rainfall, temperature, and yield. Columns such as area, area code, country, year, and element among others have been left out due to their insignificance. The data provides a world overview in terms of the yield production over the years from 1990 to 2010. The resultant dataset has around 17,864 data points for the model to train from.

4.2 4-Phase System Working

Phase 1: Collecting the Field and API Data

Initially, the system records the on-field data through the Sense nodes via LoRa and essential meteorological variables through a weather server with the help of an

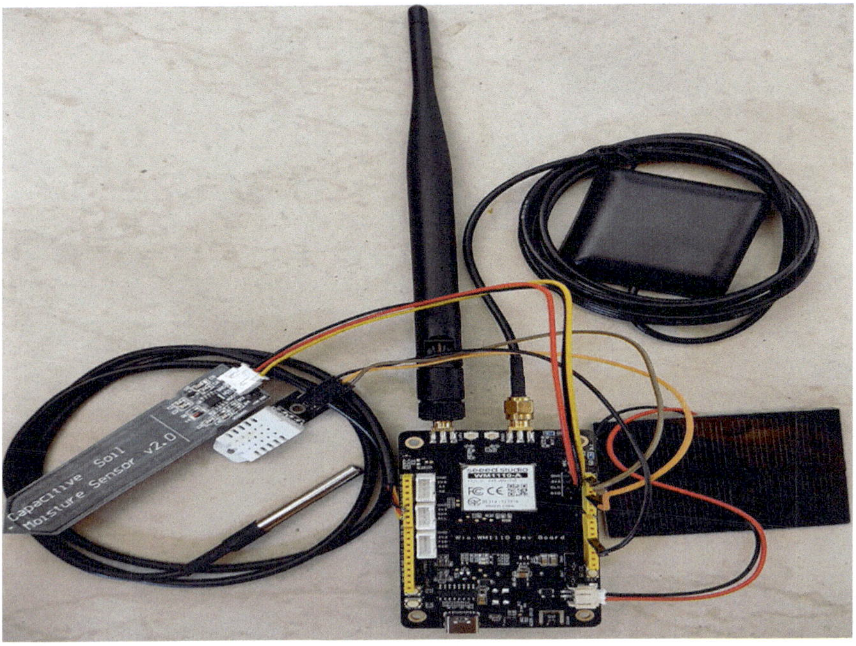

Fig. 2 Sense node

API. The Sense node uses 868 MHz LoRa frequency and GNSS (Global Navigation Satellite System) for geographical location discovery. Additionally, the node uses a capacitive soil moisture sensor, DHT22 ambient humidity and temperature sensor, and DS18B20 soil temperature sensor for reading meteorological data of the farmland.

Figure 2 shows the LoRa-based Sense Node. The node periodically sends sensor data to the Sink node via LoRa with an SF9 (Spread Frequency).

Phase 2: Soil Moisture Prediction Model.

The data that has been gathered from the field and weather API is combined and fed to the machine learning model. The ML model predicts soil moisture based on the meteorological data provided to it.

Phase 3: Yield Prediction Model.

Features such as average temperature, average rain, pesticide value, and yield value in consideration are retrieved from the dataset and are pre-processed before it further undergoes computations using the Ml model for prediction. Also, new features showing relationships between pesticide value against temperature and rainfall have been included. The dataset then created is split into 30:70, training and testing datasets. Further, the data points undergo model training and then are validated to test model performance. The model in consideration is the Random Forest and has

Fig. 3 Actuator node

shown exceptional results for the given dataset. It is to be noted that further hyper-parameter tuning may enhance the results however it is very relative to the dynamic factors that may change on the field.

Phase 4: On-Field Actuator Node.

The actuator node consists of ESP-01S which is the main microcontroller. It handles all the MQTT connections and maintains a Wi-Fi connection. The node consists of two 12 v relays that can be controlled by the MCU (Micro Controller Unit). The node is directly powered through 220 v mains. The onboard PSU converts the mains voltage to 12 v and an LM1117 which is a low dropout voltage regulator converts 12–3.3 v. 3.3 v is used to power the MCU and the maximum current supplied is 800 mA.

Figure 3 shows the water pump control node which can be connected through LoRa or Wi-Fi. Upon receiving the commands, the node acts accordingly, i.e., turn the pump on/off.

5 Result

The collected data as discussed previously in dataset details under experimentation details. It is then fed to the following models: Artificial Neural Network (ANN), Multiple Linear Regression (MLR), Decision Tree Regressor (DTR), Support Vector Regressor (SVR), and Gradient Boosting Regressor (GBR).

Table 1 Soil moisture prediction

Model	MAE	RMSE	R^2
ANN	0.421	1.217	0.947
MLR	0.445	1.237	0.945
DTR	0.437	1.493	0.920
GBR	0.386	1.147	0.953

Table 1 shows the comprehensive performance evaluation of soil moisture prediction models based on their Mean Absolute Error (MAE), Root Mean Squared Error (RMSE), and R-Squared Value. The Artificial Neural Network (ANN) shows overall better performance and can learn intricate associations from soil moisture data, allowing for the discovery and prediction of complex patterns. Its performance is a result of its ability to capture nonlinear dependencies, as seen by its high R^2 and low MAE values. On the other hand, Multiple Linear Regression (MLR) uses a simple technique that assumes a linear relationship between the target variable and the input features. MLR exhibits competitive outcomes despite its simplicity, as evidenced by its strong R^2 and low MAE values. Decision Tree Regression (DTR) is a tree-based technique that captures nonlinear correlations by recursively partitioning the dataset according to feature requirements. DTR does well, although it has a tendency to overfit to the training set, as seen by the marginally higher MAE and lower R^2 values when compared to ANN and Gradient Boosting Regression (GBR) is unique in that it uses an ensemble learning technique to create a strong prediction model by combining several weak learners. The boosting algorithm constructs decision trees in a stepwise manner while fixing mistakes from earlier iterations. With the lowest MAE and greatest R^2 values among the assessed models, this iterative method produces superior performance by improving predictive accuracy and generalization.

Figure 4 shows the scatter plot of the actual moisture versus predicted moisture of different soil moisture models stated in Table 1. It can be inferred that ANN and GBR performed better compared to MLR and DTR with better MAE and higher R-squared value, in terms of predictive accuracy and model robustness. Specifically, the GBR model stands out as the optimal choice, demonstrating the lowest MAE of 0.386 and the highest R^2 of 0.953 among all evaluated models. The observed trends not only underscore the significance of employing advanced machine learning techniques, such as ANN and GBR, in soil moisture prediction but also emphasize the need for selecting models that strike a balance between precision and generalization.

The dataset for yield prediction was used to train various models such as Linear Regression, Random Forest, Decision Tree, and Bagging Regression. The data consists of 17,864 entries representing various crops across countries further differentiated based on calendar year 1990–2010.

Table 2 shows the comprehensive performance evaluation of yield prediction models based on their Accuracy and R-squared value, and as can be seen from it, Random Forest (RF) and Bagging Regression (BR) are the models showcasing competitive performance with better accuracy and similar R-squared values, this goes to show that both the models can grasp the complex relationship between the

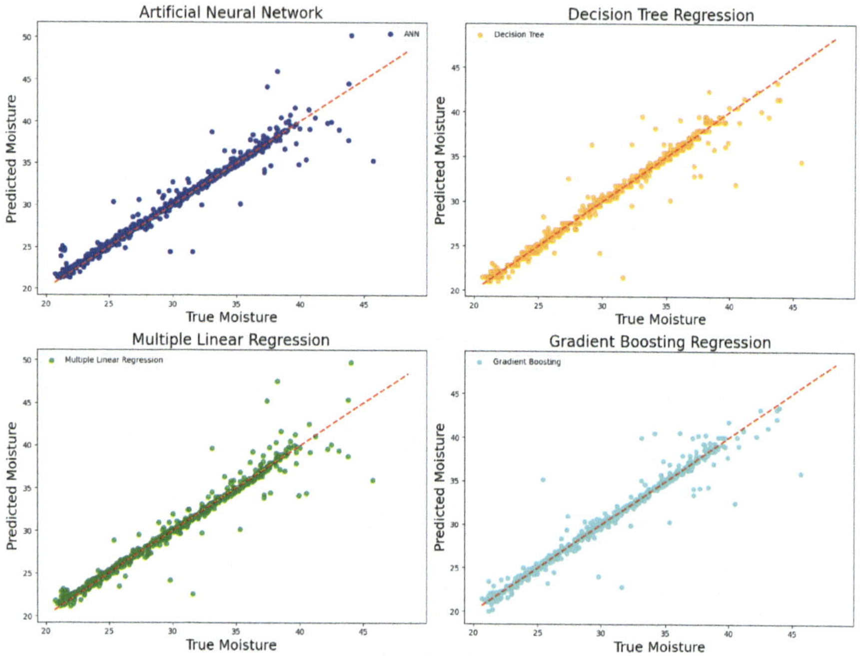

Fig. 4 Soil moisture prediction results

non-linear data. However, as observed RF has a minute improvement in accuracy of 0.02 compared to BR, the addition of other parameters such as intensity of sunlight, humidity, etc. could affect the model outcome making one model superior to the other. Decision Tree (DT) although shows good performances does lag slightly as can be seen from the results in Table 2. Linear Regression (LR) on the other hand demonstrates a lack of comprehension of the complex relationship between the data points.

Figures 5 and 6 show the evaluation of the Yield Prediction Models as stated in Table 2 based on the scatter plot of the actual yield vs predicted yield values. It can be inferred that RG and BR performed better compared to LR and DT with better spread and with maximum data points lying on the line of fitness. From Fig. 6, it can be seen clearly that LR fails to understand the complex relationship between the parameters leading to a discrete step pattern. The other three models, i.e., RG, BR, and DT seem

Table 2 Yield prediction

Model	Accuracy	R^2
Linear Regression	86.04	0.860
Random Forest	99.24	0.992
Decision Tree	98.83	0.988
Bagging Regression	99.22	0.992

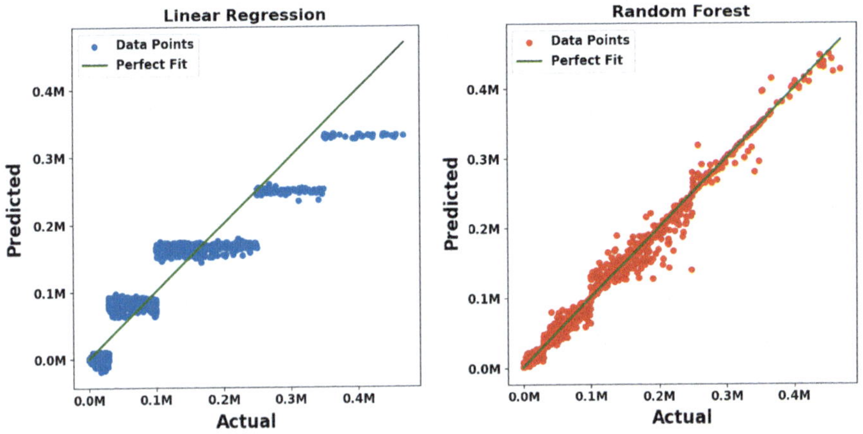

Fig. 5 Yield prediction results of LR and RF

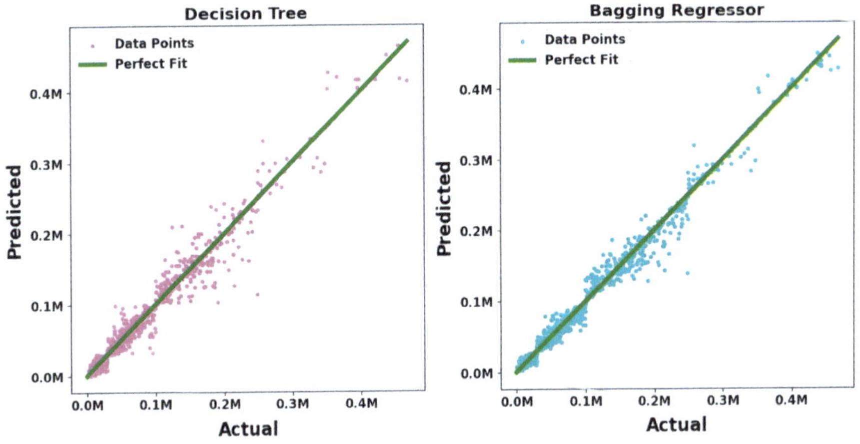

Fig. 6 Yield prediction results of DT and BR

to have a similar spread of data points but on closer observation, DT does have a spread that is distanced from the line of fitness. As for RG and BR, Table 2 sheds better light on the difference as compared to their visual plots. The trends we observed emphasize the importance of utilizing advanced machine learning techniques like RG and BR for yield prediction. Moreover, they underscore the necessity of selecting models that balance precision and generalization effectively.

A mechanism called the Spreading Factor (SF) is utilized to adjust the emission power and transmission speed based on the network circumstances surrounding the device. Thus, depending on the SF employed, performance and speed vary. SF allows for optimized power consumption and is used for radio resource management. Although the SF12 is the largest factor with the widest range, meaning that

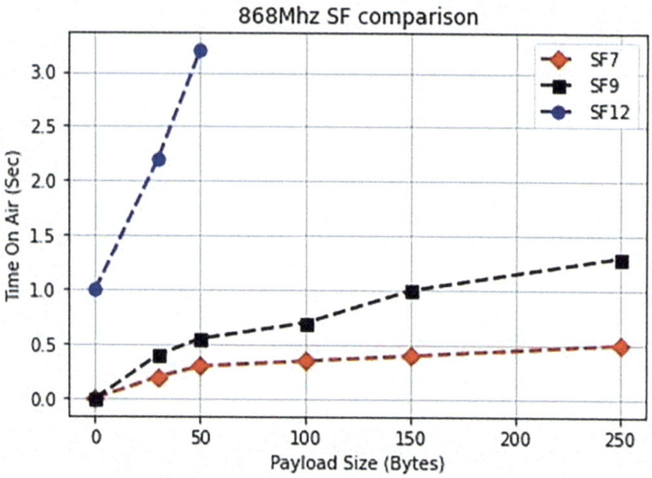

Fig. 7 Sense node LoRa SF comparison

most devices would receive the signal, using SF12 results in the slowest transmission speed and a substantial increase in power consumption. The SF7, on the other hand, has the shortest range but the fastest transmission speed and lowest power consumption. Additional SF classes include SF11, SF10, SF9, and SF8, each of which has a different transmission speed, range, and power consumption.

Figure 7 shows the amount of time taken by SF7, SF9, and SF12 to transmit different bytes of data packets from the Sense node to the Sink node. The SF12 takes longer to send data, even relatively small bytes of data. Thus, it needs more transmission power than SF7 & SF9. As a result, SF12 is less power efficient, has significant latency and reduced data transmission rates but provides a longer range than SF7 & SF9. In contrast, SF7 has a smaller range but offers high data transmission rates, minimal latency, and less power consumption. To get optimized SF the proposed system uses SF9 and has up to 6 km of range and also provides decent power efficiency.

6 Conclusions and Future Scope

The proposed architecture establishes a standard framework for building a LoRa-enabled WSN for precision agriculture. The proposed system aims to resolve issues regarding the exploitative utilization of natural resources such as water, fertilizer, and electricity. Consequently, improving constituting actions of yield output for agricultural land by providing real-time monitoring of the entire farmland via a centralized view. The implemented system considers both the live weather data as well as on-field data to predict the moisture of the soil. The entire statistics can be accessed via a dashboard that accurately represents the farmland conditions and assists farmers in managing, monitoring, and improving the agricultural practices. The system is poised in such a way that it can be deployed without resistance with only a few changes being required in the form of hyper-tuning of the ML models to further suit the conditions of the area. While there are existing solutions that demonstrate efficacy in numerous scenarios, opportunities for refinement exist to elevate the system's functionality. Employing a standardized Wireless Sensor Network (WSN) holds opportunities for optimizing the node's battery life and bolstering communication reliability within the network. Introducing diverse dashboard views could offer nuanced insights into various facets of the farm, enhancing overall operational comprehension. To enhance model accuracy, the integration of an unsupervised Machine Learning (ML) algorithm, trained on meticulously collected agricultural land data, presents a viable avenue for improvement.

References

1. Macrotrends (2023) World Population from 1950–2024. Accessed 17 December 2023. https://www.macrotrends.net/countries/WLD/world/population
2. Worldometer (2023) Countries in the world by population. Accessed 17 December 2023. https://www.worldometers.info/world-population/population-by-country/
3. Thinks T (2023) When will the world run out of food. Accessed 13 December 2023. https://thelmathinks.com/when-will-the-world-run-out-of-food/
4. Lovell R (2022) How modern food can regain its nutrients. BBC. https://www.bbc.com/future/bespoke/follow-the-food/why-modern-food-lost-its-nutrients/
5. Okolie CC, Ogundeji AA (2022) Effect of COVID-19 on agricultural production and food security: a scientometric analysis. Hum Soc Sci Commun 9(1)
6. World Health Organization (2023) Impact of Covid-19 on people livelihoods, their health and our food systems. Accessed 17 December 2023. https://www.who.int/news/item/13-10-2020-impact-of-covid-19-on-people%27s-livelihoods-their-health-and-our-food-systems
7. Langemeier M, Boehlje M (2021) Adaptation of precision agriculture technologies. Purdue University. https://ag.purdue.edu/commercialag/home/sub-articles/2021/03/adoption-of-precision-agriculture-technologies/
8. Reddy DJ, Rudra Kumar M (2021) Crop yield prediction using machine learning algorithm. In: 2021 5th international conference on intelligent computing and control systems (ICICCS). IEEE, pp 1466–1470
9. Gopal PSM, Bhargavi R (2019) A novel approach for efficient crop yield prediction. Comput Electron Agricul 165:104968
10. Singh G, Sharma D, Goap A, Sehgal S, Shukla AK, Kumar S (2019) Machine Learning based soil moisture prediction for Internet of Things based Smart Irrigation System. In: 2019 5th international conference on signal processing, computing and control (ISPCC). IEEE, pp 175–180
11. Suji Prasad SJ, Thangatamilan M, Suresh M, Panchal H, Rajan CA, Sagana B, Sharma A, Panchal T, Sadasivuni K (2022) An efficient LoKi Gunapriyashor Ra-based smart agriculture management and monitoring system using wireless sensor networks. Int J Ambient Energy 43(1):5447–5450
12. Queralta, JP, Gia TN, Zou Z, Tenhunen H, Westerlund T (2019) Comparative study of LPWAN technologies on unlicensed bands for M2M communication in the IoT: beyond LoRa and LoRaWAN. Proced Comput Sci 155:343–350

13. Chaopisit I, Kaewpraek N (2020) Low-cost remote-sensing-node based on LoRa mesh network for smart agriculture. Int J Sci Innov Technol 3(2):35–42
14. Bhoi A, Nayak RP, Bhoi SK, Sethi S, Panda SK, Sahoo KS, Nayyar A (2021) IoT-IIRS: Internet of Things based intelligent-irrigation recommendation system using machine learning approach for efficient water usage. PeerJ Comput Sci 7:e578
15. Kanuru L, Tyagi AK, Aswathy SU, Fernandez TF, Sreenath N, Mishra S (2021) Prediction of pesticides and fertilizers using machine learning and Internet of Things. In: 2021 international conference on computer communication and informatics (ICCCI). IEEE, pp 1–6
16. Bendre S, Bhosale M, Gunjal Y, Ranjan N (2017) IOT based irrigation automation and nutrients recommendation system. IJIACS 6(11):43–46
17. Akshay S, Ramesh TK (2020) Efficient machine learning algorithm for smart irrigation. In: 2020 international conference on communication and signal processing (ICCSP). IEEE, pp 867–870
18. Khutsoane O, Isong B, Abu-Mahfouz AM (2017) IoT devices and applications based on LoRa/LoRaWAN. In: IECON 2017–43rd annual conference of the IEEE industrial electronics society. IEEE, pp 6107–6112
19. Kumar Y, Nagendra J, Spandana V, Vaishnavi VS, Neha K, Devi VGRR (2020) Supervised machine learning approach for crop yield prediction in agriculture sector. In: 2020 5th international conference on communication and electronics systems (ICCES). IEEE, pp 736–741

The manufacturer's authorised representative in the EU is Springer
Nature Customer Service Centre GmbH, Europaplatz 3, 69115 Heidelberg,
Germany. If you have any concerns regarding our products, please
contact ProductSafety@springernature.com

Printed and bound by CPI Group (UK) Ltd, Croydon, CR0 4YY
01/10/2026
02225958-0011